Methods and Techniques In Plant Physiology

Methods and Techniques In Plant Physiology

Editor

Cornelio Losa

scitus academics

Methods and Techniques In Plant Physiology
Edited by **Cornelio Losa**

ISBN: 978-1-68117-543-0
Library of Congress Control Number: 2016942943

SCITUS Academics LLC,
www.scitusacademics.com
Box No. 4766, 616 Corporate Way,
Suite 2, Valley Cottage,
NY 10989

Printed in United States of America on Acid Free Paper

Preface

Plants are loved by lots of people - in our homes, on our tables as foods, and in hundreds of products we use every day. Plants have many different usages. But how do plants develop from seeds, and how do they grow? This is where plant physiology comes into play. Plant physiology is the study of how different parts of plants function. It includes many aspects of plant life, including nutrition, movement, and growth. Fundamental processes such as photosynthesis, respiration, plant nutrition, plant hormone functions, tropisms, nastic movements, photoperiodism, photomorphogenesis, circadian rhythms, environmental stress physiology, seed germination, dormancy and stomata function and transpiration, both parts of plant water relations, are studied by plant physiologists. Plant physiology includes the study of biological and chemical processes of individual plant cells. Plant cells have a number of features that distinguish them from cells of animals, and which lead to major differences in the way that plant life behaves and responds differently from animal life. This book explores how plant physiology helps us to understand the many functions and behaviors of plants.

Methods and Techniques in Plant Physiology is dedicated to physiology, biochemistry, cellular and molecular biology, genetics, biophysics, and environmental biology of plants. Techniques related to various physiological phenomenon are focus of tremendous interest and importance to plant physiologist, agronomist, horticulturist, ecologist, and biochemists.

Table of Contents

Chapter 1

Influence of Geographical Regions on Catechin and Caffeine Levels in Tea (Camellia sinensis)

Augustine Mutuku[1,2*], John Wanyoko[2], Francis Wachira[3], Samson Kamunya[2], Richard Chalo[2], Samuel Kimutai[4], Kelvin Moseti[2], Stephen Karori[1]

[1]Department of Biochemistry, Egerton University, Egerton, Kenya

[2]Tea Research Institute (TRI), Kericho, Kenya

[3]Association for Strengthening Agricultural Research in Eastern and Central Africa, Entebbe, Uganda

[4]Department of Chemistry, Egerton University, Egerton, Kenya

ABSTRACT

This study aimed at investigating the influence of different growing conditions in two regions, Timbilil and Kangaita, on the levels of catechins and caffeine. Fresh leaf samples of 60 clones cultivated in both sites were obtained, processed and assayed for levels of caffeine, total and individual catechins using Reverse Phase High Performance Liquid Chromatography (RP-HPLC). The obtained data was subjected to analysis of variance using GENSTAT-C statistical software. Safe for a few clones in the Kangaita site had significantly higher total and individual catechins than those grown in the Timbilil site. A mean epigallocatechin gallate (EGCG) content of 7.9% was observed for Kangaita clones compared to 6.7% in Timbilil. A similar trend was observed for the rest of the individual catechin fractions with a mean epicatechin gallate (ECG) of 3.3% in Kangaita and 2.5% in Timbilil, a mean epigallocatechin (EGC) of 5.2% in Kangaita compared to 5.0% in Timbilil and a mean epicatechin (EC) content of 1.5% in Kangaita compared to 1.4% in Timbilil. Similarly in the total catechins analyses, clones in Kangaita had a mean value of 18.7% compared to 16.2% observed for the Timbilil site clones. However, the Timbilil site clones had significantly high caffeine contents (mean, 4.2%) compared to the Kangaita site clones (mean, 3.9%) although with a few exceptions. Significant interactions ($P < 0.05$) between clone and region are also observed, suggesting that levels of these biomolecules are bound to differ with the growing region but the extent of variation will vary depending on the clonal genotype. The observed differences in levels of these biologically important biomolecules among the two regions shows the importance of the underlying conditions on the production of tea.

KEYWORDS

Camellia sinensis, Clonal Variations, Regional Variations, Catechins, Caffeine

1. INTRODUCTION

Tea is manufactured from the young tender leaves of the plant Camellia sinensis. The tea plants grow mainly in warm (tropical and subtropical) climates and can thrive at an altitude range of 1500 - 2700 m above mean sea levels and require at least 1270 mm of rainfall a year and prefer acidic soils [1]. Since its introduction, tea cultivation in Kenya has expanded tremendously in volume and in the tea growing regions mainly the East (Kiambu, Thika, Muranga, Meru, Nyeri, Kirinyaga and Embu) and West (Kericho, Kisii, Nyamira Bomet, and Nandi) of the Rift valley. Additionally, cultivation of the crop has expanded to include even marginal areas initially considered unsuitable for the crop [2]. There exist small to large differences in climatic and environmental conditions between the different growing regions, which consequently affects yield and quality of the resulting tea products.

Catechins and caffeine constitute the most important quality parameters in tea reference. Catechins are among the major polyphenolic compounds in the tea plant constituting up to 30% of the dry weight in a freshly picked tea leaf [3]. Catechins have been reported to be potent biomolecules responsible for the health benefits associated with regular drinking of tea [4] [5]. Studies have shown that catechins possess beneficial physiological activities such as antioxidants [6] [7], hypocholesterolemiant [8] [9], anti-atherosclerotic [10], anti-inflammatory [11], and anti-HIV activities [12].

Caffeine is highly valued for its central nervous stimulating properties [13]. It is associated with improved performance of tasks that require sustained effort and attention [14], speed of response [15], feelings of well- being and improved moods [16]. Several factors have been shown to be responsible for the tea quantity and ultimately the amount of biomolecules that are synthesized and accumulated by the plant [17]. Such factors are attributed to the environment such as climate [18] and altitude [19] of the growing region, agronomic factors such as farm management, harvesting practices and fertilizer applications and processing procedures [2]. Agronomic factors are important but considered controllable with previous studies by [20] [21] suggesting the adoption of different optimized management techniques such as fertilizer rates used, post harvesting techniques and plucking intervals that suit the different growing regions. Environmental factors have been shown to affect the levels of biomolecules synthesized and consequently the quality of the final tea product by influencing the rates of shoot growth [21] - [23]. Altitude differences affect the rates of shoot growth in tea plants, with a high altitude slowing the rate consequently increasing synthesis of biomolecules while low altitudes increase the rate of shoot growth causing a decline in synthesis of the biomolecules. Mean temperature and rainfall intensities between growing regions play an equally important role in influencing tea quality with high intensities stimulating a fast shoot growth compared to low intensities which stimulate a slower rate leading to accumulation of large amounts of the biomolecules [24] [25]. However, the extent of variations in synthesized biomolecules is depended on the clonal genotype as individual clones are unique on their levels of synthesized biomolecules. Therefore some clones are more stable and therefore less susceptible to changes in composition of biomolecules due to differences in environmental conditions [2] [22]. Despite these observations, research work detailing the extent of influence of the underlying environmental conditions on the clones grown in both the Timbilil and Kangaita regions is not well described. As the tea industry seeks to produce high quality and diversified products to expand its market outlets and satisfy different consumer needs, the need to study TRI's two main tea growing stations is highly necessary. In this study, levels of catechins and caffeine in clones grown in both sites were carried out to establish how quality parameters are influenced.

2. METHODOLOGY

2.1. Description of Study Sites

The study sites included Timbilil Estate (latitude 0°26'S, longitude 37°15'E, altitude 2020 m amsl) in the West of the Great Rift Valley and Kangaita substation (latitude 0°30'S, longitude 37°16'E, altitude 2180 m amsl) in the East of the Great Rift Valley. Kangaita experiences a weakly bimodal rainfall distribution of 2040 mm annually with peaks in April/May and October/November, while Timbilil has rains most of the year, except in the months of December, January and February when there is drought with an annual rainfall of 2175 mm. Kangaita has a mean annual temperature of 15.5°C while Timbilil has a mean annual temperature of 16.6°C.

2.2. Tea Samples

About 300 g of fresh green tea leaf samples (consisting of the two leaves and a bud) were obtained in triplicates from each of 60 tea cultivars grown in both study sites; Timbilil Estate and Kangaita substation of the TRI. These clones were selected from pre existing ongoing experimental trials in the two sites.

2.3. Sample Processing

The samples were first steamed for one to two minutes then dried in a micro-wave oven. They were then pulverized with a grinder into fine powder used in assaying for catechins and caffeine. The analysis was done using the Tea Research Institute established and published wet chemistry procedures High Performance Liquid Chromatography (HPLC) and spectrophotometric protocols.

2.4. Determination of Catechin Levels and Profiles

Approximately 0.2 ± 0.001 g of each of the sample was weighed into an extraction tube. Five milliliter of 70% methanol at 70°C was dispensed into the sample as an extraction mixture and vortexed until properly mixed. Heating of the extraction tube was continued in the water bath at 70°C for 10 min with mixing in the vortex mixer (VM-1000, Taiwan) after every 5 min. The extraction tubes were then removed from the water bath (Digital water bath LWB-122D, Korea) and allowed to cool at room temperature. The tubes were then placed in a centrifuge (HSCEN-204, MRC) at 3500 rpm for 10 minutes and the supernatant decanted into a graduated tube and the extraction procedure repeated. Both extracts were combined and the volume made up to 10 ml with cold 70% methanol. One milliliter of the sample extract was transferred into a graduated tube and diluted to 5 ml with a stabilizing solution (10% v/v acetonitrile with 500 µg/ml Ethylenediaminetetraacetic acid (EDTA) and ascorbic acid). The solution was filtered through a 0.45 µm nylon membrane filter and a 20 µl aliquot of this solution was injected into HPLC for analysis.

A high performance liquid chromatography method [26] was used to assay for the tea catechins. A Shimadzu LC 20 AT HPLC system fitted with a SIL 20A auto sampler and a SPD-20 UV visible detector with a class LC 10 chromatography workstation was used for analysis of the prepared samples. A Gemini 5 µM C6-Phenyl, 250 mm × 4.6 mm (Phenomenex, Torrance, CA, USA) separation column with a Reodyne precolumn filter disk was used. The sample was degassed before injection into the HPLC system. A gradient elution was carried out using the following solvent systems: Mobile phase A (acetonitrile/acetic acid/double distilled water-9/2/89 v/v/v) and mobile phase B (acetonitrile/acetic acid/double distilled water-80/2/18 v/v/v). The mobile phase composition for a binary gradient condition was started at 100% mobile phase A for 10 minutes then over 15 minutes a linear gradient to 60% mobile phase A, 32% mobile phase B and held at this composition for 7 minutes. The condition was reset to 100% mobile phase A and then allowed to equilibrate for 10 minutes before the next injection. The flow rate of the mobile phase was kept at 1 ml/min and the temperature in the column maintained at 35°C ± 0.5°C throughout the analyses. Identification of individual catechins was carried out by comparing the retention times and UV-absorbance of unknown peaks with peaks obtained from the mixed known catechin standards under the same conditions. The quantification of catechins was performed at 278 nm and was achieved using a caffeine standard with a calibration curve $R^2 = 0.9984$ in conjunction with the consensus individual catechin Relative Response Factor (RRF) values with respect to caffeine calculated on a dry matter basis. The total catechin content as a percentage by mass on a sample dry matter basis was given on the summation of individual catechin contents.

$$\% \text{Total catechins} = [\%ECG + \%EC + \%EGCG + \%C]$$

Since caffeine is used as a standard in identification of catechins, its levels were quantified alongside those of individual catechins by comparing the samples' retention times with that of the caffeine standard.

2.5. Data Analysis

All determinations for the levels of caffeine, total and individual catechin fractions, were done in triplicates and the data subjected to analysis of variance using GENSTAT-C statistical software packages. The least significant differences Test (LSD) was used to separate the means.

3. RESULTS AND DISCUSSION

Results shown in Table 1 revealed significant differences in the levels of catechins and caffeine between Kangaita and Timbilil sites.

3.1. Total Catechins

Kangaita clones had significantly higher ($P < 0.05$) total catechin content with a mean value of 18.7% compared to 16.2% observed for Timbilil clones. However, 10 clones namely; BBK 35, TRFK K-Purple, TRFK 301/5, TRFK 430/90, TRFK 91/1, AHP SC 31/37, TRIT 201/16, TRIT 201/44, TRIT 201/47 and TRIT 201/82 from Timbilil performed better than Kangaita clones.

3.2. Individual Catechins Fractions

3.2.1. Epigallocatechin Gallate

Clones from Kangaita region had significantly high EGCG values with a mean value of 7.9% compared to 6.7% observed in the clones from Timbilil.

However, BBK 35, TRFK K-Purple, TRFK 400/10, TRFK 430/90, AHP SC 31/37, TRIT 201/16, TRIT 201/44, TRIT 201/47, TRIT 201/50, TRIT 201/55, TRIT 201/70, TRIT 201/73, TRIT 201/75 and TRIT 201/82 from Timbilil had higher EGCG levels.

3.2.2. Epicatechin Gallate

Results obtained showed that Kangaita clones had significantly high ECG levels (mean value of 3.3%) than Timbilil clones (mean value of 2.5%) with exceptions of the clones BBK 35, AHP SC 31/37, TRIT 201/16, TRIT 201/44, TRIT 201/50, TRIT 201/73 and TRIT 201/82.

3.2.3. Epigallocatechin

Kangaita clones had a significantly high mean value of EGC (5.2%) compared to Timbilil clones (5%). However clones EPK H 81/22, TRFK 31/8, TRFK 12/12, KTDA KAG 4, AHP F7/346, MICHI (51/10/20), TRFCA SFS 150, STC M3, BBK 5, TRFK 303/577+, GW EJULU-L, BBK 35, BBK 21, AHP SC 12/28, TRFK K PURPLE, TRFK 301/1, TRFK 301/3, TRFK 301/4, TRFK 301/5, TRFK 301/6, TRFK ST 543, TRFK 306/1, TRFK 306/3, TRFK 91/1, AHP SC 31/37, TRIT 201/47 and TRIT 201/82 from the Timbilil site performed better.

3.2.4. Epicatechin

Significant difference was observed for EC levels with Kangaita clones having a high mean value of 1.5% compared to 1.4% in the Timbilil clones. However, clones KTDA KAG 4, AHP F7/346, MICHI (51/10/20), TRFCA SFS 150, STC M3, BBK 5, TRFK 303/577+, BBK 35, TRFK 301/1, TRFK 301/3, TRFK 301/4, TRFK 301/5, TRFK 301/6, TRFK 306/1 and TRIT 201/47 in Timbilil performed better.

3.2.5. Catechin (+C)

Results for the variations in simple catechins between the two regions among the studied clones showed significantly high levels in Kangaita (mean value 0.7%) compared to those in Timbilil which had a mean value of 0.6% with clones KTDA KAG 4, STC M3, TRFK 301/3, TRFK 306/1, TRFK 306/4, TRFK 91/1, AHP SC 31/37 and AHP CG28V929 performing better in Timbilil site.

3.2.6. Caffeine

Caffeine analysis results for the two regions showed an interesting pattern, different from that observed for individual catechins. A significant difference was observed with Timbilil clones containing high caffeine levels (mean value of 4.2%) compared to Kangaita clones (mean value of 3.8%) as depicted in the Table 1. However, clones EPK H 81/22, EPK C182/40, AHP F5/222, AHP F7/346, MICHI (51/10/20), AHP MNI 1/96, TRFCA SFS 150, STC M3, BBK 5, TRFK 303/577+, TRFK 54/40, BBK 21, AHP SC 12/28, TRFK-K PURPLE, TRFK 301/6, TRFK 400/1, AHP PC 81, AHP SF 186, AHP CG28V929 and TRIT 201/43 in Kangaita performed better.

There were significant interactions ($p \leq 0.05$) between the growing region and the clones, an indication that synthesized catechins and caffeine differ in regions. This is because while clonal performance is primarily genetically linked, the growing region and the underlying conditions have some level of influence on the synthesis of the biomolecules as evidenced in this study where the studied clones performed differently in the two study sites. This implies that the two factors, region and clone should be taken into consideration during tea production to ensure that high quality of the final product is achieved. From the total catechins results, several clones showed wide variations between the two regions such as EPK C12 with 22.2% in Kangaita and 16.5% in Timbilil, TRFK 100/5 with 23.9% in Kangaita and 16.7% in Timbilil. The clones in Kangaita therefore provide the suitable raw materials for production of diversified catechins-rich products such as green teas and catechin extracts. Apart from being the most biologically active biomolecules in green tea, catechins are also important precursors of black tea theaflavins and thearubigins and are therefore important quality markers. These observations of wide variations in catechins between the two regions imply that clones must be thoroughly evaluated for their performance in the intended cultivation area to test their suitability before their release. However, in clones such as TRIT 201/55, the variation in the two regions was not statistically different with 20.0% in Kangaita and 19.8% in Timbilil, an observation consistent with that of a study by Wachira et al., (2002) suggesting that irrespective of the region of production, some tea genotypes exhibit resistance to yield variations and thus quality. This observation implies that such clones with least variation between the two growing regions are more stable and thus will have little variations when grown in other tea growing regions of different climatic conditions such as Sotik, Maragwa and Embu. Utilization of such clones with adaptations to varied environmental conditions will ensure increased and consistent production of tea with uniform biomolecule composition and therefore produce teas of uniform quality. This observation is similar to a recent study [27] which found significant differences in chemical quality although based on the micronutrient level between teas from East and West of the Great Rift Valley. It is clear from the results that most of the clones in Kangaita perform better compared to those in Kericho based on their total and individual catechins. From these results, the influence of regional differences and their underlying climatic conditions on the differential synthesis and accumulation of these bioactive compounds is responsible for quality in tea. This is in agreement with previous studies [18] [21] - [23] [28] whereby factors such as seasonal climate, altitude, temperature and total rainfall density have been shown to affect the tea's yield, overall growth rate and synthesis and accumulation of the chemical compounds responsible for the quality characteristics.

Altitude has an influence on the growth of the tea plant whereby shoots of teas grown in high altitude regions grow and mature slowly and in the process accumulate high amounts of these polyphenolic catechins (quality markers) and develop a richer flavor [18] [29] [30]. This agreed with the present study in caffeine composition. However, this study differs with such observations on catechins levels in that Kangaita has a relatively lower altitude (2020 m) than Kericho (2180 m) yet the tea clones performed better implying that other factors could be contributing to the observed differences. Temperature could have contributed to this observation since Kangaita is located near Mt. Kenya which is relatively cooler throughout the year with a mean annual temperature of 15.5°C compared to 16.6°C in Timbilil. This environment induces a form of dormancy in the rate of shoot growth consequently increasing the accumulation of catechins in these clones. High temperatures on the other hand stimulate a fast shoot growth but with low levels of these biomolecules [18]. Similar results were observed in previous studies [31] which reported that elevated temperature reduced flavonoid accumulation and inhibited the gene expression of the enzymes in phenylpropanoid and flavonoid biosynthetic pathway; chalcone synthase (CHS), flavanone 3-hydroxylase (F3H) and dihydroflavonol 4-reductase (DFR) in grape berries.

Table 1. Differences in the contents (%) of catechins and caffeine between Kangaita (K) and Timbilil (T) study sites.

No.	Clone	EGC		C		CAFFEINE		EC		EGCG		ECG		TC	
		K	T	K	T	K	T	K	T	K	T	K	T	K	T
1	EPK TN 14-3	6.94	6.69	0.53	0.49	4.02	4.44	1.77	1.54	9.30	7.58	3.12	2.30	21.64	18.60
2	EPK H 81/22	6.99	7.29	0.60	0.44	4.02	4.20	2.08	2.01	7.66	6.41	2.93	1.85	20.26	17.36
3	EPK C182/40	7.83	6.83	0.62	0.57	4.00	3.90	1.38	1.38	8.00	5.65	2.35	1.64	20.85	16.08
4	EPK C12	6.64	5.62	0.84	0.55	4.24	4.11	2.03	1.41	8.90	5.63	3.78	2.01	22.19	16.51
5	TRFK 31/8	5.17	6.01	0.56	0.35	4.92	4.92	1.08	0.96	10.55	7.47	2.71	1.63	20.08	16.41
6	TRFK 12/12	4.47	5.68	1.04	0.92	5.18	5.57	1.49	1.33	9.52	6.84	4.14	2.74	20.66	17.51
7	TRFK 6/8	6.47	6.38	0.93	0.56	3.61	4.21	1.84	1.36	8.40	6.49	2.71	1.71	20.36	16.50
8	AHP S15/10	4.53	4.30	0.50	0.45	2.81	3.82	0.96	0.93	6.62	6.43	2.00	1.90	17.97	15.88
9	TRFK 108/82	7.30	6.30	0.72	0.55	4.50	4.69	1.66	1.32	8.44	6.47	2.80	1.92	20.93	16.57
10	TRFK 100/5	7.78	5.96	0.82	0.61	4.40	4.87	1.69	1.22	10.46	6.95	3.17	1.96	23.93	16.69
11	KTDA KAG 4	6.23	6.77	0.51	0.56	4.18	4.31	1.68	2.24	10.66	6.59	4.55	2.90	23.63	19.05
12	AHP F5/222	5.28	4.49	0.60	0.36	4.76	4.53	1.92	1.80	9.05	6.88	4.00	2.83	20.84	15.61
13	AHP F7/346	6.85	6.95	0.54	0.43	4.72	4.16	1.89	2.14	9.31	7.57	4.65	3.03	23.26	20.12
14	MICHI (51/10/20)	6.56	6.78	0.68	0.60	4.23	4.14	1.34	1.37	8.64	5.50	2.58	1.61	19.79	15.85

15	AHP S1/99	7.12	5.77	0.66	0.47	4.63	4.92	1.40	1.13	8.62	6.75	2.72	1.87	20.52	15.99
16	AHP MNI 1/96	6.17	5.10	0.75	0.48	5.39	4.93	1.52	1.21	10.43	7.63	3.94	2.49	22.81	16.91
17	TRFCA SFS 150	5.89	6.71	0.50	0.31	4.74	3.96	1.25	1.54	10.11	5.51	2.76	1.55	20.51	15.62
18	STC M3	3.86	4.28	0.53	0.59	4.36	4.29	0.82	0.89	9.58	6.09	2.78	1.82	17.57	13.68
19	STC M1	5.07	5.05	0.45	0.33	4.50	4.52	1.03	0.96	9.01	6.62	2.48	1.65	18.05	14.61
20	STC 5/3	6.29	4.02	0.67	0.36	3.49	3.89	1.41	0.95	9.16	7.03	2.55	2.17	20.09	14.52
21	BBK 5	4.39	5.14	0.68	0.53	4.84	3.95	0.91	0.91	9.91	7.14	2.83	1.89	18.71	17.16
22	TRFK 303/577+	5.99	6.67	0.65	0.40	3.54	2.46	1.89	1.91	8.59	5.47	2.98	1.77	20.10	16.22
23	GW EJULU-L	1.63	2.49	1.73	1.50	3.79	4.23	1.18	1.15	6.49	5.81	6.96	4.79	19.96	15.74
24	TRFK 54/40	6.41	5.66	0.57	0.38	4.37	4.35	1.22	0.99	9.78	9.44	2.89	2.61	21.56	19.07
25	BBK 35	5.48	6.75	0.68	0.78	3.33	5.60	1.19	1.35	5.68	9.41	2.20	3.07	16.41	20.78
26	BBK 21	3.76	4.31	0.70	0.61	4.33	3.73	1.34	1.35	9.34	6.14	3.98	2.21	19.12	14.34
27	AHP SC 12/28	5.27	5.91	1.36	0.81	3.99	4.10	2.21	1.67	7.33	6.76	3.30	3.00	19.68	18.14
28	TRFK K PURPLE	1.18	1.31	0.97	0.89	2.25	3.58	1.04	1.04	2.91	3.79	2.87	2.19	8.96	9.21
29	TRFK 301/1	3.51	4.10	0.95	0.82	3.79	5.51	2.09	2.72	8.64	4.50	5.08	3.16	20.27	15.30

30	TRFK 301/2	4.88	4.54	1.19	0.79	4.52	5.16	2.26	1.50	8.43	5.67	3.50	2.19	20.26	14.70
31	TRFK 301/3	2.70	3.92	0.78	0.80	3.99	5.16	1.28	1.70	9.54	6.37	4.30	3.24	18.60	16.03
32	TRFK 301/4	2.03	3.44	0.44	0.31	3.78	3.24	1.66	2.34	7.67	3.36	5.87	2.97	17.68	14.90
33	TRFK 301/5	4.77	7.00	0.64	0.51	4.38	4.84	2.79	3.75	7.17	5.78	4.84	3.47	20.22	20.52
34	TRFK 301/6	2.60	3.62	0.83	0.48	3.61	3.59	3.49	4.44	6.14	4.25	6.39	4.80	19.46	17.59
35	TRFK 400/10	5.08	3.80	0.55	0.42	3.33	5.28	1.22	0.93	7.21	9.04	2.21	2.34	18.09	16.68
36	TRFK 371/8	5.06	4.88	0.85	0.56	3.64	5.40	1.46	1.34	7.47	6.91	2.73	2.25	17.57	15.95

37	TRFK 371/3	7.69	6.58	0.94	0.60	4.68	6.44	1.93	1.33	8.64	8.49	3.12	2.49	22.34	19.50
38	TRFK 430/90	7.61	7.23	0.70	0.55	3.80	4.34	1.97	1.53	7.42	8.97	2.29	2.52	20.00	20.79
39	TRFK ST 536	4.34	3.97	0.41	0.26	2.72	3.32	1.24	1.16	6.44	4.94	1.96	1.31	14.38	11.65
40	TRFK ST 543	4.73	4.83	0.56	0.29	2.90	3.26	1.60	1.28	7.06	5.18	2.04	1.40	15.99	12.97
41	TRFK 306/1	1.68	2.05	0.80	0.83	2.04	2.67	0.92	1.10	4.39	4.09	4.55	2.68	11.04	10.76
42	TRFK 306/2	2.16	1.84	0.86	0.77	2.17	2.53	1.01	0.99	4.60	4.02	4.97	2.93	14.62	10.54
43	TRFK 306/3	2.11	2.12	0.91	0.88	2.24	2.38	1.11	1.14	4.76	4.45	5.18	4.65	14.06	13.24
44	TRFK 306/4	2.05	2.03	0.80	0.89	2.05	2.42	0.94	1.06	4.43	4.23	5.00	2.98	14.40	11.20
45	TRFK 91/1	1.25	1.94	0.58	0.73	1.06	3.13	0.53	0.68	1.47	2.82	3.20	4.90	7.03	9.62
46	AHP SC 31/37	5.48	5.97	0.59	0.77	4.31	4.77	1.15	1.42	8.28	9.75	2.52	3.13	18.03	21.04
47	TRFK 400/1e	6.53	4.13	1.06	0.43	4.84	4.13	1.88	0.92	9.33	8.29	3.97	2.15	22.78	15.93
48	AHP PC 81	6.20	4.82	0.60	0.61	3.48	4.04	1.23	1.39	7.34	6.75	2.36	2.15	20.46	15.72
49	AHP SF 186	4.70	4.52	0.42	0.37	3.97	4.33	1.04	1.05	8.27	7.35	2.36	1.73	18.24	15.02
50	AHP CG28V929	4.98	3.07	0.45	0.75	3.72	3.56	1.13	0.85	7.29	5.65	2.24	2.08	16.93	12.39
51	TRIT 201/16	5.39	4.84	0.60	0.59	2.38	3.23	1.52	1.39	4.52	6.90	1.66	2.31	15.53	16.02
52	TRIT 201/43	7.16	6.06	0.80	0.60	4.06	3.90	1.63	1.23	7.04	7.36	1.88	1.87	18.50	17.12
53	TRIT 201/44	5.91	5.84	0.84	0.75	3.13	4.02	1.53	1.47	5.28	7.92	1.74	2.34	15.31	18.32

54	TRIT 201/47	3.55	4.87	0.52	0.52	3.46	3.99	1.05	1.08	8.37	10.73	2.66	2.60	16.16	19.81
55	TRIT 201/50	7.27	5.71	0.86	0.68	3.53	4.04	1.66	1.24	7.01	7.45	1.98	2.08	18.78	17.17
56	TRIT 201/55	5.60	5.34	0.79	0.64	3.74	4.33	1.90	1.33	8.47	9.35	3.21	3.08	19.97	19.75
57	TRIT 201/70	6.70	4.95	0.55	0.53	3.33	3.74	1.51	1.28	7.92	9.39	2.50	2.61	19.83	18.76
58	TRIT 201/73	5.87	4.64	0.63	0.61	2.61	4.23	1.07	1.01	5.50	7.81	1.45	1.90	16.86	15.97
59	TRIT 201/75	6.63	3.65	0.65	0.46	3.87	4.21	1.65	0.98	9.70	9.89	2.98	2.70	21.60	17.68
60	TRIT 201/82	6.13	6.15	0.68	0.54	2.81	3.35	1.39	1.09	5.39	7.12	1.40	1.64	15.00	16.54
Grand mean		5.16	4.96	0.72	0.59	3.75	4.15	1.50	1.41	7.73	6.68	3.21	2.46	18.67	16.16
CV (%)		10.2		10.7		5.2		5.7		5.1		8.6		5.0	
LSD (5% Level)	Region	0.11		0.08		0.04		0.02		0.08		0.05		0.18	
	Clone	0.59		0.01		0.24		0.10		0.43		0.28		0.99	

EGC-Epigallocatechin, C-Catechin, CAFF-Caffeine, EC-Epicatechin, EGC-Epigallocatechin gallate, ECG-Epicatechin gallate, TC-Total catechins.

Differences in the amounts of rainfall received in the two regions could have played a role in the variation of catechin composition. Kangaita has a mean annual rainfall of 2040 mm with peaks in the months of April/May and October/November while Timbilil receives 2175 mm of rainfall spread through the year except in the dry season (December-February). A high precipitation like in Timbilil stimulates a fast shoot growth rate resulting in high yields but producing low quality teas compared to the relatively low precipitation rate in Kangaita, an observation supporting an earlier study [25]. It is worth noting that whereas a slow rate of shoot growth contributes to a high synthesis and accumulation of catechins, the mean productivity decreases owing to the fact that development of pluckable shoots takes much longer thus reducing plucking intensity [24] [29] [32]. It's important to note that clonal genetic differences also affects their individual ability to absorb nutrients even under similar agronomic practices and this consequently affects their ability to synthesize and accumulate these bioactive molecules such as catechins [33] [34]. This explains why some clones in Kangaita had relatively low total and individual catechin content compared to the same clones grown in Timbilil even when the conditions in Kangaita seemingly impacted positively on the catechin levels in majority of the clones.

Caffeine analysis showed that growing conditions in Timbilil favored the synthesis and accumulation of caffeine more than Kangaita. A typical example in the variation was observed in clones such as BBK 35 with caffeine levels of 3.3% in Kangaita and 5.6% in Timbilil and clone TRFK 400/10 with 3.3% in Kangaita and 5.28% in Timbilil. Like catechins, synthesis and accumulation of caffeine has been shown to be affected by climatic factors [35]. Previous studies [36] [37] have revealed that a high relative humidity stimulates caffeine synthesis in tea plants, observations supported by this study. Timbilil has a relatively high annual rainfall (2175 mm) compared to 2040 mm in Kangaita. Consequently, the high mean temperature in Timbilil (16.6°C) seems to increase synthesis of caffeine in the tea clones compared to those in Kangaita with relatively low mean temperature (15.5°C). This can be explained by the fact that caffeine synthesis by plants is usually in response to adverse climatic conditions such as the relatively high temperature in Timbilil. The high caffeine levels in clones grown in Timbilil could also be attributed to the high altitude of 2180 m compared to the relatively low altitude of 2020 m in Kangaita. This observation on the influence of altitude on caffeine synthesis concurs with a study by Owuor et al., (1990a) in which they found that a high altitude significantly influenced caffeine synthesis in tea plants by slowing down the rates of shoot growth consequently increasing their accumulation.

The differences in the climatic conditions (rainfall intensity and mean annual temperature) between the two regions are not so wide but significant differences in the levels of catechins and caffeine were observed. This observation corroborates a previous study [2] and refutes earlier assumptions stating that larger climatic differences were necessary for significant quality differences to be observed and that a superior clone selected in one location maintains its desirable attributes within the country.

4. CONCLUSION

The obtained results imply that a high performing clone in one region should be checked for its stability in new regions to ascertain if it maintains relatively high levels of the desired biomolecules. This is desirable since it could lead to development of region-specific clones. Additionally, Kangaita region, favoring synthesis of relatively high quantities of most biomolecules should be considered for growing teas for manufacture of high quality teas or for extraction of biomolecules for use in pharmaceutical and cosmetic industries.

ACKNOWLEDGEMENTS

This study was supported by the Tea Research Institute (TRI) of Kenya and published with permission from the Director.

REFERENCES

1. Rolfe, J. and Yvonne, C. (2003) Camellias: A Practical Gardening Guide. Timber Press, New York.

2. Wachira, F.N., Ng'etich, W., Omolo, J. and Mamati, G. (2002) Genotype x Environment Interactions for Tea Yields. Euphtica, 127, 289-296.

3. Lu, M.J., Chu, S.C., Yan, L. and Chen, C. (2009) Effect of Tannase Treatment on Proteintannin Aggregation and Sensory Attributes of Green Tea Infusion. LWT-Food Science and Technology, 42, 338-342.

4. Cabrera, C., Gimenez, R. and Lopez, C.M. (2003) Determination of Tea Components with Antioxidant Activity. Journal of Agriculture and Food Chemistry, 51, 4427-4435.

5. Amie, D., Amie, D.D., Beslo, D. and Trinajstie, N. (2003) Structure-Radical Scavenging Activity Relationships of Flavonoids. Croatia Chemica Acta, 76, 55-61.

6. Hara, Y. (2006) Prophylactic Functions of Tea Catechins. In: Jain, W.K., Siddiqi, M.A. and Weisburger, J.H., Eds., Protective Effects of Tea on Human Health, CABI Publishing, Wallingford, 16-24.

7. Frei, B. and Higdon, J.V. (2003) Antioxidant Activity of Tea Polyphenols in Vivo: Evidence from Animal Studies. Journal of Nutrition, 133, 3275-3284.

8. Boschmann, M. and Thielecke, F. (2007) The Effects of Epigallocatechin-3-Gallate on Thermogenesis and Fat Oxidation in Obese Men: A Pilot Study. Journal of the American College of Nutrition, 26, 389-395.

9. Yang, T.T. and Koo, M.W. (2000) Chinese Green Tea Lowers Cholesterollevel through an Increase in Fecal Lipid Excretion. Life Sciences, 66, 411-423.

10. Chyu, K.Y., Babbidge, S.M., Zhao, X., Dandillaya, R., Rietveld, A.G., Yano, J., Dimayuga, P., Cercek, B. and Shah, P. K. (2004) Differential Effects of Green Tea-Derived Catechin on Developing versus Established Atherosclerosis in Apolipoprotein E-Null Mice. Circulation, 109, 2448-2453.

11. Karori, S., Wachira, F., Wanyoko, J. and Ngure, R. (2008) Different Types of Tea Products Attenuate Inflammation Induced in Trypanosome Brucei Infected Mice. Parasitology International, 57, 325-333.

12. Khan, N. and Mukhtar, H. (2007) Tea Polyphenols for Health Promotion. Life Sciences, 81, 519-533.

13. Davis, J.M., Zhao, Z., Stock, H.S., Mehl, K.A., Buggy, J. and Hand, G.A. (2003) Central Nervous System Effects of Caffeine and Adenosine on Fatigue. American Journal of Physiology-Regulatory, Integrative and Comparative Physiology, 284, R399-R404.

14. Bonnet, B.H. and Arand, D.L. (1994) The Use of Prophylactic Naps and Caffeine to Maintain Performance during a Continuous Operation. Ergonomics, 37, 1009-1020.

15. Ruijter, J., De Ruiter, M.B. and Snel, J. (2000) The Effects of Caffeine on Visual Selective Attention to Color: An ERP Study. Psychophysiology, 37, 427-439.

16. Mumford, G.K., Evans, S.M., Kaminski, B.J., Preston, K.L., Sannerud, C.A. and Griffiths, R.R. (1994) Discriminant Stimulus and Subjective Effects of Theobromine and Caffeine in Humans. Psychopharmacology, 115, 1-8.

17. Kamunya, S.M., Wachira, F.N., Owuor, P.O., Pathak, R.S., Wanyoko, J.K., Sharma, R.K. and Muoki, R.C. (2010) Quantitative Genetic Parameters for Yield, Drought Tolerance and some Quality Traits in Tea (Camellia sinensis (L.) O. Kuntze). Agricultural Science Research Journal, 1, 53-65.

18. Wei, K., Wang, L.Y., Zhou, J., He, W., Zeng, J.M., Jiang, Y.W. and Cheng, H. (2011) Catechin Contents in Tea (Camellia sinensis) as Affected by Cultivar and Environment and Their Relation to Chlorophyll Contents. Food Chemistry, 125, 44-48.

19. Chen, W.-Q., Zhao. X.-L., Wang, D.-L., Li, S.-T., Hou, Y., Hong, Y. and Cheng, Y.-Y. (2010) Effects of Epigallocatechin-3-Gallate on Behavioral Impairments Induced by Psychological Stress in rats. Experimental Biology and Medicine, 235, 577-583.

20. Owuor, P.O., Kamau, D.M. and Jondiko, E.O. (2009) Responses of Clonal Tea to Location of Production and Plucking Intervals. Food Chemistry, 115, 290-296.

21. Owuor, P.O., Kamau, D.M. and Jondiko, E.O. (2010) The Influence of Geographical Area of Production and Nitrogenous Fertilizer on Yields and Quality Parameters of Clonal Tea. Journal of Food, Agriculture and Environment, 8, 682-690.

22. Owuor, P.O., Wachira, F.N. and Ng'etich, W.K. (2010) Influence of Region of Production on Relative Clonal Tea Quality Parameters in Kenya. Food Chemistry, 119, 1168-1174.

23. Owuor, P.O., Obanda, M., Nyirenda, H.E. and Mandala, W.L. (2008) Influence of Region of Production on Clonal Black Tea Chemical Characteristics. Food Chemistry, 108, 363-271.

24. Anandacoomaraswamy, A., De Costa, W.A., Shyamalie, H.W. and Campbell, G.S. (2000) Factors Controlling Transpiration of Mature Field-Grown Tea and Its Relationship with Yield. Agriculture and Forest Meteorology, 103, 375-386.

25. Odhiambo, H.O., Owuor, P.O., Othieno, C.O. (1988) Factors Affecting Tea Quality: I. Agronomic Practices: A Review. Tea, 9, 36-40.

26. Kilel, E.C., Faraj, A.K., Wanyoko, J.K., Wachira, F.N. and Mwingirwa, V. (2013) Green Tea from Purple Leaf Coloured Tea Clones in Kenya—Their Quality Characteristics. Food Chemistry, 141, 769-775.

27. Omwoyo, W.N., Owuor, P.O., Ongeri, D.M. and Kamau, D.M. (2014) Effect of Genotypes in Different Environments on Micronutrient Content of Black Tea. Journal of Tea Science, 4, 2-6.

28. Owuor, P.O., Obaga, S.O. and Othieno, C.O. (1990) The Effects of Altitude on the Chemical Composition of Black Tea. Journal of Science of Food and Agriculture, 50, 9-17.

29. Squire, G.R., Obaga, S.M. and Othieno, C.O. (1993) Altitude, Temperature and Shoot Production of Tea in Kenyan Highlands. Experimental Agriculture, 29, 107-120.

30. Pruess, J. and Harney, J. (2006) Tea Cuisine: A New Approach to Flavoring Contemporary and Traditional Dishes. Globe Pequot Press, Guilford.

31. Mori, K., Sugaya, S. and Gemma, H. (2005) Decreased Anthocyanin Biosynthesis in Grape Berries Grown under Elevated Night Temperature Condition. Scientia Horticulturae, 105, 319-330.

32. Obaga, S.O. and Ng'etich, W.K. (1989) Thermal-Time and Growth of Shoots of Tea Clones. Tea, 10, 80-89.

33. Yemane, M., Chandravanshi, B.S. and Wondimu, T. (2008) Levels of Essential and Non-Essential Metals in Leaves of the Tea Plant (Camellia sinensis L.) and Soil of the Wushwush Farms, Ethiopia. Food Chemistry, 107, 1236-1243.

34. Wanyoko, J.K. and Njuguna, C.K. (1983) Variation of Foliar Nutrients of Some Kenyan Tea and Their Relationship with Yield. Tea, 4, 37-43.

35. Suzuki, T., Ashihara, H. and Waller, G.R. (1992) Purine and Purine Alkaloid Metabolism in Camellia and Coffea Plants. Phytochemistry, 31, 2575-2584.

36. Wang, L.Y., Wei, K., Jiang, Y.W., Cheng, H., Zhou, J., He, W. and Zhang, C.C. (2011) Seasonal Climate Effects on Flavanols and Purine Alkaloids of Tea (Camellia sinensis L.). European Food Research and Technology, 233, 1049-1055.

37. Lee, J.E., Lee, B.J., Chung, J.O., Hwang, J.A., Lee, S.J., Lee, C.H. and Hong, Y.S. (2010) Geographical and Climatic Dependencies of Green Tea (Camellia sinensis) Metabolites: A1H NMR-Based Metabolomics Study. Journal of Agricultural and Food Chemistry, 58, 10582-10589.

Chapter 2

Effects of Nitrogen Addition on the Mixed Litter Decomposition in Stipa baicalensis Steppe in Inner Mongolia

Wenya Li[1,2,3], Wenchao Yu[1,2,3], Long Bai[1], Hongmei Liu[2,3], Dianlin Yang[1,2,3*]

[1]Department of Horticulture, Shenyang Agricultural University, Shenyang, China

[2]Agro-Environmental Protection Institute, Ministry of Agriculture, Tianjin, China

[3]Key Laboratory of Original Agro-Environment Quality of Ministry of Agriculture and Tianjin Key Laboratory of Agro-Environment and Safe-Product, Tianjin, China

ABSTRACT

During the past two centuries, global changes (i.e., enhanced nitrogen deposition) have exerted profound effects on ecological processes of steppe ecosystems. We used litterbag method and mixed litters of three different plant species tissues (Stipa baicalensis: Sb, Leymus chinensis: Lc and Artemisia frigid: Af), endemic to Stipa baicalensis Steppe, and measured the mass loss of mixtures over 417 days under the N addition treatment. We studied the effect of N addition (N0: no N addition; N15: 1.5 g $N/m^2 \cdot a$; N30: 3.0 g $N/m^2 \cdot a$; N50: 5.0 g $N/m^2 \cdot a$; N100: 10.0 g $N/m^2 \cdot a$; N150: 15.0 g $N/m^2 \cdot a$) on the rate of mixed litter decomposition and nutrient dynamics change. The decomposition constant (k) of leaf mixtures was higher than that of root mixtures. The k values of leaf mixed combinations were 0.880 (Sb + Lc), 1.231 (Lc + Af), 1.027 (Sb + Lc + Af), respectively. The k value of stem was 0.806 (Lc + Af) and the root mixed combinations were 0.665 (Sb + Lc), 0.979 (Lc + Af) and 1.164 (Sb + Lc + Af), respectively. The results indicated that N addition had significantly effect on the mixed litter decomposition and nutrient releasing. The rate of plant tissues litter decomposition had different response to N addition. In the context of N addition, litter decomposition rate and nutrient dynamics were changed by synthetic effect of decaying time, specie types and N addition dose. Our findings suggested that prairie plants may adapt to environmental change by adjusting litter quality, thus retaining the stability of the steppe ecosystem.

KEYWORDS

Grassland, Litter Decomposition, Mixed Effect, N Addition, Mass Loss, Nutrient Dynamics

1. INTRODUCTION

As an important phenomenon of global change, atmospheric nitrogen deposition leads to a series of ecological problems, and affects elements circulation on broad spatial scales potentially [1] [2]. Natural grassland, as the largest green vegetable layers of terrene, is easily affected by the global change, meanwhile, it has important functions in maintenance of global carbon, nitrogen balance and water cycling [3] [4]. Plant litters are produced by the vegetable metabolism, and litter decomposition is a key process of nutrients cycling and energy flowing in the steppe ecosystem and plays a critical role in plant productivity and ecosystem carbon storage [5] [6]. Soil nutrient has long been considered as predominant factor which affects the dynamic change process of litter decomposition, therefore nitrogen deposition can affect the rate of litter decomposition directly and indirectly, and eventually influence nutrient cycling and productivity of steppe ecosystem [7] [8].

Litter decomposition keeps material transformation and nutrient cycling operation expedite [9], has irreplaceable impact on carbon and nitrogen circulation in ecosystems [10], and has caused an extensive concern of ecologists, especially in steppe ecosystem. The process of litter decomposition contains leaching, mechanical disruption, transformation of organic material, digestion of soil animals and enzymatic action of soil microbies. It is controlled by many factors including climate, the microenvironment around the litter, decomposer activity and litter quality [11] - [13]. Litter quality parameters, especially initial N content, have correlation with the rate of litter mass loss. Aerts [14] manifested that initial chemistry differences among litters, especially nitrogen and lignin concentrations, are primarily controlled from early to late stages of decomposition, respectively. In the complexity of real ecosystems, mixed litter from different species with varying litter quality and plant structure changes the chemical environment and physical character could drive interaction among adjacent litters from different species during decomposition. Gartner et al. [15] stated that litter mixtures often increased mass loss and had greater nutrient concentrations than predicted from single-species dynamics, meanwhile enhanced nutrient content could reflect less nutrient loss and enhanced import from surrounding litter and soil. Mixtures decaying can influence the litter decomposition rate potentially, may deviate from expectation values calculated as the additive mean of component litters decaying alone, and have important effect on litter mass loss dynamic, nutrient content, decomposers abundance and activity. In the process of litter decomposition, the microenvironment of mixtures decomposition (physical and chemical environment) differs from single-species litter decomposition environment [16] [17]. Due to leaching or biological effects, specific nutrients or secondary metabolites of litter transfer among different types influence the rate of litter decomposition and nutrient dynamic process. Almost 70% of the literatures, both domestic and international, on litter mixtures show mixed-litter decomposition in accordance with a non-additive model, but have inconsistent conclusions: synergistic or antagonistic and is insensitive to changes in mixtures [18] [19]. Gatner et al. [12] summarized that 76% mixed litters performed non-additive effect during the litter decaying among 123 kinds of specie types. Mixed litter could predict litter decomposition in natural ecosystem more accurately, and is of great significance on the biogeochemical cycles.

So far, due to agriculture activities, power plant and vehicle emissions, large areas have excessive nitrogen deposition, have side effects on aeschynomenous phytocoenosium and sensitive species components, which are far higher than thresholds of ecosystem (Carly et al. 2012) [20]. It is estimated that global nitrogen deposition amount was 103 Tg/a (1 Tg = 1012 g) in 1990, three times of 1860, and was expected to reach 195 Tg/a by 2050 [21]. In the late 80s, international start to pay close attention to the impact of global change (i.e. nitrogen deposition) on the dynamic change process of litter decomposition [22]. Atmospheric nitrogen deposition to the soil surface can improve the soil N content and then promote the plant growth and impact litter production and decaying rate directly. Numerous studies of nitrogen addition control tests verified simulated nitrogen deposition could reduce vegetation species richness and change the dominant species of settlement, it can also lead to the loss of rare species and the succession of species [23] [24], finally these will affect the rate of litter decomposition. Stevens et al. [25] found each additional 0.25 $g/m^2 \cdot a$ nitrogen deposition would reduce a plant species within 4 m^2 in 68 acid grasses. Xia et al. [26] stated the gramineae was more sensitive to nitrogen and easier to lose than grass. Therefore, nitrogen deposition leads to the decrease of species richness and different functional group of plants (i.e. legumes and grasses) transform, which will affect litter decomposition and cause long term chronic effects on the nutrient cycling in the ecosystem.

The continued increasing of nitrogen deposition could change the litter chemical constituents to affect the rate of litter decomposition and nutrient dynamics [27]. Previous simulation experiments of nitrogen deposition showed there were three cases of litter decomposition in response to nitrogen deposition: Vestgarden [28] demonstrated nitrogen deposition accelerate litter decomposition, especially in some N limited sites, while others found negative effects, especially high levels nitrogen deposition areas [29] [30], or have no significant effects [31]. Some research showed that litter decomposition showed different response to nitrogen from initial phase of accelerate to the later stages of inhibit. It looks like nitrogen deposition could decrease the decomposers composition and activity [32], or N element with some decomposes substance forming harder degradable polymeric substances, thus inhibit litter decomposition [33]. It implied that the effects of nitrogen deposition on litter decomposition depended on decomposition time, site condition, species types and levels of nitrogen addition. The inconsistency in the relationships with stimulated nitrogen deposition and litter decomposition, hence, it is necessary to carry out relative investigation, especially on the effect of nitrogen deposition on mixed litter decomposition and nutrient dynamic change.

The Stipa baicalensis steppe, an endemic grassland type in Central Asia, is one of the representative meadow steppe types. It is mainly located in Songliao Plain and the eastern part of Mongolia Plateau forest steppe in China [34], in which nutrient availability is a limiting factor of the vegetation growth. Therefore, Stipa baicalensis steppe is an ideal experimental platform to carry out the study of temperate grassland response to global change. The studies abroad of mixed litter decomposition are mainly concentrated on temperate forest ecosystem, while domestic studies on mixed litter decomposition in steppe ecosystem merely restrict to Wang et al. (2000) [35] in three different steppes (Meadow steppe, Leymus chinensis Steppe and Stipa grandis Steppe), Liu et al. [36] in Stipa krylovii Roshev. Steppe. Chen et al. [17] in the typical grassland of Inner Mongolia and Zhang et al. [30] in Hulun Buir Meadow Steppe, while there remains lack of the study about mixed litter decomposition in Stipa baicaleninsis Steppe and potential impact of nitrogen deposition on nutrient cycling in steppe ecosystem. This paper will compare and analyze the effect of N addition, simulated nitrogen deposition experiment, on the mixed litter decomposition rate of mixed litter and nutrient dynamic changes of different species tissues litter (leaf, stem and root) among Stipa baicalensis (perennial bunch grass, Sb), Leymus chinensis (perennial rhizomatous grass, Lc) and Artemisia frigid (small semishrub, Af), which are the main grasses in the Stipa baicalensis steppe ecosystem. The main objective of this study is to assess the effect of N addition on litter decomposition in Stipa baicalensis steppe by comparing three species mixed litter decomposition rates, and the chemical composition change to adapt global change, which influences ecosystem nutrient cycling. Our study postulated N addition would have significant influence on mixed litter decomposition, and three different species with contrasting initial chemical composition and functional groups have diverse response to N addition. We aimed to provide data support for nutrient cycle in the steppe ecosystem in the context of global changes, and serve theoretical basis and scientific guidance rational utilization of natural grasslands, biodiversity conservation and degraded grassland restoration.

2. MATERIALS AND METHODS

2.1. Study Area

The study area (119°35'E - 119°41'E, 48°27'N - 48°35'N, 760 - 770 m above sea level) was located in Ewenke County, Hulun Buir, Inner Mongolia. It was fenced to protect against domestic animals. The site had an annual average temperature of −1.6 degrees Celsius and an annual average rainfall of 328.7 mm. The annual accumulated temperature (≥0 degrees Celsius) and annual evaporation was 2567.5 degrees Celsius and 1478.8 mm, respectively. The study area has a northern Mid-temperate continental monsoon climate. The soil at the site was Dark Chestnut. The plant community was composed of Stipa baicalensis as the dominant species, as well as Leymus chinensis, Artemisia frigida, Achnatherum sibiricum, Serratula yamatsutanna, Carex pediformis, Filifolium sibicum, Koeleriacr istata, Pocockia ruthenica, Carex duriuscula, and Cleistogenes squarrosa.

2.2. Experimental Design

We chose smooth terrain and a vegetation representative area as an experiment site in Stipa baicalensis grassland. From 2010, we sprayed ammonium nitrate (NH_4NO_3) twice, at June 15th and July 15th, respectively. We designed six N addition controlled levels: 0, 1.5, 3.0, 5.0, 10.0, 15.0 g N/m^2·a, abbreviate to N_0, N_{15}, N_{30}, N_{50}, N_{100} and N_{150} respectively, with 3 repetitions and in total of 18 blocks were get, each block was 8 m × 8 m at intervals of 2 m, using a randomized complete block design [37]. N concentration was calculated by the percentage of months' average precipitation in annual total precipitation multiplied by N addition levels. NH_4NO_3 was dissolved in 8 L water (This corresponded to an additional 1.0 mm rainfall as annual rainfall), and sprayed uniformly on the each plot. CK was sprayed equal water to avoid the error of external water in each block. The same N concentration was used in each block per year.

2.3. Litter Preparation

We collected the litter of S. baicalensis, L. chinensis and A. frigid outside of the experiment treatment sites, and air dried the samples to constant mass. The naturally dried samples were divided into leaf, stem and root and cut into 10 cm fragments. We used the litter-bag technique with a nylon mesh bag, which was used to quantify litter decomposition [14]. We placed leaf, stem and root litter into 15 cm × 25 cm of 0.2 mm litterbags. Each bag was filled with 8 g of single litter and the equivalent amount of mixed litter by three plant species tissues (Table 1). We prepared three bags for each species litter combination, and in total of 45 litterbags were placed on every treatment block in 29th August 2011, using metal pins to fix litterbags on the ground surface to prevent being blown away by the wind. 810 litterbags were retrieved randomly after litter decaying 295d, 356d and 417d respectively. The retrieved materials were cleaned from soil particles and picked out other plants, roots and shoots. The cleaned materials were then oven-dried for 48 h at 70 degree Celsius. The dry litter materials were weighed to determine the remaining mass.

The oven dried litter was ground and sieved through a 1mm mesh and analyzed for C, N, P, K and lignin contents. Organic carbon content was determined using the potassium dichromate oxidation-concentrated sulfuric acid external heating method. For the determination of N, P and K content powdered samples were digested in a H_2SO_4-H_2O_2 solution. Total N was determined using Auto Analyzer 3 [38]. Phosphorus (P) was determined by ammonium molybdate colorimetry [39]. Potassium was determined by flame photometer method [40]. Lignin content was determined by ultraviolet spectrophotometer method [41].

2.4. Statistical Analyses

The remaining mass loss of the three species plant tissues litter was calculated by the following equation: R (%) = $M_t/M_0 \times 100\%$. Where R is litter mass remaining, M_t is the measured litter mass remaining in the litterbag at any time (t) of decomposition; M_0 is the mass of initial litter before decomposition.

Equal amounts of fragment litter from the three different species were mixed together in the litterbags, thus the expected mass remaining (M_e) can be calculated using the equation [42] : $M_e = [M_A/M_A + M_B + M_C] \times R_A + [M_B/M_A + M_B + M_C] \times R_B + [M_C/M_A + M_B + M_C] \times R_C$. Where R_A, $R_B$and R_C is the mass remaining in the single species litterbag of species A, B and C respectively; M_A, M_B and M_C is the initial litter mass of each fragment from the three species in the mixture.

All statistical analyses were performed by SPSS 16.0. The data of litter decomposition were subjected to analysis of variance (ANOVA), and three-way variance was used to test the effect of N addition level, species type and decomposing time on litter decomposition. T-test was used to test differences between the observed and expected mass remaining of mixed litter.

Table 1. Litter quality combinations.

Plant tissues	Richness level	Percentage of each litter quality (%)	Litter composition
Leaf	1	100	Sb, Lc, Af
	2	50	Sb + Lc, Lc + Af
	3	33	Sb + Lc + Af
Stem	1	100	Lc, Af
	2	50	Lc + Af
Root	1	100	Sb, Lc, Af
	2	50	Sb + Lc, Lc + Af
	3	33	Sb + Lc + Af

Note: Sb, Stipa baicalensis; Lc, Leymus chinensis; Af, Artemisia frigid. "1" represents only one species in one bag, "2" represents two different species were mixed in one bag, "3" represents three different species were mixed in one bag.

3. RESULTS

3.1. Chemical Parameters of Initial Litter

Comparing the different plant tissues litter of S. baicalensis, L. chinensis and A. frigid, the initial chemical contents were different (Table 2). For the different tissue litters of each plant, the N content of leaf litter was significantly higher than stem and root litter, but C/N lower than stem and root litter ($P < 0.05$). There was no significant difference of lignin content between tissue litters of three plants. For the different plants litter, the leaf litter of A. frigid had the highest C, N, P, K contents and lowest C/N, followed by L. chinensis, S. baicalensis was opposite, and three plant leaf litters had similar lignin content. The stem litter of A. frigid had lowest C, N, P, K contents than L. chinensis. The root litter of L. chinensis had highest C contents than S. baicalensis and A. frigid, but had lowest N concentrations.

3.2. Effects of N Addition on the Mixed Litter Mass Losing

Without N addition treatment, after 417d, the leaf mixture litter residual rate was 79.8%, 80.2%, 80.2% (Sb + Lc, Lc + Af, Sb + Lc + Af) respectively, stem mixture litter residual rate was 85.6% (Lc + Af), root mixture litter residual rate was 82.1%, 79.5%, 80.3% (Sb + Lc, Lc + Af, Sb + Lc + Af) respectively. After N addition treatment, as time goes by, the mixture litter residual rate presented the tendency of continuous declination, from the day of 356, the mass losing became faster until the day of 417 (Figure 1).

3.3. Effects of N Addition on the Rate of Mixed Litter Decomposition

Under the six levels of nitrogen addition treatments, the mixed litter remaining mass of different plant tissues were on the decline with the litter continuous decomposing (Table 3). For the leaf mixed litter, the remaining mass (%) under nitrogen addition treatments were higher than N0 at 295d after decaying. When decaying to 417d, the mixed litter remaining mass of plant tissues turned to opposite side, N addition accelerated mixtures decomposing except for N150. For the stem mixed litter, the remaining mass of the litters under nitrogen addition treatments lower than N0. During root mixed litter decomposing, the remaining mass of the litters under nitrogen addition treatments higher than N0. In general, the plant tissues litter remaining mass of three grasses, with different initial chemical composition, had different response to N addition, but no consistent increasing or decreasing were found with elevated levels of N addition. N150 significantly inhibited leaf mixtures mass loss, while other levels showed different degrees of promotion. N30 significantly accelerated stem mixtures mass loss, but N100 and N150 were opposite. N addition significantly inhibited root mixtures, only N100 and N150 transitorily promoted L. chinensis and A. frigid (Lc + Af) root mixtures decomposing ($P < 0.05$).

Control plot with non-nitrogen addition could simulate the natural condition to predict the rate of mixed litter decomposition and non-additive effect of mixed plant tissues litter (Table 3). After 417d, we found the leaf and root litter of S. baicalensis, L. chinensis and A. frigid (Sb + Lc + Af) showed synergistic effect, which the observed residual rate was lower than expected values 3.3% and 4.1%, respectively. L. chinensis and A. frigid (Lc + Af) stem mixtures , S. baicalensis and L. chinensis (Sb + Lc) root mixtures showed antagonism effect, which the observed residual rate was higher than expected values 4.0% and 5.1%, respectively. N addition significantly affected mixed effect, and different mixture combination had various responses to N addition.

Three-way ANOVA showed the effect of N addition on mixed litter decomposition rate changed with the decomposition time (Table 4). Decomposing time, species types and N addition all had significantly influence on the rate of aboveground mixed-litter decomposition ($P < 0.01$), and had interaction effect except for decomposing time and species types. Due to belowground litter undergo different decayed microenvironment, specie types played dominant role in controlling belowground mixed-litter decomposing under N addition treatment.

Table 2. Initial chemical compositions of three litter from various plant species (mg/g).

Chemical composition		Carbon	Nitrogen	Phosphorus	Potassium	C/N	Lignin
Stipa baicalensis	Leaf	436.9 ± 13.3 a	14.4 ± 0.2 a	0.89 ± 0.2 a	10.8 ± 0.2 a	30 ± 0.6 a	119.6 ± 6.3 a
	Root	349.9 ± 82.0 b	8.9 ± 0.8 b	0.63 ± 0.2 a	4.9 ± 1.0 b	50 ± 6.3 b	117.3 ± 1.2 a
Leymus chinensis	Leaf	451.7 ± 2.4 a	16.5 ± 1.2 a	0.89 ± 0.1 a	11.9 ± 0.4 a	27 ± 2.1 b	120.7 ± 7.6 a
	Stem	420.5 ± 36.7 a	7.9 ± 0.6 b	0.83 ± 0.1 a	10.2 ± 0.5 b	53 ± 5.1 a	119.3 ± 5.0 a
	Root	396.2 ± 40.3 a	7.9 ± 0.7 b	0.63 ± 0.1 b	4.9 ± 1.0 c	50 ± 8.8 a	113.0 ± 6.23 a
Artemisia frigida	Leaf	460.7 ± 75.2 a	21.1 ± 0.6 a	1.78 ± 0.2 a	18.5 ± 1.6 a	22 ± 1.3 b	121.8 ± 5.4 a
	Stem	354.9 ± 54.0 ab	7.2 ± 1.3 c	0.68 ± 0.1 b	10.2 ± 0.3 c	49 ± 10.4 a	111.9 ± 1.9 a
	Root	313.0 ± 39.2 b	9.0 ± 0.3 b	0.67 ± 0.1 b	14.0 ± 0.1 b	35 ± 7.9 b	109.4 ± 1.0 a

Note: Values within the same column with different small letters mean significant difference at P = 0.05 level, the same as follows.

Table 3. The effect of N addition on different tissues mixture litter remaining mass (%) after decaying 417 d.

N addition levels		Leaf			Stem		Root	
		Sb + Lc	Lc + Af	Sb + Lc + Af	Lc + Af	Sb + Lc	Lc + Af	Sb + Lc + Af
	Obs	59.7 ± 1.05	53.2 ± 0.68	57.4 ± 0.41	65.6 ± 0.67	66.0 ± 0.30	57.5 ± 0.51	54.5 ± 0.68
N_0	Exp	63.8 ± 2.05	58.0 ± 2.13	60.8 ± 1.78	61.6 ± 0.99*	60.9 ± 1.17*	54.5 ± 2.38	58.6 ± 2.29
	k	0.880 a	1.231 bc	1.027 c	0.806 b	0.665 c	0.979 c	1.164 a
	Obs	58.8 ± 0.89	52.3 ± 1.10	54.8 ± 0.35	65.5 ± 0.61	63.3 ± 0.40	59.1 ± 0.41	60.6 ± 0.56
N_{15}	Exp	61.3 ± 1.13	55.6 ± 0.63*	57.6 ± 0.74*	61.6 ± 1.19*	61.7 ± 1.98	57.1 ± 1.23	60.0 ± 1.26
	k	1.008 a	1.347 ab	1.181 b	0.774 b	0.791 a	0.951 cd	0.821 bc
	Obs	56.0 ± 0.27	55.1 ± 0.43	54.4 ± 0.22	61.8 ± 0.31	65.1 ± 1.02	60.2 ± 0.29	61.2 ± 0.82
N_{30}	Exp	60.3 ± 0.56*	53.8 ± 0.42	56.7 ± 0.58*	60.9 ± 1.51	68.6 ± 0.74*	64.1 ± 0.57*	66.1 ± 0.58*
	k	1.125 b	1.158 c	1.077 c	0.945 a	0.651	0.917 d	0.806 c
	Obs	60.0 ± 0.77	50.9 ± 0.62	57.9 ± 0.20	66.8 ± 0.14	67.2 ± 0.52	60.0 ± 0.29	59.2 ± 1.53
N_{50}	Exp	61.2 ± 1.12	52.6 ± 2.11	55.9 ± 1.72	61.0 ± 0.86*	61.7 ± 2.78	55.6 ± 1.46*	59.0 ± 0.25
	k	0.922a	1.449 a	1.048 c	0.757 b	0.668 c	0.995 c	0.861 b
	Obs	62.1 ± 0.72	55.8 ± 0.44	52.1 ± 0.54	65.6 ± 0.54	63.9 ± 0.26	56.3 ± 0.16	60.5 ± 0.25
N_{100}	Exp	59.4 ± 1.04	55.1 ± 0.91	56.2 ± 0.70*	61.3 ± 0.83*	58.7 ± 3.03	56.8 ± 2.08	58.6 ± 2.31
	k	0.944 a	1.076 cd	1.396 a	0.818 b	0.746 ab	1.095 b	0.743 d
	Obs	60.6 ± 0.31	59.9 ± 0.16	61.9 ± 0.30	65.4 ± 0.66	67.9 ± 0.44	53.0 ± 0.17	61.6 ± 0.72
N_{150}	Exp	61.6 ± 0.48	54.5 ± 0.59*	57.4 ± 0.56*	63.8 ± 1.29	59.0 ± 1.52*	54.3 ± 1.50	55.4 ± 1.92*
	k	0.848 a	0.946 d	0.810 d	0.812 b	0.690 bc	1.232 a	0.724 d

Note: Values with star means significant difference between Obs and Exp at P = 0.05 level. Obs means the values of experiment observed, Exp means the values calculated means of mixtures component of single litter from three plants. k was litter decomposition constant, and the litter letter in the same line behind the values mean significant difference among six nitrogen addition treatments.

3.4. Effects of N Addition on the Nutrient Dynamic Change of Mixed Litter Decomposition

Due to the difference of plant tissues litter types, the response of mixed litter nutrient dynamics to N addition were different, and the response of aboveground and belowground litter was opposite (Table 5 and Table 6). At the early stage of litter decomposing to 295d, N addition significantly inhibited leaf and stem C element release, oppositely N addition accelerated root C element release ($P < 0.05$). At the end to 417d, the C element releasing of different mixed combination performed differently under the N addition treatments. Individual nitrogen levels accelerated N element releasing of leaf mixed litter, N_{50}, N_{100} and N_{150} significantly enhanced stem mixed litter the enrichment capacity of N element, retarded root mixed litter N element releasing. The ability of P element releasing of mixed litter was more sensitive than N element. High levels of nitrogen addition significantly inhibited P element releasing of aboveground mixed litter, N_{15} and N_{30}accelerated S. baicalensis, L. chinensis and A. *frigid* (Sb + Lc + Af) leaf mixtures and L. chinensis, A. frigid (Lc + Af) stem mixtures P element releasing. Comprehensively, N addition was incline to inhibit each plant tissues mixed litter nutrient releasing, especially high N addition levels. For the aboveground mixtures, the nutrient releasing of mixed combination of S. baicalensis and L. chinensis (Sb + Lc), which was the same family plant, was less sensitive than the mixed combination with A. frigid. Compared with aboveground mixtures, the nutrient dynamics of belowground mixtures was more readily affected by N addition.

Three-way ANOVA showed N addition had significant effect on mixed litter nutrient releasing (Table 6). Dynamic change process of C element similar to the tendency of litter residual rate could be regarded as the principal part of litter decomposition, affected by decomposing time, species types and N addition simultaneously.

4. DISCUSSION

In comparison with the condition of single litter decomposition, the mixtures of various specie types or initial chemical composition, altered litter quality and spatial structure, increase the complexity of the litter decomposition, and the change of decomposition environment had influence on the rate of mixed litter decomposition and nutrient dynamics

[43] [44]. The studyon mixed litter decomposition of four kinds of functional groups at the initial stage found that mixed effects had important influence on litter decaying rate, which was not the simple additive mass loss arithmetic from single litter [45]. Previous studies on mixed litter decomposition stated that there would be synergism effect [18], antagonism effect [46] and additive effect [47] during the mixed litter decomposition. In our study, at the end to 417d, the rate of individual mixed litter decomposition was deviated from its components single litter decomposition. Compared the actual residual rate of mixed litter decomposition with its expected values, we found the leaf and root litter of S. baicalensis, L. chinensis and A. frigid (Sb + Lc + Af) showed synergistic effect, and L. chinensis and A. frigid (Lc + Af) stem mixtures and S. baicalensis and L. chinensis (Sb + Lc) root mixtures showed antagonism effect, while other combinations showed additive effect. Meanwhile, there was a slight tendency of the remaining two leaf mixed litter combination (Sb+Lc and Lc+Af) showed synergism effect, L. chinensis and A. frigid (Lc + Af) root mixtures showed slightly antagonism effect. Whether this trend would reverse as the decomposing time increasing or had a significant enhancement to perform non-additive effect still needed long-term tracking and investigation. Antagonism or additive effects could be responsible for the inhibit effect of transfer secondary metabolism substance (i.e. phenolic) between adjacent mixed litters [48]. In our study, as the duration of litter decomposition, mixed effects would in flux. Under the interaction of adjacent single litters, mixed litter would conserve, enrich or release nutrient element to adapt the environmental changes, and thus maintain the stability of the entire ecosystem. N addition had no significant effect on the litter decomposition because that N addition may have a long-term impact on litter decomposition.

In the similar studies on N addition experiments, Liu et al. [49] found N addition had significant effect on the rate of mixed litter decomposition, while Zhang et al. [30] found no significant effect. In our previous speculation, we found N addition affected the rate of mixed litter decomposition significantly. The different conclusions were due to the discrepancy of N addition does. Zhang et al. [30] was focus on low N addition dose, they were 0, 1.0 and 2.0 $g/m^2 \cdot a$ nitrogen addition levels, which was significantly lower than our study. Liu et al. [16] designed three N addition levels on typical steppe which was 8.0, 16.0, 32.0 $g/m^2 \cdot a$, respectively. In addition, the impact of N addition on decomposition rate was affected by decomposition phase, experimental field condition, soil nitrogen availability and natural factors. We found significant interaction effect between species and N addition on the rate of mixed litter decomposition, which would change over decomposing time. Aboveground and belowground litter decomposition rate had different response to species types and N addition. The promote effect of N addition on aboveground mixed litter decomposition rate was stronger than belowground, whilst belowground was more sensitive to specie types. Due to aboveground and belowground litter were subject to different decomposition microenvironment. Therefore they may experience different decomposing patterns and exhibit opposite decomposition result. We found that N addition accelerated leaf and stem mixed litter decaying except N_{150}, significantly inhibited leaf mixtures decaying. High dose of N addition would promote S. baicalensis and L. chinensis (Sb + Lc) and L. chinensis and A. frigid (Lc + Af) root mixtures decaying, but N addition retarded S. baicalensis, L. chinensis and A. frigid (Sb + Lc + Af) root mixtures decaying. The effect of N addition on decomposition were relate to some aspects of litter chemistry [50]. Previous studies suggested litter initial chemical composition could better predict litter decomposition rate such as N content, C/N and lignin/N [51] - [53]. The higher initial N content may lead to faster litter decomposition rate, excessive lignin content would retard litter decay [53]. Some prior studies elucidated that N addition was resistant to decomposition may produce microbial decomposers synthesize phenolic compound or break down litter lignin and other polyphenolic compounds into compounds that react with inorganic N to decomposition [54]. Hobbie et al. [55] found residual rate of litter was strongly influenced by the interaction of plant species and N addition, which was the same to our study. N addition altered decomposer community composition in steppe ecosystem [30]. He et al. (2010) [56] demonstrated N addition decreased microbial abundance and microbial activities, high level of N addition could disrupt balanced relationships in plant-microbial symbiosis system. Tanya et al. [10] manifested decomposer organisms with low functional diversity slowed the litter decomposition and the cycling of carbon and nitrogen. N addition, increased soil nitrogen availability, had effect on root mixed litter decomposition directly. Therefore, in the further trial, we would measure decomposer communities and relative impact factors to explore the mechanism of effect of N addition on mixed litter decomposition.

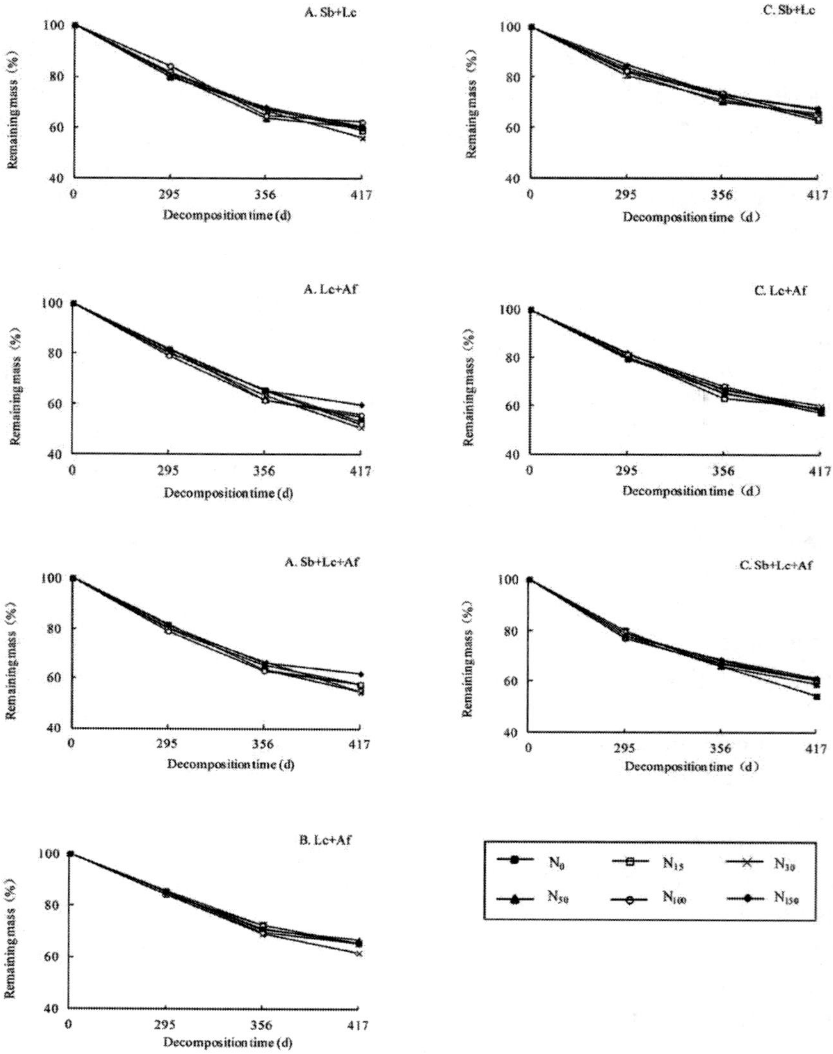

Figure 1. The effects of N addition on mixed litter remaining mass (%) after decaying to 417d. Sb, S. baicalensis; Lc, L. chinensis; Af, A. frigid. A, Leaf; B, Stem; C, Root.

In the natural environment, mixed litter with different physical characters of shape, hardness and specific surface area had higher heterogeneity and beneficial environment to hold on more decomposers. Besides physical characters, nutrient substances and secondary metabolism transferred through the connection effect of leaching and fungal hyphae caused litter decomposition rate changing (Salamanca et al. 1998). It is established that the litter contained phenol and lignin in compound form decreasing the decomposition of the mixed litter [57]. Different physical and chemical properties were responsible for various litter decomposition rate, nutrient dynamics and had different response to N addition. Previous reports had proposed that faster decomposition rates and nutrient turnover rate in mixed litters were due to the transfer of nutrients between the litters and the slower decomposition was due to the release of inhibitory compounds [48]. Recent study [10] confirmed the existence of nitrogen transformation from nitrogen-fixing litter species type to fast decomposing litter species type. The shift of nutrient between mixtures may change the litter decomposition, even affect the carbon cycling in ecosystem.

Table 4. Three-way ANOVA for decomposition rate of mixed-litter under N addition after decomposing 295 d, 356 d and 417 d.

Source of variation	*df*	F	*P*
Aboveground			
Time (T)	2	53.24	<0.01
Specie types (S)	3	8.36×10^3	<0.01
N addition (Ni)	5	1.52×10^3	<0.01
T*S	6	2.43	0.27
T*Ni	10	13.97	<0.01
S*Ni	15	935.56	<0.01
T*S*Ni	30	4.55	<0.01
Belowground			
Time (T)	2	0.16	0.854
Specie types (S)	2	295.93	<0.01
N addition (Ni)	5	12.57	<0.01
T*S	4	6.38	<0.01
T*Ni	10	1.88	0.052
S*Ni	10	30.60	<0.01
T*S*Ni	20	0.96	0.514

Table 5. The nutrient releasing rate of belowground mixture after decaying 417 d.

Species	Fertilization level	Carbon	Nitrogen	Phosphorus
Leaf				
	N0	50.8 ± 2.5 a	31.2 ± 0.4 a	38.1 ± 1.5 a
	N15	40.8 ± 3.8 a	31.9 ± 1.4 a	34.1 ± 3.6 a
Sb + Lc	N30	49.6 ± 2.6 a	33.0 ± 0.4 a	38.0 ± 1.0 a
	N50	44.7 ± 2.0 a	35.1 ± 3.3 a	34.4 ± 1.6 a
	N100	42.4 ± 1.6 a	31.2 ± 1.4 a	30.9 ± 3.0 a
	N150	46.7 ± 2.8 a	30.7 ± 1.3 a	30.5 ± 2.4 a
	N 0	55.4 ± 2.2 a	30.6 ± 1.6 ab	53.9 ± 2.6 d
	N15	50.8 ± 1.8 a	32.1 ± 2.8 b	49.7 ± 3.6 cd
Lc + Af	N30	57.8 ± 4.1 a	29.8 ± 0.4 ab	43.9 ± 5.1 bc
	N50	53.2 ± 2.4 a	34.7 ± 1.5 b	50.1 ± 1.1 cd
	N100	52.3 ± 1.1 a	30.3 ± 1.9 ab	36.8 ± 1.0 ab
	N150	51.7 ± 1.8 a	25.3 ± 0.9 a	31.9 ± 1.3 a
	N0	51.3 ± 1.3 b	37.1 ± 0.5 bc	37.8 ± 0.6 bc
	N15	53.3 ± 4.3 b	33.9 ± 1.4 b	50.1 ± 5.9 c
Sb + Lc + Af	N30	57.2 ± 1.0 b	34.6 ± 1.7 b	48.0 ± 5.9 c
	N50	51.0 ± 1.8 b	35.6 ± 0.9 b	32.1 ± 0.8 ab
	N100	55.7 ± 1.6 b	40.5 ± 0.9 c	40.2 ± 1.5 bc
	N150	43.3 ± 0.9 a	26.0 ± 1.6 a	24.1 ± 4.1 a
Stem				
	N0	37.9 ± 2.8 a	-5.5 ± 2.3 bc	42.6 ± 0.9 b
	N15	40.0 ± 4.6 a	-4.1 ± 2.8 bc	46.6 ± 1.2 b
Lc + Af	N30	44.0 ± 4.1 a	-0.5 ± 3.0 c	45.2 ± 1.1 b
	N50	37.7 ± 5.0 a	-10.7 ± 0.4 ab	34.5 ± 1.3 a
	N100	31.9 ± 2.9 a	-16.3 ± 2.9 a	30.3 ± 4.2 a
	N150	33.1 ± 4.0 a	-13.9 ± 3.2 a	33.6 ± 0.9 a
Root				
	N0	52.2 ± 2.3 ab	22.8 ± 1.8 b	58.0 ± 0.5 bc
	N15	55.2 ± 1.8 b	33.5 ± 3.2 c	59.1 ± 0.3 c
Sb + Lc	N30	56.8 ± 2.9 b	17.7 ± 3.3 b	57.3 ± 1.7 bc
	N50	48.2 ± 0.9 a	18.4 ± 2.3 b	48.7 ± 1.2 a
	N100	56.5 ± 1.4 ab	19.1 ± 0.9 b	56.6 ± 1.7 bc
	N150	54.1 ± 1.9 ab	15.2 ± 4.4 a	54.6 ± 1.0 b
	N0	50.0 ± 2.2 a	23.3 ± 2.9 a	33.3 ± 0.9 b
	N15	46.5 ± 5.5 a	21.3 ± 1.2 a	25.0 ± 1.0 a
Lc + Af	N30	49.2 ± 3.6 a	24.3 ± 7.5 a	31.3 ± 4.7 ab
	N50	48.2 ± 5.6 a	21.3 ± 3.7 a	42.6 ± 1.7 ab
	N100	45.6 ± 1.2 a	19.8 ± 3.0 a	32.2 ± 2.6 c
	N150	48.0 ± 1.9 a	17.1 ± 2.9 a	36.8 ± 2.0 bc
	N0	57.4 ± 1.4 b	30.6 ± 2.8 c	50.9 ± 3.5 b
	N15	53.9 ± 3.7 ab	24.0 ± 1.3 bc	46.2 ± 3.6 b
Sb + Lc + Af	N30	55.1 ± 1.3 ab	25.0 ± 2.8 bc	47.1 ± 2.3 b
	N50	52.9 ± 4.4 ab	26.4 ± 3.1 bc	44.0 ± 2.7 ab
	N100	47.4 ± 3.7 a	20.9 ± 3.5 b	40.3 ± 1.9 b
	N150	48.5 ± 1.5 ab	12.1 ± 0.6 a	36.2 ± 2.2 a

Note: Values were mean ± SE, litter letter in the same line behind the values for each litter mean significant difference at $P < 0.05$ level.

Table 6. Three-way ANOVA for C, N and P releasing rate of mixed-litter under N addition after decomposing 417 d.

Source of variation	*df*	Carbon		Nitrogen		Phosphorus	
		F	*P*	F	*P*	F	*P*
Aboveground							
Time (T)	2	954.343		1.80×10^3		423.303	
Specie types (S)	3	129.325		968.98		17.649	
N addition (Ni)	5	5.415		5.628		17.465	
T*S	6	1.999	0.067	62.782		16.951	
T*Ni	10	1.281	0.242	3.875		7.692	
S*Ni	15	1.521	0.099	2.464		2.914	
T*S*Ni	30	0.853	0.690	4.116		3.856	
Belowground							
Time (T)	2	369.592		944.323		812.960	
Specie types (S)	2	14.953		19.938		283.533	
N addition (Ni)	5	2.801	0.019	4.469		6.786	
T*S	4	0.763	0.550	2.052	0.090	10.441	
T*Ni	10	3.064		6.440		3.084	
S*Ni	10	3.171		1.020	0.429	5.663	
T*S*Ni	20	2.261		2.949		4.119	

Note: Vacancy in the table means had significant difference at $P < 0.01$ level.

In this study, we found the nutrient dynamic process of each plant tissues litter had different response to N addition. From the chemical composition perspective, the mixture combination within similar initial chemical composition was inclined to negative effect, while ones within discrepant initial chemical composition (i.e. L. chinensis and A. frigid leaf litters) were inclined to positive or non-additive effect. Litter initial chemical composition was the key factor for the mixed effects, and the difference of initial chemical composition between adjacent mixtures determined mixed effect direction. Litter with high nutrient contents could fix and support more microorganisms on the surface, and promoted the nutrient turnover rate of low litter quality [58]. The litter decomposition and nutrient turnover rate were deviated from single litter decomposition by the nutrient transfer and fungal hyphae [48], mixed litters decomposition dynamics were affected by mixed effect, and more sensitive to N addition, and then reacted corresponding variational mechanism to adaptation environmental change. From physical perspective, mixed litter of different plant families, compared to the same plant families' mixtures, was prone to releasing nutrient elements. Therefore, the richness of different plant families may give rise to litters variance response to N addition. But in our study, we only chose S. baicalensis (Gramineae), L. chinensis (Gramineae) and A. frigid (Composite), and did not consider more different families plant litters. It has been suggested that primarily chemical and physical interactions between litter species are the factors in non-additive effects [36]. In addition, identifying the key species or key properties (e.g. secondary chemistry) seems decisive for predicting the process of mixed litter decomposition. Longer-term study is needed to further investigate the effect of litter quality diversity and plant family richness on decomposition.

In our study, N addition had significant effect on litter nutrient dynamics and had interaction effect with decomposing time and specie types. N addition inhibited different plant tissues mixed litter nutrient releasing rate, especially high N addition level. P element releasing rate was more sensitive to N addition than C and N element, showed continuous releasing over litter decomposing. Due to soil N content replenished, litter could get more availability nitrogen, but limited the availability of P element [59]. Vivanco et al. [60] suggested N addition accelerated mixed litter decomposition rate and nutrient releasing, and converted non-additive effects into additive effect in temperate forest of South America. Qi et al. [61] stated that N addition was not direct factor to restrict litter decomposition, but to the growth of microbial and influenced litter decomposition rate and nutrient release indirectly. The effect of N addition on litter nutrient dynamics had relationship with decomposing time scale, the succession in decomposers species and community structure would be more complex and efficient, and had higher demand for nitrogen with exogenous nitrogen increasing (Berg and Staaf 1980) [39]. In our study, high levels N addition had inhibited effect on mixed litter nutrient releasing. N addition prolonged the nutrient releasing time and was conducive to the retention of nutrient and organic matter in the ecosystem. N addition, increased nitrogen input in soil and enhanced soil fertility directly, could provide more carbon and nutrient for microbial in the short term, which was advantageous to the grassland ecosystem balance, especially to recover degraded grassland. Excessive nitrogen reduced community structure and activity of decomposer and weaken decompose capacity for decaying litter of disintegrators. Hence, the inhibited effect of N addition or nitrogen deposition increased caused the decrease of nutrient turnover in long-term, in the lack of nutrient adverse to plant growth and microbial activity, and impedes the positive succession of ecosystem. As the main carbon source, litter decomposition provided plentiful nutrient for microbial activity. N addition, simulated nitrogen deposition increasing, accelerated the rate of leaf, stem mixed litters, S. baicalensis and L. chinensis (Sb + Lc) root mixed litter and L. chinensis and A. frigid (Lc + Sb) root mixed litter, but inhibited litters nutrient releasing rate. The possible explanation could be that litter quality of mixed litters in trial could not sheer meet decomposer microbial activity as absorbed nutrient from environment and reduced litter nutrient releasing [61]. Abby et al. [62] stated that the litter quality change associated with N deposition led to increasing litter decomposition. Knorr [63] indicated that N could affect the litter decomposition rate by altering the initial quality and decomposer community in chronic and high N deposition areas. As the process of litter decomposition, mixed litters mediate autologous chemical composition and improve litter quality constantly, conducive to microbial activity and promote litter decomposition, whilst N addition promoted the adaption mechanism of litter decomposition to environmental change.

5. CONCLUSION

In conclusion, our results demonstrated N addition had significantly effect on mixed litter decomposition rate and nutrient dynamics of plant tissues litter from three plants. We found mixed litter decomposition was influenced by the interaction of decomposing time, species types and N addition. N addition could alter mixed effects among different plant tissues litter, correspondingly different litter quality may lead to different response to N addition. N addition simulated nitrogen deposition tended to inhibit litter nutrient releasing and influenced the retention of nutrient in Stipa baicalensis steppe in Inner Mongolia.

ACKNOWLEDGEMENTS

The authors gratefully acknowledged the financial supported by the National Natural Science Fund of China (Project 31170435), "Twelfth five-year" national science and technology plan projects funding (2012 bad13b07), Central public welfare basic scientific research institutes were special funds (Agro-Environmental Protection Institute, Ministry of Agriculture, Tianjin) and Key laboratory of original Agro-environment quality of ministry of Agriculture and Tianjin Key laboratory of Agro-environment and Safe-product funding.

REFERENCES

1. Jiang, Z.B., Cheng, J.M. and Chen, A. (2013) Assessment of Vegetative Ecological Characteristics and the Succession Process during Three Decades of Grazing Exclusion in a Continental Steppe Grassland. Ecological Engineering, 57, 162-169.

2. Chang, Y.H., Liu, X.J., Li, K.H., Lv, J.L. and Song, W.B. (2012) Research Progress in Atmospheric Nitrogen Deposition. Arid Zone Research, 29, 972-979.

3. Liu, J., Wu, J.G., Liu, F.Q. and Han, X.G. (2013) Quantitative Assessment of Bioenergy from Crop Stalk Resources in Inner Mongolia, China. Applied Energy, 93, 305-318.

4. Su, J.Q., Li, X.R., Hui, R., Zhao, Y. and Liu, Y.M. (2013) Effects of Nitrogen Deposition on Species Diversity and Community Composition of Herbaceous Vegetation in Desert Steppe. Acta Botanica Boreali-Occidentalia Sinica, 32, 795-801.

5. Carrera, A.L. and Bertiller, M.B. (2013) Combined Effects of Leaf Litter and Soil Microsite on Decomposition Process in Arid Rangeland. Journal of Environmental Management, 114, 505-511.

6. Tu, L.H., Hu, H.L., Hu, T.X., Zhang, J., Zou, S.H. and Dai, H.Z. (2012) Response of Betula luminifera Leaf Litter Decomposition to Simulation Nitrogen Deposition in Rainy Area of West China. Chinese Journal of Plant Ecology, 36, 99-108.

7. Liu, Y.S., Pan, Q.M., Zheng, S.X., Bai, Y.F. and Han, X.G. (2012) Intra-Seasonal Precipitation Amount and Pattern Differentially Affect Primary Production of two Dominant Species of Inner Mongolia Grassland. Acta Oecologica, 44, 2-10.

8. Geng, Y.B. and Shi, J.J. (2012) Influencing Factors of Grassland Litter Decomposition and Nutrient Release and Accumulation. Progress in Geography, 31, 655-663.

9. Scowcroft, P.G. (1997) Mass and Nutrient Dynamics of Decaying Litter from Passiflora mollissima and Selected Native Species in a Hawaiian Montane Rain Forest. Journal of Tropical Ecology, 13, 407-426.

10. Handa, I.T., Aerts, R., Berendse, F., Berg, M.P., Bruder, A., Butenschoen, O., Chauvet, E., Gessner, M.O., Jabiol, J., Makkonen, M., McKie, B.G., Malmqvist, B., Peeters, E.T.H.M., Scheu, S., Schmid, B., van Ruijven, J., Vos, V.C.A. and H&aumlttenschwiler, S. (2014) Consequences of Biodiversity Loss for Litter Decomposition across Biomes. Nature, 509, 218-221.

11. Meentemeyer, V. (1978) Macroclimate and Lignin Control of Hardwood Leaf Litter Decomposition Dynamics. Ecology, 59, 465-472.

12. Gartner, T.B. and Cardon, Z.G. (2004) Decomposition Dynamics in Mixed-Species Leaf Litter. Oikos, 104, 230-246.

13. Patricia, I.A., Laura, Y. and Amy, T.A. (2012) Do Soil Organisms Affect Aboveground Litter Decomposition in the Semi-Arid Patagonian Steppe, Argentina? Oecologia, 168, 221-230.

14. Aerts, R. (1997) Climate, Leaf Litter Chemistry and Leaf Litter Decomposition in Terrestrial Ecosystems: A Triangular Relationship. Oikos, 79, 439-449.

15. Gartner, T.B. and Cardon, Z.G. (2008) Site of Leaf Origin Affects How Mixed Litter Decomposes. Soil Biology and Biochemistry, 38, 2307-2317.

16. Liu, P. (2006) Characteristics and Responses to Changes in Environmental Conditions of Litter Decomposition in Plant Species of Stipa krylovii Roshev. Steppe. Institute of Botany, the Chinese Academy of Sciences, Beijing.

17. Chen, J., Li, Y. and Huang, J.H. (2011) Decomposition of Mixed Litter of Four Dominant Species in an Inner Mongolia Steppe. Chinese Journal of Plant Ecology, 35, 9-16.

18. Hector, A., Beale, A.J., Minns, A., Otway, S.J. and Lawton, J.H. (2000) Consequence of the Reduction of Plant Diversity for Litter Decomposition: Effects through Litter Quality and Microenvironment. Oikos, 90, 357-371.

19. Smith, C.V. and Bradford, A.M. (2003) Do Non-Additive Effects on Decomposition in Litter-Mix Experiments Result from Differences in Resource Quality between Litters? Oikos, 102, 235-242.

20. Stevens, C.J., Dise, N.B., Mountford, J.O. and Gowing, D.J. (2004) Impact of Nitrogen Deposition on the Species Richness of Grasslands. Science, 303, 1876-1879.

21. Galloway, J.N., Dentener, F.J., Capone, D.G., Boyer, E.W., Howarth, R.W., Seitzinger, S.P., Asner, G.P., Cleveland, C.C., Green, P.A., Holland, E.A., Karl, D.M., Michaels, A.F., Porter, J.H., Townsend, A.R. and Voosmarty, C.J. (2004) Nitrogen Cycles: Past, Present, and Future. Biogeochemistry, 70, 153-226.

22. Norby, R.J. and Cotrufo, M.F. (1998) Global Change: A Question of Litter Quality. Nature, 396, 17-18.

23. Bai, Y.F., Wu, J.G., Christopher, M.C., Shahid, N., Pan, Q.M., Huang, J.H., Zhang, L.X. and Han, X.G. (2010) Tradeoffs and Thresholds in the Effects of Nitrogen Addition on Biodiversity and Ecosystem Functioning: Evidence from Inner Mongolia Grasslands. Global Change Biology, 16, 358-372.

24. Cleland, E.E. and Harpole, W.S. (2010) Nitrogen Enrichment and Plant Communities. Annals of the New York Academy of Science, 1195, 46-61.

25. Stevens, C.J., Smart, S.M., Henrys, P.A., Maskell, L.C., Cheffings, C.M., Whitfield, C., Gowing, D.J.G., Rowe, E.C., Dore, A.J. and Emmett, B.A. (2012) Terricolous Lichens as Indicators of Nitrogen Deposition: Evidence from National Records. Ecological Indicators, 20, 196-203.

26. Xia, J.Y. and Wan, S.Q. (2008) Global Response Patterns of Terrestrial Plant Species to Nitrogen Addition. New Phytologist, 179, 428-439.

27. Ma, C., Dong, S.F. and Mo, J.M. (2012) Response of Litter Decomposition to Changes of Litter Input in a Pine Forest of Dinghushan. Ecology and Environmental Sciences, 21, 647-653.

28. Vestgarden, L.S. (2001) Carbon and Nitrogen Turnover in the Early Stage of Scots Pine (Pinus sylvestris L.) Needle Litter Decomposition: Effects of Internal and External Nitrogen. Soil Biology & Biochemistry, 33, 465-474.

29. Magill, A.H. (2000) Dissolved Organic Carbon and N Relationships in Forest Litter as Affected by N Deposition. Soil Biology and Biochemistry, 32, 603-613.

30. Zhang, C.H., Li, S.G., Zhang, L.M., Xin, X.P. and Liu, X.R. (2013) Effects of Species and Low Does Nitrogen Addition on Litter Decomposition of Three Dominant Grasses in Hulun Buir Meadow Steppe. Journal of Resources and Ecology, 4, 20-26.

31. Carreiro, M.M., Sinsabaugh, R.L., Repert, D.A. and Perkhurst, D.F. (2000) Microbial Enzyme Shifts Explain Litter Decay Responses to Simulated N Deposition. Ecology, 81, 2359-2365.

32. Johansson, O., Palmqvist, K. and Olofsson, J. (2012) Nitrogen Lichen Community Changes through Differential Species Responses. Global Change Biology, 18, 2626-2635.

33. Berg, B. and Staaf, H. (1980) Decomposition Rate and Chemical Changes in Decomposing Needle Litter of Scots Pine: Influence of Chemical Composition. Ecological Bulletin, 32, 373-390.

34. Yang, D.L., Han, G.D., Hu, Y.G. and Wuyungerile (2006) Effects of Grazing Intensity on Plant Diversity and Aboveground Biomass of Stipa baicalensis Grassland. Chinese Journal of Ecology, 25, 1470-1475.

35. Wang, Q.B., Li, L.H., Bai, Y.F. and Xing, X.R. (2000) Effects of Simulated Climate Change on the Decomposition of Mixed Litter in Three Steppe Communities. Acta Phyytoecologica Sinica, 24, 674-679.

36. Liu, P., Huang, J.H., Han, X.G., Sun, J.X. and Zhou, Z.Y. (2007) Non-Additive Effects of Litter Mixtures on Decomposition and Correlation with Initial Litter N and P Concentrations in Grassland Plant Species of Northern China. Biology and Fertility of Soils, 44, 211-216.

37. Tilman, D. and Wedin, D. (1991) Plant Traits and Resource Reduction for Five Grasses Growing on a Nitrogen Gradient. Ecology, 72, 685-700.

38. Zhang, Y.L., Xu, A.M., Shang, H.B. and Ma, A.S. (2006) Determination Study of Total Nitrogen in Soil and Plant by Continuous Flow Analytical System. Journal of Northwest Sci-Tech University of Agriculture and Forestry (Natural Science Edition), 34, 128-132.

39. Murphy, J. and Riley, J.P. (1962) A Modified Single Solution Method for Determinantion of Phosphate in Natural Waters. Analytica Chimica Acta, 26, 31-36.

40. Bao, S.D. (2000) Soil Analysis. 3rd Edition, China Agriculture Press, Beijing.

41. Li, J., Cheng, Z., Yang, X.L., Li, S., Gu, M., Wan, S.W., Zhang, W.J. and Chen, J.K. (2006) Determination of Lignin Content in Tiny Panax ginseng by UV Spectrophotometry. Journal of Chinese Medicinal, 29, 239-241.

42. Wardle, D.A. (1997) Biodiversity and Plant Litter: Experimental Evidence Which Does Not Support the View That Enhanced Species Richness Improves Ecosystem Function. Oikos, 79, 247-258.

43. Liao, L.P., Ma, Y.Q., Wang, S.L., Gao, H. and Yu, X.J. (2000) Decomposition of Leaf Litter of Chinese Fir in Mixture with Major Associated Broad-Leafed Plantation Species. Acta Phytoecologica Sinica, 24, 27-33.

44. Xiong, Y., Xu, G.Q. and Wu, L. (2012) Progress on Non-Additive Effects of Mixed Litter Decomposition. Evironmental Science & Technology, 35, 56-60.

45. Wardle, D.A., Bonner, K.I. and Nicholson, K.S. (1999) Biodiversity and Plant Litter Experimental Evidence Which Does Not Support the View That Enhanced Species Richness Improves Ecosystem Function. Oikos, 79, 147-158.

46. Fylea, J.W. and Fyles, J.H. (1993) Interaction of Douglas Fir with Red Alder and Salal Foliage Litter during Decomposition. Canadian Journal of Forest Research, 23, 358-361.

47. Scowcroft, P.G. (1997) Mass and Nutrient Dynamics of Decaying Litter from Passiflora mollissima and Selected Native Species in a Hawaiian Montane Rain Forest. Journal of Tropical Ecology, 13, 407-426.

48. Chapman, K., Whittaker, J.B. and Heal, O.W. (1988) Metabolic and Faunal Activity in Litter of Tree Mixtures Compared with Pure Stand. Agriculture Ecosystem Environment, 24, 33-40.

49. Liu, P., Huang, J.H., Han, X.G., Sun, J.X. and Zhou, Z.Y. (2006) Differential Responses of Litter Decomposition to In-creased Soil Nutrients and Water between Two Contrasting Grassland Plant Species of Inner Mongolia, China. Applied Soil Ecology, 34, 266-275.

50. Hobbie, S.E. (2008) Nitrogen Effects on Decomposition: A Five-Year Experiment in Eight Temperate Sites. Ecology, 89, 2633-2644.

51. Swift, M.J., Heal, O.W. and Anderson, J.M. (1979) Decomposition in Terrestrial Ecosystem. University of California Press, Berkley.

52. Cotrufo, M.F. (1994) Decomposition of Tree Leaf Litters Grown under Elevated CO2: Effect of Litter Quality. Plant and Soil, 163, 121-130.

53. Berg, B. and McClaugherty, C. (2003) Plant Litter: Decomposition, Humus Formation, Carbon Sequestration. Springer, Berlin.

54. Davidson, E.A., Chorover, J. and Dail, D.B. (2003) A Mechanism of Abiotic Immobilization of Nitrate in Forest Ecosystems: The Ferrous Wheel Hypothesis. Global Change Biology, 9, 228-236.

55. Hobbie, S.E. (2005) Contrasting Effects of Substrate and Fertilizer Nitrogen on the Early Stages of Litter Decomposition. Ecosystems, 8, 644-656.

56. He, Y.T., Qi, Y.C., Dong, Y.S., Peng, Q., Xiao, S.S. and Liu, X.C. (2010) Advances in the Influence of External Nitrogen Input on Soil Microbiological Characteristics of Grassland Ecosystem. Advances in Earth Science, 25, 877-885.

57. Berg, B. and Ekboh, G. (1991) Litter Mass-Loss Rates and Decomposition Patterns in Some Needle and Leaf Litter Types. Long-Term Decomposition in a Scots Pine Forest. VII. Canadian Journal of Botany, 69, 1449-1456.

58. Victoria, C.S. and Mark, A.B. (2003) Do Non-Additive Effects on Decomposition in Litter-Mix Experiments Result from Differences in Resource Quality between Litters? Oikos, 102, 235-242.

59. Mo, J.M., Xue, J.H. and Fang, Y.T. (2004) Litter Decomposition and Its Responses to Simulated N Deposition for the Major Plants of Dinghushan Forests in Subtropical China. Acta Ecologica Sinica, 24, 1413-1420.

60. Vivanco, L. and Austin, A.T. (2011) Nitrogen Addition Stimulates Forest Litter Decomposition and Disrupts Species Interaction in Patagonia, Argentina. Global Change Biology, 17, 1963-1974.

61. Qi, H., Liu, Y. and Guo, J.X. (2012) Decomposition of Leymus chinensis and Phragmites communis Litter Responses to Simulated Warming and Nitrogen Application in Songnen Grassland. Journal of Northeast Normal University (Natural Science Edition), 44, 114-119.

62. Abby, G.S., Edith, B.A., Thomas, M. and Michael, F.A. (2007) Impacts of Anthropogenic N Additions on Nitrogen Mineralization from Plant Litter in Exotic Annual Grasslands. Soil Biology and Biochemistry, 39, 24-32.

63. Knorr, M., Frey, S.D. and Curtis, P.S. (2005) Nitrogen Additions and Litter Decomposition: A Meta-Analysis. Ecology, 86, 3252-3257.

Chapter 3

PP2A Phosphatase as a Regulator of ROS Signaling in Plants

Moona Rahikainen [1], Jesús Pascual [2], Sara Alegre [1], Guido Durian [1] and Saijaliisa Kangasjärvi [1,*]

[1]Department of Biochemistry, Molecular Plant Biology, University of Turku, FI-20014 Turku, Finland

[2]Plant Physiology Lab, Organisms and Systems Biology, Faculty of Biology, University of Oviedo, 33006 Oviedo, Spain

ABSTRACT

Reactive oxygen species (ROS) carry out vital functions in determining appropriate stress reactions in plants, but the molecular mechanisms underlying the sensing, signaling and response to ROS as signaling molecules are not yet fully understood. Recent studies have underscored the role of Protein Phosphatase 2A (PP2A) in ROS-dependent responses involved in light acclimation and pathogenesis responses in *Arabidopsis thaliana*. Genetic, proteomic and metabolomic studies have demonstrated that trimeric PP2A phosphatases control metabolic changes and cell death elicited by intracellular and extracellular ROS signals. Associated with this, PP2A subunits contribute to transcriptional and post-translational regulation of pro-oxidant and antioxidant enzymes. This review highlights the emerging role of PP2A phosphatases in the regulatory ROS signaling networks in plants.

KEYWORDS

plant; reactive oxygen species (ROS); signaling; protein phosphorylation; Protein Phosphatase 2A (PP2A)

1. INTRODUCTION

Plants are frequently challenged by changing weather conditions and the presence of detrimental biotic stress factors, including microbial pathogens and insect herbivores. To cope with the ever-changing natural circumstances, plants have evolved mechanisms to sense and respond to the external factors to optimize their fitness. To achieve this, stress-exposed plant tissues must decide whether it is advisable to sacrifice the tissue through a tightly regulated process of programmed cell death, or whether it is more beneficial to invest in energy-consuming acclimation processes in order to ensure seed production. The "choice" is the outcome of signal integration from multiple sources, and deploys signaling pathways elicited by reactive oxygen species (ROS) in different cellular compartments [1,2]. Analysis of cross-communication among different pathways is a modern trend in plant biology [3]. This review focuses on the emerging role of PP2A phosphatases as a regulator of antioxidant activities and ROS signaling in plants.

During the past decade, intensive research has demonstrated that ROS signaling is a common factor in biotic and abiotic stress responses. In biotic interactions, recognition of external factors by plasma membrane receptor kinases leads to activation of plasma membrane NADPH oxidases causing an ROS burst in the apoplast, with a consequent activation of phosphorylation-relay cascades that trigger the first line of defense gene expression in the nucleus [4,5,6]. Abiotic stresses in turn are generally considered to involve alterations in organellar redox biology and ROS signaling, albeit the exact mechanisms remain unsolved [2]. Various combinations of light stress, heat and drought are, however, well known to promote increased formation of ROS, and the photosynthetic machinery is generally considered the major site of this ROS production [7,8,9,10]. While harsh abiotic stress conditions may cause oxidative damage to the photosystems [11,12], studies have demonstrated that ROS-induced signals also essentially regulate the expression of nuclear-encoded genes responsible for defensive measures and antioxidant functions, thereby triggering acquired stress resistance in aboveground organs [10,13, 14, 15,16].

More recently, the role of organellar ROS signaling in plant biotic interactions has also been recognized [17,18]. Recognition of a presence of a pathogen in the apoplast is rapidly relayed into the chloroplasts, where Ca^{2+}-dependent signaling interactions are needed to trigger additional signaling events, which further modulate immunity-related gene expression in the nucleus ([19,20], Figure 1). The importance of chloroplast ROS burst in plant immunity is also reflected by a recent report by Zabala *et al.* [21], who showed that the bacterial pathogen *Pseudomonas syringae*secretes pathogen effectors that are targeted to chloroplasts, where they prevent the organellar ROS burst by modulating the function of the photosynthetic electron transfer chain.

Even though it is evident that plants can sense the chemical nature (superoxide, hydrogen peroxide, singlet oxygen and hydroxyl radical) and sub-cellular localization of ROS, and trigger defense reactions accordingly, the signaling networks determining the acclimation response remain poorly understood [22]. Even so, it is clear that evolution of mechanisms to sensitively respond to ROS as signaling molecules has provided significant competitiveness to plants in different ecological niches.

Since redox regulation is a key component in plant acclimation to varying light conditions, particularly in high light stress, light-dependent changes in gene expression may partially overlap with those induced by pathogens [2]. Besides light intensity, the duration of the photoperiod has proved to be a key component that modulates both the availability of ROS [23] as well as plant responses to elevated ROS levels [24,25,26]. Yet another layer of cross-communication is commonly observed as acquired cross-tolerance. Plants pre-exposed to high levels of light, for example, display increased resistance to virulent strains of the bacterial hemibiotrophic pathogen*Pseudomonas syringae* or green peach aphid (*Myzus persicae*) [27,28]. Such cross-tolerance commonly associates with enhanced production of ROS and consequent oxidative signaling [29]. To keep the oxidative signals in balance, plants deploy a multi-layered antioxidant network, including enzymes such as catalases (CAT), superoxide dismutases (SOD), ascorbate peroxidases (APX), peroxiredoxins (PRX) and thioredoxins

(TRX) as well as the low-molecular-weight antioxidants ascorbate and glutathione, which maintain ROS balance and metabolic homeostasis in different cellular compartments [30].

Beside antioxidant activities, post-translational regulation that modulates the activation state of signaling proteins is an important level of regulation in ROS signaling. Reversible phosphorylation of signaling proteins is an evolutionarily conserved mechanism that drives phosphorylation-relay cascades and their signaling interactions in animals and plants. Controlled activities among mitogen-activated protein kinases (MPKs) and Ca^{2+}-dependent protein kinases (CDPKs, or CPKs in *Arabidopsis thaliana*) are well known for their importance in cellular stress signaling [31,32]. Here, we focus on the emerging role of type 2A protein phosphatases in these interactions.

2. PP2A PHOSPHATASE AS A REGULATORY ENZYME IN PLANT STRESS

Trimeric type 2A protein phosphatases, composed of a catalytic subunit C, scaffold subunit A and regulatory subunit B, are evolutionarily conserved signaling components that regulate stress signaling in both animals and plants [33]. The regulatory B subunit is commonly referred to as the "specificity unit" that determines the target specificity of the trimeric PP2A holoenzyme [33,34]. The *Arabidopsis thaliana* genome contains five different genes encoding C subunits, three genes for the A subunits and 17 genes encoding the variable regulatory B subunits, which are further divided into B, B′ and B″ families [35]. The trimeric holoenzyme compositions therefore provide extensive variability and versatility for PP2A in regulatory networks. Besides the regulatory B subunit, PP2A function is further regulated by reversible methylation of a conserved C-terminal leucine residue [36]. Moreover, both animals and plants possess regulatory components that further modulate the phosphatase activity as has been shown e.g., for the regulatory TAP46 protein in *Arabidopsis* [37].

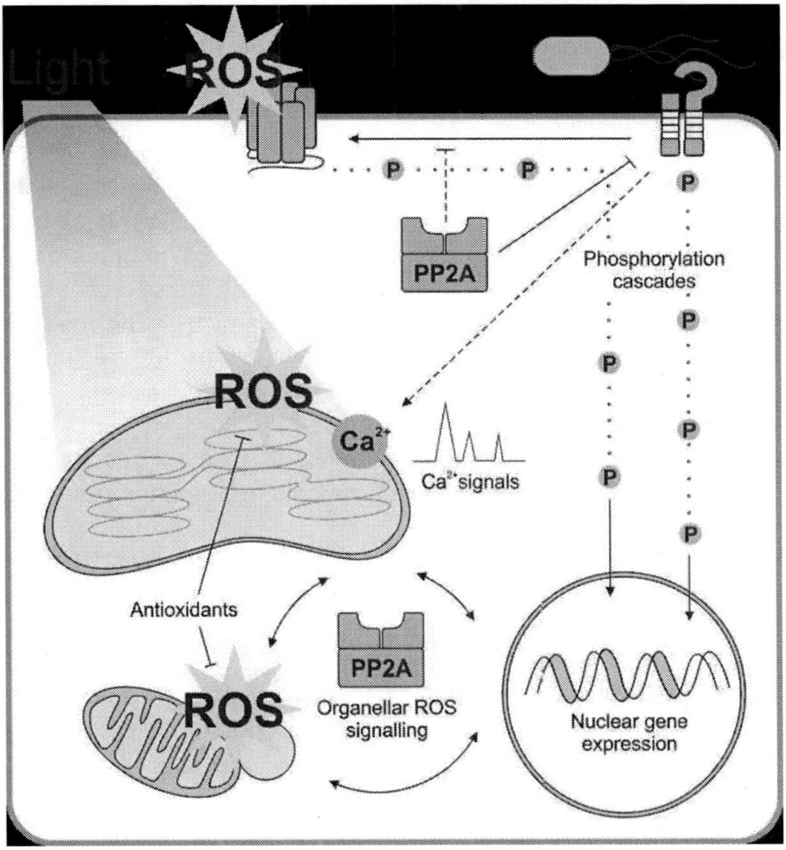

Figure 1. Protein Phosphatase 2A as a regulator of ROS signaling in plant leaves. In plant biotic interactions, recognition of conserved microbial structures, such as bacterial flagellin, by plasma membrane receptor kinases leads to NADPH oxidase-driven ROS burst in the apoplast and activation of phosphorylation-relay cascades that trigger defence gene expression in the nucleus. Abiotic stresses, such as light stress, in turn promote alterations in organellar redox biology and ROS signaling, which also modulate nuclear gene expression and stress resistance in green plant tissues. Recognition of attempted infection is also rapidly relayed into the chloroplasts, where Ca^{2+}-dependent signaling interactions are needed to trigger additional signaling events, which further modulate immunity-related gene expression in the nucleus. PP2A protein phosphatase limits the activity of the plasma membrane receptor kinase BAK1, and, consequently, the flagellin-induced ROS burst into the apoplast. Additionally, PP2A modulates photo-oxidative stress responses by controlling organellar ROS signaling and the abundance of alternative oxidases in leaf mitochondria.

Recent studies have assigned functions for PP2A in plant stress signaling in several different model crop species. The potato PP2A catalytic subunit StPP2Ac2 and the tomato subunits LePP2Ac1 and LePP2Ac2 mediate jasmonic acid signaling after wounding [38], while the tomato subunits LePP2Ac1, LePP2Ac2 and LePP2Ac3 operate in cold stress [38]. Studies have also shown that the rice PP2A catalytic subunit OsPP2A-1-5 as well as the potato subunits StPP2Ac1, StPP2Ac2a, StPP2Ac2b and StPP2Ac3 become transcriptionally upregulated under salinity stress [38,39,40]. Moreover, in wheat, TaPP2Ac-1 acts as a positive regulator of salt stress responses [41]. Most of the research efforts have focused on unraveling the roles of the catalytic C-subunits, whereas little is known about the regulatory B-subunits. In wheat, B″α subunit has been shown to positively regulate lateral root formation under osmotic stress conditions [42]. Identification and characterization of other PP2A subunits in crop species will help in interpreting how PP2A regulates abiotic stress responses in useful plants.

In the model plant *Arabidopsis thaliana*, a recessive mutation of the catalytic subunit PP2A-C2 leads to enhanced sensitivity to abscisic acid (ABA), suggesting that this subunit is a negative regulator of ABA-mediated responses [43]. The scaffold subunit PP2A-A1, or ROOTS CURL IN NAPHTHYLPHTHALAMIC ACID1 (RCN1), in turn acts as a positive regulator in ABA-related pathways, and *rcn1* mutants additionally display pleiotropic phenotypes in other phytohormone signaling pathways as well [44,45]. PP2A-A3 has been identified as an important element in the low-temperature signaling pathway in *Arabidopsis*, inhibiting the cold adaptative response [46]. Regarding the regulatory B-subunits, B″α and B″β have been shown to modulate plant isoprenoid biosynthesis pathway, B″α playing a distinct role in plants under salt stress [47]. The regulatory subunit PP2A-B′γ, in turn, seems to be present in heterotrimers involved in heat stress response [48]. A *pp2a-b′γζ* double mutant, deficient in two highly similar subunits B′γ and B′ζ showed increased resistance towards photo-oxidative stress under a stress-combination of high light, elevated temperature and limited water availability [49]. Hence, different PP2A subunits, presumably operating in different trimeric PP2A holoenzymes, mediate both positive and negative regulation of abiotic stress responses, allowing delicate fine-tuning of responses under a range of environmental challenges.

PP2A has also been shown to function in plant-biotic interactions. Catalytic PP2A subunits of tobacco have been related to effector-triggered immunity (ETI) [50], a defense response which commonly includes a hypersensitive response, a particular type of cell death in plants. In*Arabidopsis*, the subunit B′γ negatively regulates defense against green peach aphids and is required for transcriptional and post-translational control of immune processes [28,51]. Moreover, mutants deficient in B′γ and B′θ were shown to display lowered susceptibility towards the bacterial pathogen *Pseudomonas syringae pv. tomato* [52]. Mechanistic insights into PP2A-dependent regulation were recently provided by Segonzac *et al.* [53], who showed that a trimeric PP2A protein phosphatase composed of the catalytic subunit C4, the scaffold subunit A1 and regulatory B subunits B′η or B′ζ binds and limits the autophosphorylation, and hence activity, of the plasma membrane receptor kinase BAK1. These protein interactions also negatively regulate the flagellin-induced ROS burst into the apoplast [53]. Consequently, *pp2a-c4* and *pp2a-a1* knock-out lines are more resistant to the virulent bacterial pathogen *Pseudomonas syringae pv. tomato* [53]. Taken together, PP2A subunits respond to and mediate various stress signaling effects in different plant species, but the molecular mechanisms are only starting to emerge.

PP2A mutant lines with deficiencies in the regulation of stress responses may show simultaneous changes both in hormone and ROS signaling. This stems from the fact that ROS signaling and hormonal signaling pathways are tightly connected, not only in eliciting stress resistance but also in determining developmental processes [54]. ROS act as secondary messengers in many hormone-signaling pathways, but, in the case of stress signaling, ROS work also upstream by eliciting hormonal signals. In *Arabidopsis*, PP2A-B′γ has been shown to be a negative regulator of organellar ROS accumulation and the consequent salicylic acid (SA) signaling elicited by these ROS [55]. In ABA signaling, PP2A-B′α, B′β and B′δ are direct interactors of the protein kinase OPEN STOMATA 1 (OST1), which is a major component of ABA signaling in guard cells and has been shown to regulate apoplastic ROS production through phosphorylation of the NADPH oxidase RbohF [56,57]. However, the physiological role of PP2A in these interactions remains unclear. Altogether, further studies are needed to pinpoint the specific regulatory nodes in the stress signaling networks governed by PP2A.

3. PP2A-B′Γ AS A REGULATOR OF ROS SIGNALING AND CELL DEATH

Attempts to understand the role of PP2A in stress resistance led to the identification of PP2AB′γ as a putative component in light acclimation and defense signaling in *Arabidopsis* [51]. When grown under 50% humidity and moderate light intensity, *pp2a-b′γ* mutant displayed premature yellowing and constitutive activation of salicylic acid-(SA-) and jasmonic acid-(JA-) dependent pathogenesis responses, whereas growth under high levels of light abolished the yellowing phenotype and also the immune reactions become alleviated. Notably, even slight variations in the growth conditions, such as growth under 65% humidity at 200 $\mu mol \cdot m^{-2} \cdot s^{-1}$ [50] was sufficient to rescue the yellowing *pp2a-b′γ* mutant phenotype, which makes it a tricky tool for analysis of PP2A mediated phenotypes by mutant approaches. The yellowing *pp2a-b′γ* mutant phenotype was accompanied by elevated foliar ROS levels and eventually cell death under moderate light intensities, but significant alterations in the redox status of the main antioxidants ascorbate and glutathione were not observed [51,55,58].

Recently, by taking advantage of selective reaction monitoring (SRM) mass spectrometry, Konert*et al.* [58] showed that *pp2a-b′γ* may partially circumvent the constitutive ROS accumulation through a feedback loop where ROS induce up-regulation of the mitochondrial bypass pathways employing alternative oxidases AOX1A and AOX1D. AOX1A has been extensively studied and shown to operate in diverting excess reducing equivalents and hence in minimizing the formation of ROS within the photosynthetic and mitochondrial electron transfer chains [59,60]. The other AOX isoforms have remained less well characterized and their importance in stress responses is not well understood. Shuttling of redox-active intermediates between the organelles and the cytoplasm has a very likely impact on the formation of ROS in different cellular compartments and hence modulate redox signaling effects within the entire cell [59,61].

To promote organellar ROS signaling effects and to stabilize the *pp2a-b′γ* mutant phenotype, Li*et al.* [55] took advantage of the *catalase 2* (*cat2*) mutant as a tool to trigger oxidative signaling in the*pp2a-b′γ* mutant background. CATALASE 2 (CAT2) is the major antioxidant enzyme that quenches photorespiratory H_2O_2 in peroxisomes, and hence provides a highly informative system to study factors involved in intracellular ROS signaling [62]. Intriguingly, however, SA-induced cell death of *cat2* is conditioned by day length and becomes observable only under long-day conditions, even though oxidative stress in the *cat2* mutant prevails also under short photoperiods [24,25,63]. Recently, Li *et al.* [55] showed that, under short photoperiods, the ROS-induced cell death becomes suppressed through pathways that require the activity of PP2AB′γ. This conclusion was supported by proteomic analysis, which showed that PP2A-B′γ is required to control ROS-induced changes in the abundance and phosphorylation of the key pathogenesis-associated marker proteins PATHPGENESIS-RELATED PROTEIN 2 (PR2) and PATHOGENESIS-RELATED PROTEIN 5 (PR5) in short-day conditions [55].

Even though PP2A-B′γ is required to control intracellular oxidative stress responses and associated cell death in *Arabidopsis* [55], analysis of PP2A-B′γ and the highly similar PP2A-B′ζ in stress responses revealed that deficiencies in these PP2A subunits lead to alleviation of biotic stress responses and induction of photoprotective mechanisms and enhanced tolerance against abiotic stress [28,49]. Hence, PP2A-B′γ and PP2A-B′ζ appear to mediate opposing effects on cell death regulation and impact the decision between ROS-induced cell death and acclimation [49]. However, the underlying mechanisms remain to be established, and hence what determines the threshold is not well understood. It has been suggested that the ultimate outcome of ROS-induced responses, which may lead to activation of a strictly controlled cascade of programmed cell death or elicit protective mechanisms and acclimation, becomes determined by opposing actions of pro-cell death and anti-cell death signaling mechanisms [22]. Intriguingly, the outcomes of the genetic interactions in the *cat2 pp2a-b′γ* double mutant and in the *pp2a-b′γζ* are very different. While salicylic acid signaling and cell death are promoted in the *cat2 pp2a-b′γ* double mutant that experiences oxidative stress [55], in the double mutant combination *pp2a-b′γζ,* protective photo-oxidative stress responses are activated at an increased level, resulting in an enhanced activation of light acclimation [49]. Signaling in plant cells is characterized by interplay between different signal transduction pathways, often also such diverse ones as biotic and abiotic stress response pathways. PP2A-B′γ and PP2A-B′ζ control intracellular ROS homeostasis and signaling and thereby influence the plants different physiological outcomes, comprising cell death, growth and acclimation upon environmental perturbations.

4. PP2A AS A REGULATOR OF ANTIOXIDANT ACTIVITIES

Proteomic studies suggested that PP2A-B′γ is required to control the abundance of antioxidant enzymes, such as the chloroplastic copper/zinc SOD 2 (CSD2) and monodehydroascorbate reductase 2 (MDAR2), as well as the mitochondrial AOX1A and 1D, all of which have well-known roles in the maintenance of cellular ROS homeostasis [51,55,58]. The AOX1A gene is a target for ANAC013-mediated transcriptional activation in response to mitochondrial stress signals [64]. ANAC013 is an important regulator of mitochondrial retrograde regulation/signaling, since it controls a set of genes that are common and robust candidates for mitochondrial retrograde regulation under different types of stresses (denoted Mitochondrion Dysfunction Stimulon Genes) [64]. Indeed, Giraud *et al.* [65] showed that AOX1A is required for full tolerance against combined effects of light stress and drought.

While the *pp2a-b'γ* mutant may partially circumvent ROS accumulation through enhanced quenching of redox equivalents by mitochondrial AOX activity [58], transcript profiling of high light acclimated *pp2a-b'γζ* double mutants did not show up-regulation of AOX1A gene expression [49]. Instead, *pp2a-b'γζ* double mutants displayed increased transcript abundance for the H_2O_2scavenging enzyme ascorbate peroxidase 2 (APX2), heat shock transcription factor A3 (HSFA3) and the heat shock proteins HSP18.2, HSP21 and HSP22, which are transcriptionally co-regulated targets of HSFs [49]. The APX2 gene is transcriptionally highly responsive to a number of factors, including reduction state of the photosynthetic electron transfer chain, ABA signaling, chloroplastic and apoplastic ROS, and metabolic signals exemplified by accumulation of phosphoadenosine 5'-phosphate (PAP) in stress-exposed leaves [13,66,67,68]. Jung *et al.* [69] suggested, that HSFA3 would modulate APX2 expression under high levels of light and high temperature but would not mediate signals arising from reduction of the plastoquinone pool in the photosynthetic thylakoid membranes. A good candidate for mediating the HSFA3-dependent activation of APX2 is H_2O_2, which is generated in both heat- and light-induced stresses [70]. Through a yet unidentified mechanism, PP2A-B'γ is required to control the extent of photo-oxidative stress responses, exemplified here by enhanced expression of APX2 in *Arabidopsis* leaves.

5. CONCLUSIONS

PP2A phosphatases appear to control multiple targets in the stress signaling networks in plants. Such modulation of different signaling nodes is likely to involve PP2A phosphatases with different heterotrimeric compositions. Besides resolving the identity of the regulatory PP2A holoenzymes and the functional redundancy among individual PP2A subunits, the key research questions still to be resolved include elucidation of the PP2A target phosphoproteins and their functional significance and co-operation in the stress resistance in plants.

ACKNOWLEDGMENTS

This work was supported by Academy of Finland Center of Excellence project 271832 and University of Turku Doctoral Programme in Molecular Life Sciences. Guido Durian was supported by Academy of Finland post-doctoral fellowship (289687). Jesús Pascual was supported by a fellowship from the Formación de Profesorado Universitario (FPU) program (AP2010-5857, Ministry of Education, Spain).

AUTHOR CONTRIBUTIONS

Saijaliisa Kangasjärvi conceived the idea; Moona Rahikainen, Jesús Pascual, Sara Alegre, Guido Durian and Saijaliisa Kangasjärvi designed and wrote the paper.

REFERENCES

1. Mullineaux, P.; Ball, L.; Escobar, C.; Karpinska, B.; Creissen, G.; Karpinski, S. Are diverse signaling pathways integrated in the regulation of *Arabidopsis* antioxidant defence gene expression in response to excess excitation energy? *Phil. Trans. R. Soc. B* **2000**, *355*, 1531–1540

2. Sierla, M.; Rahikainen, M.; Salojärvi, J.; Kangasjärvi, J.; Kangasjärvi, S. Apoplastic and chloroplastic redox signaling networks in plant stress responses. *Antioxid. Redox Signal.* **2013**, *18*, 2220–2239.

3. Allahverdiyeva, Y.; Battchikova, N.; Brosché, M.; Fujii, H.; Kangasjärvi, S.; Mulo, P.; Mähönen, A.P.; Nieminen, K.; Overmyer, K.; Salojärvi, J.; *et al.* Integration of photosynthesis, development and stress as an opportunity for plant biology. *New Phytol.* **2015**, *208*, 647–655.

4. Asai, T.; Tena, G.; Plotnikova, J.; Willmann, M.R.; Chiu, W.L.; Gomez-Gomez, L.; Boller, T.; Ausubel, F.M.; Sheen, J. MAP kinase signaling cascade in *Arabidopsis* innate immunity. *Nature* **2002**, *415*, 977–983.

5. Zhang, J.; Shao, F.; Li, Y.; Cui, H.; Chen, L.; Li, H.; Zou, Y.; Long, C.; Lan, L.; Chai, J.; Chen, S.; Tang, X.; Zhou, J.M. A Pseudomonas syringae

effector inactivates MAPKs to suppress PAMP-induced immunity in plants. *Cell Host. Microb.* **2007**, *1*, 175–185.

6. Savatin, D.V.; Bisceglia, N.G.; Marti, L.; Fabbri, C.; Cervone, F.; de Lorenzo, G. The Arabidopsis NUCLEUS-AND PHRAGMOPLAST-LOCALIZED KINASE1-related protein kinases are required for elicitor-induced oxidative burst and immunity. *Plant Physiol.* **2014**,*165*, 1188–1202.

7. Hideg, É.; Kálai, T.; Vass, I. Photoinhibition of photosynthesis *in vivo* results in singlet oxygen production detection via nitroxide-induced fluorescence quenching in broad bean leaves. *Biochemistry* **1998**, *37*, 11405–11411.

8. Foyer, C.H.; Noctor, G. Redox sensing and signaling associated with reactive oxygen in chloroplasts, peroxisomes and mitochondria. *Physiol. Plant.* **2003**, *119*, 355–364.

9. Karpinski, S.; Szechynska-Hebda, M. Cellular light memory, photo-electrochemical and redox retrograde signaling in plants. *Biotechnologia* **2012**, *93*, 27–39.

10. Karpinski, S.; Szechyńska-Hebda, M.; Wituszyńska, W.; Burdiak, P. Light acclimation, retrograde signaling, cell death and immune defences in plants. *Plant. Cell Environ.* **2013**, *36*, 736–744.

11. Suorsa, M.; Järvi, S.; Grieco, M.; Nurmi, M.; Pietrzykowska, M.; Rantala, M.; Kangasjärvi, S.; Paakkarinen, V.; Tikkanen, M.; Jansson, S.; *et al.* PROTON GRADIENT REGULATION5 is essential for proper acclimation of *Arabidopsis* photosystem I to naturally and artificially fluctuating light conditions. *Plant Cell* **2012**, *24*, 2934–2948.

12. Tyystjärvi, E. Photoinhibition of Photosystem II. *Int. Rev. Cell Mol. Biol.* **2013**, *300*, 243–303.

13. Karpinski, S.; Reynolds, H.; Karpinska, B.; Wingsle, G.; Creissen, G.; Mullineaux, P.M. Systemic signaling and acclimation in response to excess excitation energy in *Arabidopsis.Science* **1999**, *284*, 654–657.

14. Karpinski, S.; Gabrys, H.; Mateo, A.; Karpinska, B.; Mullineaux, P.M. Light perception in plant disease defence signaling. *Curr. Opin. Plant Biol.* **2003**, *6*, 390–396.

15. Pogson, B.J.; Woo, N.S.; Förster, B.; Small, I.D. Plastid signaling to the nucleus and beyond.*Trends Plant Sci.* **2008**, *13*, 602–609.

16. Kangasjärvi, S.; Tikkanen, M.; Durian, G.; Aro, E.-M. Photosynthetic light reactions—An adjustable hub in basic production and plant immunity signaling. *Plant Physiol. Biochem.***2014**, *81*, 128–134.

17. Trotta, A.; Rahikainen, M.; Konert, G.; Finazzi, G.; Kangasjärvi, S. Signalling crosstalk in light stress and immune reactions in plants. *Phil. Trans. R. Soc. B* **2014**, *369*.

18. Stael, S.; Kmiecik, P.; Willems, P.; van der Kelen, K.; Coll, N.S.; Teige, M.; van Breusegem, F. Plant innate immunity—Sunny side up? *Trends Plant Sci.* **2015**, *20*, 3–11.

19. Nomura, H.; Komori, T.; Uemura, S.; Kanda, Y.; Shimotani, K.; Nakai, K.; Furuichi, T.; Takebayashi, K.; Sugimoto, T.; Sano, S.; *et al.* Chloroplast-mediated activation of plant immune signaling in *Arabidopsis*. *Nat. Commun.* **2012**, *3*, 926.

20. Caplan, J.L.; Kumar, A.S.; Park, E.; Padmanabhan, M.S.; Hoban, K.; Modla, S.; Czymmek, K.; Dinesh-Kumar, S.P. Chloroplast stromules function during innate immunity. *Dev. Cell* **2015**,*34*, 45–57.

21. Zabala, M.; Littlejohn, G.; Jayaraman, S.; Studholme, D.; Bailey, D.; Lawson, T.; Tillich, M.; Licht, D.; Bölter, B.; Delfino, L.; *et al.* Chloroplasts play a central role in plant defence and are targeted by pathogen effectors. *Nat. Plants* **2015**, *1*, 15074.

22. Mullineaux, P.M.; Baker, N.R. Oxidative stress: Antagonistic signaling for acclimation or cell death? *Plant Physiol.* **2010**, *154*, 521–525.

23. Lepistö, A.; Rintamäki, E. Coordination of plastid and light signaling pathways upon development of *Arabidopsis* leaves under various photoperiods. *Mol. Plant* **2012**, *5*, 799–816.

24. Chaouch, S.; Queval, G.; Vanderauwera, S.; Mhamdi, A.; Vandorpe, M.; Langlois-Meurinne, M.; Van Breusegem, F.; Saindrenan, P.; Noctor, G. Peroxisomal hydrogen peroxide is coupled to biotic defense responses by ISOCHORISMATE SYNTHASE1 in a daylength-related manner. *Plant Physiol.* **2010**, *153*, 1692–1705.

25. Queval, G.; Issakidis-Bourguet, E.; Hoeberichts, F.A.; Vandorpe, M.; Gakière, B.; Vanacker, H.; Miginiac-Maslow, M.; van Breusegem, F.; Noctor, G. Conditional oxidative stress responses in the *Arabidopsis* photorespiratory mutant cat2 demonstrate that redox state is a key modulator of daylength-dependent gene expression, and define photoperiod as a crucial factor in the regulation of H_2O_2-induced cell death. *Plant J.* **2007**, *52*, 640–657.

26. Perez-Perez, J.M.; Esteve-Bruna, D.; Gonzalez-Bayon, R.; Kangasjärvi, S.; Caldana, C.; Hannah, M.A.; Willmitzer, L.; Ponce, M.R.; Micol, J.L. Functional redundancy and divergence within the *Arabidopsis* RETICULATA-RELATED gene family. *Plant Physiol.* **2013**,*162*, 589–603.

27. Mühlenbock, P.; Szechynska-Hebda, M.; Plaszczyca, M.; Baudo, M.; Mateo, A.; Mullineaux, P.M.; Parker, J.E.; Karpinska, B.; Karpinski, S. Chloroplast signaling and LESION SIMULATING DISEASE1 regulate crosstalk between light acclimation and immunity in*Arabidopsis*. *Plant Cell* **2008**, *20*, 2339–2356.

28. Rasool, B.; Karpinska, B.; Konert, G.; Durian, G.; Denessiouk, K.; Kangasjärvi, S.; Foyer, C.H. Effects of light and the regulatory B-subunit composition of protein phosphatase 2A on the susceptibility of *Arabidopsis thaliana* to aphid (*Myzus persicae*) infestation. *Front. Plant Sci.***2014**.

29. Bostock, R.M. Signal crosstalk and induced resistance: Straddling the line between cost and benefit. *Annu. Rev. Phytopathol.* **2005**, *43*, 545–580.

30. Foyer, C.H.; Noctor, G. Redox regulation in photosynthetic organisms: Signaling, acclimation, and practical implications. *Antioxid. Redox Sign.* **2009**, *11*, 861–905.

31. Boudsocq, M.; Willmann, M.R.; McCormack, M.; Lee, H.; Shan, L.; He, P.; Bush, J.; Cheng, S.H.; Sheen, J. Differential innate immune signaling via Ca^{2+} sensor protein kinases. *Nature***2010**, *464*, 418–422.

32. Vogel, M.O.; Moore, M.; König, K.; Pecher, P.; Alsharafa, K.; Lee, J.; Dietz, K.J. Fast retrograde signaling in response to high light involves metabolite export, MITOGEN-ACTIVATED PROTEIN KINASE6, and AP2/ERF transcription factors in *Arabidopsis*. *Plant Cell* **2014**, *26*, 1151–1165.

33. Uhrig, R.G.; Labandera, A.-M.; Moorhead, G.B. *Arabidopsis* PPP family of serine/threonine protein phosphatases: Many targets but few engines. *Trends Plant Sci.* **2013**, *18*, 505–513.

34. Janssens, V.; Longin, S.; Goris, J. PP2A holoenzyme assembly: In cauda venenum (the sting is in the tail). *Trends Biochem. Sci.* **2008**, *33*, 113–121.

35. Farkas, I.; Dombradi, V.; Miskei, M.; Szabados, L.; Koncz, C. *Arabidopsis* PPP family of serine/threonine phosphatases. *Trends Plant Sci.* **2007**, *12*, 169–176.

36. Wu, G.; Wang, X.; Li, X.; Kamiya, Y.; Otegui, M.S.; Chory, J. Methylation of a phosphatase specifies dephosphorylation and degradation of activated brassinosteroid receptors. *Sci. Signal.* **2011**.

37. Hu, R.; Zhu, Y.; Shen, G.; Zhang, H. TAP46 plays a positive role in the ABSCISIC ACID INSENSITIVE5-regulated gene expression in *Arabidopsis*. *Plant Physiol.* **2014**, *164*, 721–734.

38. País, S.M.; González, M.A.; Téllez-Iñón, M.T.; Capiati, D.A. Characterization of potato (*Solanum tuberosum*) and tomato (*Solanum lycopersicum*) protein phosphatases type 2A catalytic subunits and their involvement in stress responses. *Planta* **2009**, *230*, 13–25.

39. Yu, R.M.; Zhou, Y.; Xu, Z.F.; Chye, M.L.; Kong, R.Y. Two genes encoding protein phosphatase 2A catalytic subunits are differentially expressed in rice. *Plant Mol. Biol.* **2003**,*51*, 295–311.

40. Yu, R.M.; Wong, M.M.; Jack, R.W.; Kong, R.Y. Structure, evolution and expression of a second subfamily of protein phosphatase 2A catalytic subunit genes in the rice plant (*Oryza sativa* L.). *Planta* **2005**, *222*, 757–768.

41. Xu, C.; Jing, R.; Mao, X.; Jia, X.; Chang, X. A wheat (*Triticum aestivum*) protein phosphatase 2A catalytic subunit gene provides enhanced drought tolerance in tobacco. *Ann. Bot.* **2007**,*99*, 439–450.

42. Liu, D.; Li, A.; Mao, X.; Jing, R. Cloning and characterization of TaPP2AbB″-α, a member of the PP2A regulatory subunit in wheat. *PLoS ONE* **2014**, *9*, e94430.

43. Pernas, M.; Garcia-Casado, G.; Rojo, E.; Solano, R.; Sanchez-Serrano, J.J. A protein phosphatase 2A catalytic subunit is a negative regulator of abscisic acid signaling. *Plant. J.***2007**, *51*, 763–778.

44. Blakeslee, J.J.; Zhou, H.-W.; Heath, J.T.; Skottke, K.R.; Barrios, J.A.; Liu, S.Y.; DeLong, A. Specificity of RCN1-mediated protein phosphatase 2A regulation in meristem organization and stress response in roots. *Plant Physiol.* **2008**, *146*, 539–553.

45. Cutler, S.R.; Rodriguez, P.L.; Finkelstein, R.R.; Abrams, S.R. Abscisic acid: Emergence of a core signaling network. *Annu. Rev. Plant Biol.* **2010**, *61*, 651–679.

46. Luo, J.H.; Shen, G.X.; Yan, J.Q.; He, C.X.; Zhang, H. AtCHIP functions as an E3 ubiquitin ligase of protein phosphatase 2A subunits and alters plant response to abscisic acid treatment. *Plant J.* **2006**, *46*, 649–657.

47. Leivar, P.; Antolin-Llovera, M.; Ferrero, S.; Closa, M.; Arro, M.; Ferrer, A.; Boronat, A.; Camposa, N. Multilevel control of *Arabidopsis* 3-hydroxy-3-methylglutaryl coenzyme A reductase by protein phosphatase 2A. *Plant Cell* **2011**, *23*, 1494–1511.

48. Latorre, K.; Harris, D.M.; Rundle, S.J. Differential expression of three *Arabidopsis* genes encoding the B′ regulatory subunit of protein phosphatase 2A. *Eur. J. Biochem.* **1997**, *245*, 156–163.

49. Konert, G.; Rahikainen, M.; Trotta, A.; Durian, G.; Salojärvi, J.; Khorobrykh, S.; Tyystjärvi, E.; Kangasjärvi, S. Subunits B′γ and B′ζ of protein phosphatase 2A regulate photo-oxidative stress responses and growth in *Arabidopsis thaliana*. *Plant Cell Environ.* **2015**, *38*, 2641–2651.

50. He, X.; Anderson, J.C.; del Pozo, O.; Gu, Y.Q.; Tang, X.; Martin, G.B. Silencing of subfamily I of protein phosphatase 2A catalytic subunits results in activation of plant defense responses and localized cell death. *Plant J.* **2004**, *38*, 563–577.

51. Trotta, A.; Wrzaczek, M.; Scharte, J.; Tikkanen, M.; Konert, G.; Rahikainen, M.; Holmström, M.; Hiltunen, H.M.; Rips, S.; Sipari, N.; *et al.* Regulatory subunit B′γ of protein phosphatase 2A prevents unnecessary defense reactions under low light in *Arabidopsis*. *Plant Physiol.***2011**, *156*, 1464–1480.

52. Kataya, A.R.A.; Behzad, H.; Cathrine, L. Protein phosphatase 2A regulatory subunits affecting plant innate immunity, energy metabolism, and flowering time—Joint functions among B′η subfamily members. *Plant Signal. Behav.* **2015**, *10*, 37–41.

53. Segonzac, C.; Macho, A.P.; Sanmartín, M.; Ntoukakis, V.; Sánchez-Serrano, J.J.; Zipfel, C. Negative control of BAK1 by protein phosphatase 2A during plant innate immunity. *EMBO J.* **2014**, *33*, 1–11.

54. Xia, X.-J.; Zhou, Y.-H.; Shi, K.; Zhou, J.; Foyer, C.H.; Yu, J.-Q. Interplay between reactive oxygen species and hormones in the control of plant development and stress tolerance. *J. Exp. Bot.* **2015**, *66*, 2839–2856.

55. Li, S.C.; Mhamdi, A.; Trotta, A.; Kangasjärvi, S.; Noctor, G. The protein phosphatase subunit PP2A-B′γ is required to suppress day length-dependent pathogenesis responses triggered by intracellular oxidative stress. *New Phytol.* **2014**, *202*, 145–160.

56. Sirichandra, C.; Gu, D.; Hu, H.-C.; Davanture, M.; Lee, S.; Djaoui, M.; Valot, B.; Zivy, M.; Leung, J.; Merlot, S.; *et al.* Phosphorylation of the *Arabidopsis* AtrbohF NADPH oxidase by OST1 protein kinase. *FEBS Lett.* **2009**, *583*, 2982–2986.

57. Waadt, R.; Manalansan, B.; Rauniyar, N.; Munemasa, S.; Booker, M.A.; Brandt, B.; Waadt, C.; Nusinow, D.A.; Kay, S.A.; Kunz, H.-H.; *et al.* Identification of open stomata1-interacting proteins reveals interactions with sucrose non-fermenting1-related protein kinases2 and with type 2A protein phosphatases that function in abscisic acid responses. *Plant Physiol.***2015**, *169*, 760–779.

58. Konert, G.; Trotta, A.; Kouvonen, P.; Rahikainen, M.; Durian, G.; Blokhina, O.; Fagerstedt, K.; Muth, D.; Corthals, G.L.; Kangasjärvi, S. Protein phosphatase 2A (PP2A) regulatory subunit B′ gamma interacts with cytoplasmic ACONITASE 3 and modulates the abundance of AOX1A and AOX1D in *Arabidopsis thaliana*. *New Phytol.* **2015**, *205*, 1250–1263.

59. Rhoads, D.M.; Umbach, A.L.; Subbaiah, C.C.; Siedow, J.N. Mitochondrial reactive oxygen species. Contribution to oxidative stress and interorganellar signaling. *Plant Physiol.* **2006**,*141*, 357–366.

60. Strodtkoetter, I.; Padmasree, K.; Dinakar, C.; Speth, B.; Niazi, P.S.; Wojtera, J.; Voss, I.; Do, P.T.; Nunes-Nesi, A.; Fernie, A.R.; *et al.*

Induction of the aox1d isoform of alternative oxidase in *A. thaliana* T-DNA insertion lines lacking isoform AOX1A is insufficient to optimize photosynthesis when treated with antimycin A. *Mol. Plant* **2009**, *2*, 284–297.

61. Vandenabeele, S.; Vanderauwera, S.; Vuylsteke, M.; Rombauts, S.; Langebartels, C.; Seidlitz, H.K.; Zabeau, M.; van Montagu, M.; Inzé, D.; van Breusegem, F. Catalase deficiency drastically affects gene expression induced by high light in *Arabidopsis thaliana*. *Plant J.* **2004**, *39*, 45–58.

62. Mhamdi, A.; Queval, G.; Chaouch, S.; Vanderauwera, S.; van Breusegem, F.; Noctor, G. Catalase function in plants: A focus on *Arabidopsis* mutants as stress-mimic models. *J. Exp. Bot.* **2010**, *61*, 4197–4220.

63. Queval, G.; Neukermans, J.; Vanderauwera, S.; van Breusegem, F.; Noctor, G. Day length is a key regulator of transcriptomic responses to both CO_2 and H_2O_2 in *Arabidopsis*. *Plant Cell Environ.* **2012**, *35*, 374–387.

64. De Clercq, I.; Vermeirssen, V.; van Aken, O.; Vandepoele, K.; Murcha, M.W.; Law, S.R.; Inzé, A.; Ng, S.; Ivanova, A.; Rombaut, D.; *et al.* The membrane-bound NAC transcription factor ANAC013 functions in mitochondrial retrograde regulation of the oxidative stress response in *Arabidopsis*. *Plant Cell* **2013**, *25*, 3472–3490.

65. Giraud, E.; Ho, L.H.; Clifton, R.; Carroll, A.; Estavillo, G.; Tan, Y.F.; Howell, K.A.; Ivanova, A.; Pogson, B.J.; Millar, A.H.; Whelan, J. The absence of alternative oxidase1a in *Arabidopsis*results in acute sensitivity to combined light and drought stress. *Plant Physiol.* **2008**, *147*, 595–610.

66. Fryer, M.J.; Ball, L.; Oxborough, K.; Karpinski, S.; Mullineaux, P.M.; Baker, N.R. Control of Ascorbate Peroxidase 2 expression by hydrogen peroxide and leaf water status during excess light stress reveals a functional organisation of *Arabidopsis* leaves. *Plant J.* **2003**, *33*, 691–705.

67. Estavillo, G.M.; Crisp, P.; Pornsiriwong, W.; Wirtz, M.; Collinge, D.; Carrie, C.; Giraud, E.; Whelan, J.; David, P.; Javot, H.; *et al.* Evidence for a SAL1-PAP chloroplast retrograde pathway that functions in drought and high light signaling in *Arabidopsis*. *Plant Cell* **2011**, *23*, 3992–4012.

68. Rossel, J.B.; Wilson, I.W.; Pogson, B.J. Global changes in gene expression in response to high light in *Arabidopsis*. *Plant Physiol.* **2012**, *130*, 1109–1120.

69. Jung, H.-S.; Crisp, P.A.; Estavillo, G.M.; Cole, B.; Hong, F.; Mockler, T.C.; Pogson, B.J.; Chory, J. Subset of heat-shock transcription factors required for the early response of*Arabidopsis* to excess light. *Proc. Natl. Acad. Sci. USA* **2013**, *110*, 14474–14479.

70. Volkov, R.; Panchuk, I.I.; Mullineaux, P.M.; Schöffl, F. Heat stress-induced H_2O_2 is required for effective expression of heat shock genes in *Arabidopsis*. *Plant Mol. Biol.* **2006**, *61*, 733–746.

Chapter 4

Impact of Cell Wall Composition on Maize Resistance to Pests and Diseases

Rogelio Santiago *, Jaime Barros-Rios † and Rosa A. Malvar

Misión Biológica de Galicia (CSIC), Pontevedra E-36080, Spain Current address: Department of Plant Physiology, Umeå Plant Science Centre, Umeå University, Umeå 90187, Sweden

ABSTRACT

In cereals, the primary cell wall is built of a skeleton of cellulosic microfibrils embedded in a matrix of hemicelluloses and smaller amounts of pectins, glycoproteins and hydroxycinnamates. Later, during secondary wall development, *p*-coumaryl, coniferyl and sinapyl alcohols are copolymerized to form mixed lignins. Several of these cell wall components show a determinative role in maize resistance to pest and diseases. However, defense mechanisms are very complex and vary among the same plant species, different tissues or even the same tissue at different developmental stages. Thus, it is important to highlight that the role of the cell wall components needs to be tested in diverse genotypes and specific tissues where the feeding or attacking by the pathogen takes place. Understanding the role of cell wall constituents as defense mechanisms may allow modifications of crops to withstand pests and diseases.

KEYWORDS

maize (*Zea mays* L.); pests and diseases; cell wall; cellulose; arabinoxylans; hydroxycinnamates; lignin; ferulic acid; borer; maize weevil; ear and stalk rot; rust; leaf blight

1. INTRODUCTION

Plants rely on their cell walls to provide shape and strength to cells, to glue cells together and to give rigidity to the whole plant [1]. The composition of plant cell walls varies significantly from one cell type to another, one species to another, and between accessions within species [2]. There is a special overview of the structure and synthesis of plant cell walls in Carpita and McCann, 2000 [3]. The growing cell wall consists mainly of cellulose microfibrils embedded in a matrix of hemicelluloses, pectins, proteins and phenolics [3,4]. For instance, a typical dicot wall contains 30% cellulose, 30% hemicellulose, 35% pectin and 1%–5% structural protein. By contrast, a typical grass species contains, 25% cellulose, 55% hemicellulose (arabinoxylans, xyloglucan and, in some tissues, mixed-linked glucans) and only 10% pectin [5]. The xylans are substituted with arabinose and glucuronic acid side chains, and are called arabinoxylan (AX) or glucuronoarabinoxylans (GAX) [6]. A proportion of the arabinose side chains are further substituted with phenolic acid residues, notably ferulic (4-hydroxy-3-methoxycinnamate), *p*-coumaric (4-hydroxycinnamate) and 4-*O*-methylglucuronic acids [7]. The arabinoxylans in grasses serve a major structural role by binding to cellulose microfibrils and becoming oxidatively cross-linked with each other and with lignin by phenolics [8,9]. Later in plant development, the lignin polymer becomes a significant constituent of the cell wall (Figure 1).

Plants anti-nutritional defenses against insects can occur as both pre-ingestion to limit food supply, and post-ingestion to reduce nutrient value to the attacking insect. In the pre-ingestion phase, host plants can limit food supplies to insects via physical barriers such as the cell wall fortification. Lignin and other phenolics can strengthen cell walls against digestion and therefore can be anti-nutritional [48,49]. Plant cell wall components are important factors dictating plant resistant to mechanical injury and the tearing action of mandibles [50,51]. On the other hand, diverse cell wall components have been implicated in disease resistance by a number of researchers [52–55]. Lignin and the presence of phenols in host cell walls act as a physical barrier against fungal penetration [55,56]. In the current review, we describe the role of key cell wall components as barriers that limit insects feeding and diseases progression (Table 1). The research is focused on the maize (*Zea mays* L.) cultivation as a model of monocotyledon crop.

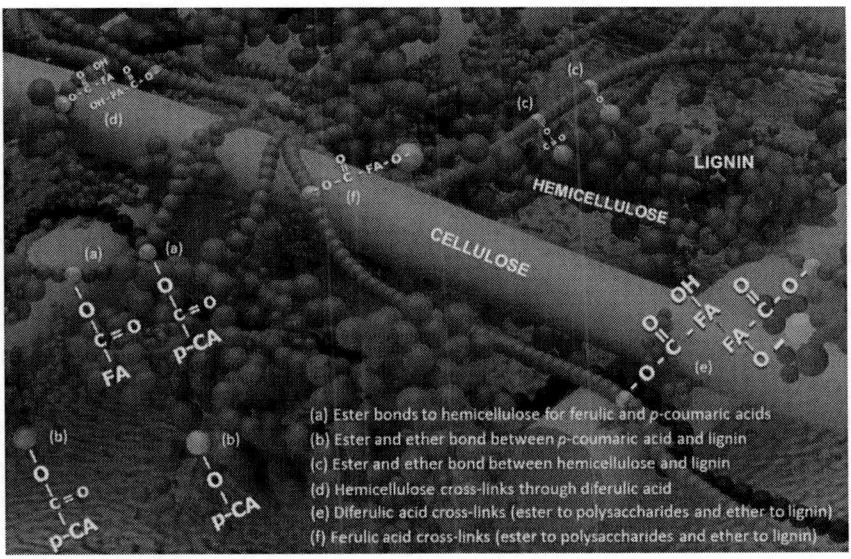

Figure 1. Representation of the secondary cell wall structure in maize. Adapted from Barros-Rios *et al.* [10].

2. CELL WALL POLYSACCHARIDES

Cellulose is the most abundant polymer in nature. It is the main component of the plant cell wall, highly stable and insoluble in water, is synthesized by plasma membrane protein complexes, and is deposited directly on the wall in a directional way [57]. It is a linear homopolymer composed of β (1→4) linked D-glucose units. Cellobiose [glucose dimer linked by β (1→4)] is the repeating unit of cellulose. In contrast, the term hemicellulose refers to a diverse class of polysaccharides which differ from the cellulose because this fraction is amorphous and easily soluble or hydrolysable in alkaline or acid solutions. Hemicellulose is tangled with cellulose microfibrils by hydrogen bonds giving the latter a greater inaccessibility [58]. In maize the most abundant polysaccharides are GAX, considered substitutes of pectins [59], and β-glucans with mixed connections. The basic structure of GAX is a β (1→4)-D-xylan backbone with substitutions of arabinose, glucuronic acid and acetic acid.

Most of the literature on the role of cell wall polysaccharides in pests and diseases resistance is based on the study of fibers. Fiber is composed largely of cellulose, hemicellulose and lignin, which are the primary components of plant cell walls [60]. Neutral detergent fiber (NDF) is the most common measure of fiber; it covers most of the structural components in the cell walls (*i.e.*, cellulose, hemicellulose and lignin). On the other hand, acid detergent fiber (ADF) refers solely to the insoluble fiber within a plant cell (cellulose and lignin). In this section we present research concerning the fibers as a bulk (including acid detergent lignin (ADL)), while a more specific role of the lignin polymer in resistance is discussed in section 4.

Fibers can disturb feeding insects from both the nutritional and physical point of view. High levels of fiber can increase the bulk density of the diet, and thereby, larvae are unable to ingest adequate amounts of nutrients and water. Besides, tissues with high fiber content are tougher and more resistant to the cutting or chewing action of the mandibles. In addition, some studies pointed out the role of cell walls polysaccharides as barriers to the fungal invasion.

European corn borer (ECB), *Ostrinia nubilalis*, Hübner, is a tunneling insect that bores into maize stems causing physical damage, disruption of nutrient and water flow, stalk lodging and grain damage, leading to extensive yield losses. In temperate regions it has one to three generations per year. Damage to the plant results from leaf feeding in the whorl (first generation), leaf sheath, tassel and stalk tunneling (first, second and third generations), and ear damage (second and third generations). Whorl, sheath and collar tissues of 22 inbred lines and 3 full-sib families were evaluated for first and second generation ECB damage, and ADF, lignin and silica concentration were quantified [11]. For the first generation ECB the role of lignin in the resistance was shown, while for the second generation ECB the silica was the only significant trait related with resistance.

Table 1. Summary of literature attending to the role of the cell wall composition on maize resistance to pests and diseases.

Pest or disease	Tissue/organ	Cell wall component	References
Ostrinia nubilalis (ECB)	whorl tissue and leaf sheath, leaf blade and stalk	NDF, ADF and lignin	[11,12]
ECB	stalks and leaf sheaths	crude fiber, cellulose, hemicellulose, NDF, ADF, NDF adjusted by ADL, ADL, lignin and silica	[13–18]
ECB	immature and mature leaf blade and leaf sheath, rind, node, and pith	crude fibre and hydroxycinnamates	[19–21]
Diatraea grandiosella (SWCB) and *Spodoptera frugiperda* (FAW)	whorl leaves and leaves	hemicellulose, cellulose, lignin	[22–25]
SWCB and *Diatraea saccharalis* (SCB)	leaves	hydroxycinnamates	[26]
Sesamia nonagrioides (MCB)	stalk, pith, and leaf-sheaths	hydroxycinnamates	[27–31]
MCB and ECB	pith and rind	cell wall polysaccharides, lignin and hydroxycinnamates	[32]

Sithophilus zeamais (MW)	grain and pericarp	protein and hydroxycinnamates	[33–37]
Fusarium graminearum	pith	Hydroxycinnamates lignin	[38,39]
F. graminearum	grain, pericarp and aleurone	hydroxycinnamates	[40,41]
F. graminearum	silks	hemicellulose	[42]
Fusarium verticillioides	pericarp	hydroxycinnamates	[43]
Aspergillus flavus	grain	fiber and carbohydrates	[44]
A. flavus	cob	lignin content and composition	[45]
Puccinia polysora	leaves	NDF and ADF	[46]
Helminthosporium maydis	leaves	hydroxycinnamates	[47]

This initial study was focused on the analysis of diverse genotypes, but it is interesting to analyze the relationships between the cell wall polysaccharides and resistance within the same gene pool. Buendgen and coauthors [13] studied if five cycles of recurrent selection for first and second generation ECB resistance in the BS9 maize population resulted in correlated changes in ADF, NDF, cellulose, hemicellulose, lignin, ash, and N concentration in the whorls, leaf sheaths, and stalks. No appreciable changes in NDF, ADF, lignin, ash, or N concentrations in the whorls were associated with the first generation ECB resistance. Traits like the hydroxamic acid 2,4-dihydroxy-7-methoxy-1,4- benzoxazin-3-one (DIMBOA) were more effective on this relationship [19,61]. However, the authors highlight the highly significant linear responses observed in the leaf sheath for NDF, ADF, cellulose, and lignin over cycles of selection in BS9, suggesting a genetic association of these constituents with the second generation ECB resistance. Similar results were obtained when stalk tissues were analyzed. Furthermore, Bergvinson and coworkers [19], within the same recurrent selection material, concluded that feeding on mature leaf, leaf sheath, and stalk tissues appears to be affected, among other factors, by fiber.

Two maize populations, WFISIHI and WFISILO, which were developed for high and low concentrations, respectively, of indigestible leaf-sheath constituents (ADF, lignin, and silica) were evaluated by Buendgen *et al.* [13] in relation to ECB resistance. The comparison of tissue composition between WFISIHI and WFISILO suggests that, contrary to the results in BS9, structural carbohydrate, lignin, and ash appeared to have little effect on ECB resistance. As the authors remark, the genetic backgrounds of BS9, WFISIHI, and WFISILO differ greatly. Resistance mechanisms in maize can differ depending on genotype analyzed [62]. Subsequently, Ostrander and Coors [12] found that after two cycles of selection WFISILO and WFISIHI have diverged further for all compositional traits except whorl ADF, and WFISILO C2 was significantly more susceptible to first and second generation ECB infestations than WFISIHI C2. In the same way, selection for high NDF and lignin concentrations in BS9 C2 produced materials with greater second generation ECB resistance than selection for low concentration. New evidence of the relationship between the fibers and the ECB second generation resistance was shown in a study conducted by Beeghly and coauthors [14]. Significant phenotypic and genetic correlations were found between ECB tunneling and fiber components. Overall, we can note an association between the plant cell wall composition and ECB resistance (mainly in the second generation), although the genetic backgrounds influence this relationship.

In an incidental way, Martin *et al.* [15] carried out a divergent selection program for stalk strength by using a rind penetrometer. After twelve cycles of divergent selection stalk strength was decreased and increased in the high and low directions of selection, respectively. Selection for high rind penetrometer resistance was effective at providing resistance to second-generation ECB damage as well as resistance to stalk lodging. Interestingly, crude fiber, cellulose and lignin concentration increased in the high direction of selection, suggesting that stalk composition was important in providing rind penetrometer resistance, stalk lodging resistance, and ECB resistance.

More recently, four maize inbred lines were selected for their consistent resistance (EP39 and A509) or susceptibility (EP47 and EP42) to corn borers across multiple evaluations [63,64]. The resistant group showed higher total cell wall material in the pith, particularly those components abundant in secondary walls (glucose, xylose, and lignin), suggesting that cell wall thickness may be an important barrier that corn borers must mechanically rupture in order to gain access to the cytoplasmic nutrients contained within the pith tissues [32].

Multiple regression equations to predict ECB damage were based on the concentration of cell wall constituents in most of the previous studies. This approach is effective when the independent variables are not highly correlated. Whenever independent variables are highly collinear, their relative importance cannot be determined using either ordinary least-squares or biased regression methods since the data are inadequate for this purpose [65]. Cardinal and coauthors [66] found that the cell wall components were highly correlated in a recombinant inbred line population (RIL) obtained from the cross of maize inbred lines B73 and B52. Therefore, the authors used a direct approach: the comparisons of quantitative trait loci (QTL) detected for cell wall components and ECB tunneling in the same maize population. NDF, ADF, and ADL levels were analyzed in stalk and leaf-sheath. The results showed that most of the QTL for ECB resistance (10/13) were at QTL positions for one or more cell wall components. Seven of the ten QTL associated with both traits contributed to the negative correlations observed between these traits, while relatively few QTL effects were positively correlated. This suggests that although cell wall components contribute to ECB resistance in this population, other mechanisms and likewise other genes are involved [16].

Genotypic correlations between ECB stalk tunnel and cell wall components were not significant in RIL population developed from the cross of inbred lines B73 and DE811. Nevertheless, clustering of QTL for ECB damage and cell wall components was observed and negative genotypic correlations between both traits were observed at some loci. So, ECB resistance may be associated with a subset of the QTL observed for cell wall components and ADF in particular [17]. Comparing both studies, only three of the nine QTL for ECB resistance in the RILs of B73 × B52 were linked (within 25 cM) to QTL in the RILs of B73 × DE811 [16,18], what showing once again that the relationship found seems to be depend on the genetic materials assessed.

In Spain and other Mediterranean countries, a major insect pest of maize is the Mediterranean corn borer (MCB), *Sesamia nonagrioides* Lefèbvre [67,68]. Like other borers the main importance resides in the second generation larvae that feed on the pith during plant development causing weakening and lodging of the maize plants. Up to 95% of plants may be damaged and yield losses may reach 30% [69,70]. Three QTLs for MCB stalk tunneling detected in the RIL population EP42 × EP39 by Ordás and coauthors [71] were close to QTLs for stalk strength or cell wall constituents identified in various studies, suggesting that genes involved in the synthesis of cell wall compounds could be good candidates for resistance to corn borers. Two QTLs for MCB resistance were found in the RIL population B73 × Mo17 [72]. The QTL for stalk tunneling in the region of bin 9.04 overlapped with the stalk tunneling QTL detected in crosses of B73 with B52, De811 and Mo47, which are three inbred lines derived from diverse genetic backgrounds [73]. The region was also associated with leaf sheath and stalk fiber and, moreover, the allele associated with highest fiber was associated with lowest ECB tunneling [16,17].

Southwestern corn borer (SWCB), *Diatraea grandiosella* Dyar. and Fall armyworm (FAW),*Spodoptera frugiperda* J.E. Smith, are additional pests of maize. These insects attack plants in both vegetative and reproductive stages. In relation with the cell wall composition, Hedin and coauthors [22] demonstrated that resistant inbreds to SWCB have higher hemicellulose and crude fiber content than susceptible inbred lines, and moreover, crude fiber was negatively correlated with larval feeding damage. Additionally, the relationship between hemicellulose and FAW resistance was also observed [23,24]. Subsequent research showed that leaves of resistant hybrids to FAW presented higher amounts of hemicellulose than leaves of susceptible genotypes [25]. Hemicellulose and protein levels appear to be associated with FAW resistance, but not with SWCB resistance. Instead, relative amounts of cellulose were negatively correlated with leaf feeding by both insects. Amounts of ash, fat, and lignin did not appear to be associated with resistance to these pests [25].

The role of the cell wall polysaccharides in disease resistance has been scarcely studied. Mideros and coauthors (2012) [44] investigated the relation between *Aspergillus flavus* Link colonization and different kernel traits in a set of 19 inbred lines. *A. flavus* is the most common causal agent of aflatoxin contamination, aflatoxin B1 being the most potent naturally occurring carcinogen known [74]. They found significant correlations of aflatoxin content with fiber, ash, carbohydrates, and seed weight. Fiber and ash content may correlate with the thickness of pericarp and aleurone layers, reflecting barriers to fungal entrance.

Of the three rust pathogens that occur on corn worldwide, southern rust of corn (SRC), caused by *Puccinia* polysora Underw., has been reported as the most destructive [75]. Higher ADF and NDF contents were found in the leaves of a resistant inbred line to *P. polysora* than in a susceptible, though note that only two genotypes were analyzed [46].

Fusarium graminearum Schwabe is one of the most important plant pathogens infecting several cereal species including maize, wheat, and barley, in cold and temperate regions [76,77]. The major economic importance comes from the contamination of grains with mycotoxyns, mainly zearalenone and deoxynivalenol (DON). In maize, both the ear and the stalk can be infected [76,78]. Ester-bound ferulic acid decreased in the silks tissues after inoculation with *Fusarium graninearum*, probably due to degradation of hemicellulose by hydrolytic enzymes produced by the fungus [42]. Differences in the release of ferulates in resistant and susceptible genotypes suggest the influence of hemicelluloses in *F. graminearum* resistance [42].

3. CELL WALL BOUND HYDROXYCINNAMATES

The most common hydroxycinnamates found in a wide range of grasses are ferulates (FA) and *p*-coumarates (pCA). Cell-wall hydroxycinnamates are derived from the phenylpropanoid pathway, which originates from phenylalanine and tyrosine. Though structurally related, they seem to have different functional roles within the cell wall. FA is the major hydroxycinnamate derivative in young grass cell walls. Maize cell walls can contain up to 5% ferulate monomers plus dimers [79]. Ferulic acid is primarily esterified to arabinosyl residues of arabinoxylan chains, and feruloylated arabinoxylans are later cross-linked to G units of lignins via ether bonds. *p*-Coumaric acid is mainly esterified to the γ-position of the side chains of S lignin units [80,81] and lignified maize cell walls can contain up to 3% *p*-coumarate [79].

It is suspected that FA is a very important components not only for the biology of the cell wall, but also for its structure because they can be coupled by peroxidase- or laccase-mediated oxidative coupling to form a variety of dimers, cross-linking polysaccharide chains [1,82,83]. In addition, the ability of ferulic acid to participate in ester linkages and phenol coupling reactions confers it with the ability to covalently attach polysaccharide with lignin. The resultant product is a ferulate-polysaccharide-lignin complex bonded through ester-ether linkages [8,84]. Thus, FA deposition in the wall may not only lead to cell wall cross-linking during plant growth and development, but may also regulate the non-random pattern of lignin formation within the wall [9,85,86]. On the other hand, the role of pCA in cell wall matrices is not fully understood. It appears the majority of pCA remains unincorporated other than its attachment to monolignols via an ester linkage. Therefore, pCA does not function as a cross-linking agent between wall matrix polymers, at least between different lignin polymers or between lignin and polysaccharides. It has been suggested that pCA may function as radical transfer agent to aid in the formation of sinapyl alcohol and lignin radicals [87,88]. There are reports of pCA being ester linked to arabinoxylans just as for FA, however this level is much lower (1:15) compared to FA [89]. Incorporation of pCA into grass walls appears to be primarily associated with the lignin fraction. Several studies have shown a positive relationship between pCA incorporation into cell wall matrices and levels of lignin formation [90–92].

Accumulation of FA follows the elongation pattern, with up to 70% of alkali-labile ferulate deposition that occurs during secondary wall formation and lignification [93]. Deposition of esterified and etherified ferulate derivatives continue at a high rate after cell elongation has ceased [94]. FA content plateaus as the walls mature [90]. The accumulation of pCA is a relevant indicator of lignin deposition [95,96]. In addition, it indicates a higher proportion of secondary wall in tissue [79].

Although pCA also form cell wall cross-links by the formation of radically or photochemically formed dimers, ferulate derivatives are the quantitatively most important cross-links in the plant cell wall. The first FA dimer (DFA) discovered was the 5-5′-dehydrodimer, isolated after alkaline hydrolysis of the cell wall polysaccharides [97]. Since then, a number of dehydrodimers have been identified and characterized in different grass species. The major forms are: 5-5′-DFA: (*E*,*E*)-4-4′-dehydroxy-5-5′-dimethoxy-3′-bicinnamic acid; 8-0-4′-DFA: (*Z*)-β-4-hydroxy-3-methoxycinnamic acid; 8-5′-DFA: (*E*-*E*)-4-4′-dihydroxy-3,5′-dimethoxy-β,3′-bicinnamic acid; 8-5′-C (benzofuran form):*trans*-5-((*E*)-2-carboxyvinyl))-2-(4-hydroxy-3-methoxyphenyl)-7-methoxy-2,3- dihydrobenzofuran-3-carboxylic acid, and 8-8-DFA (4,4′-dihydroxy-3,3′-dimethoxy-β,β′-bicinnamic acid) [98]. FA and DFAs contents have been estimated to be 25 g kg^{-1} in maize stems [99]. More recently, specific ferulate dehydrotrimers and tetramers have been isolated and identified in a variety of plant tissues [100–103]. Formation of intramolecular versus intermolecular polysaccharide crosslinks is a key question to be answered in the future.

Thus, hydroxycinnamate mediated cross-links are able to connect cell wall polymers, especially polysaccharides to each other but also to lignin and to proteins [9,104]. Plant cell wall cross-linking strengthens the cell wall [105] and is involved in the regulation of the cell wall extensibility and cessation of elongation processes [106–109]. Bonding contributes, as many studies have shown, to the low digestibility of most grass species even by ruminants and, in part, to the recalcitrance of grass tissues to direct degradation by simple mixtures of cellulases and xylanases [56,79,110–115].

There is some minor chemical disruption of the plant cell wall attending to the insect attack, for instance, majority of the plant cell wall is indigestible to lepidopteran [51,116]. Some studies with insects used cellulose as an indigestible nutrient diluter [117,118], but if the effect of the cell wall is that of a barrier to the easily digested contents, then the arrangement of the cell wall components may be more important than the absolute amount of cellulose, hemicellulose, *etc.* Plant nitrogen is a major factor contributing to larval growth [119], however, it does not matter how much nitrogen the plant contains if it is inaccessible. We conclude, once again, that the cell wall matrix appears to be a barrier that insects must mechanically rupture to assimilate nutrients. In accordance with this idea, the link between cell wall cross-linking and the insect resistance was first suggested by Fry (1986) [120]. Since then, various studies have pointed out the role of hydroxycinnamates in pests and diseases resistance [19,27,32,40]. A working theory has proposed that susceptibility of maize to pests and diseases depends on these hydroxycinnamates contents [121,122].

Concerning to ECB resistance, increased levels of FA monomer have been quantified in epidermal cell walls of leaves in maize resistant inbred lines [20], while the content of FA dimers in leaves was negatively correlated across genotypes with the leaf damage [21]. The mechanism of resistance suggests early cell wall fortification through phenolic cross-linking of hemicellulose by DFAs, increasing the hardness of the leaf tissue. Additional evidence was provided by DFAs increases in diverse feeding tissues over cycles of selection to ECB resistance in the Iowa Stiff Stalk Synthetic BS9 [19]. In addition, both DFA and pCA were negatively related with damage parameters in the rind, node and pith [19].

The related insect species SWCB and sugarcane borer (SCB), *Diatraea saccharalis* (Fabricius), are among the most important lepidopteran pests affecting maize production in Central and South America [123]. Later generations damage the plant mainly by stalk boring and tunneling, resulting in indirect yield losses. Resistance to these borers could be sensitive to early cell wall fortification by phenolic cross-linking [26]. The study showed that both cell wall bound pCA and FA were significantly and negatively correlated to leaf feeding damage. Significant negative correlations were also found between DFA content and leaf feeding damage by both insects.

Multiple studies with maize genotypes that differ in their resistance to the MCB noted the role of hydroxycinnamates in the resistance to this borer pest [27–30,32]. Monomers, pCA and FA, and diverse isomers of DFA, were identified in different feeding tissues. The amount of all these compounds in the pith was correlated with the resistance level in the genotypes, with the resistant inbreds having the highest concentrations [27]. Negative correlations were observed between the larvae weight and the DFAs concentration in the leaf-sheath [29]. Subsequent investigations reinforced the effect of hydroxycinnamates in the borer resistance. Higher concentrations of total DFAs were associated with shorter tunnel length and lower numbers of larvae per stem over cycles of selection to MCB resistance in maize synthetic EPS12 [30]. Barros-Rios and coworkers (2011) found that the stem tunneling by ECB and MCB was negatively correlated with total DFAs, 8-5′-DFA and pCA esters [32]. Additionally, it was demonstrated that the DFAs in the pith of the basal section of the maize internode could facilitate MCB larval feeding inside the stem in particular synthetic varieties [28]. Nowadays, Santiago and coworkers has divided the research topic in two branches: evaluation of corn borer resistance in a divergent selection program for DFAs concentration, and study of the variation in the hydroxycinnamates contents after borer attack. Maize populations selected for the highest total DFAs (third cycle of selection for high DFAs (CH3)) had 31% higher total DFAs than maize populations selected for the lowest total DFAs (third cycle of selection for low DFAs

(CL3)) [124]. The gain indicates that DFA deposition in maize pith was modified during the selection program and provides unique plant material to validate the genes potentially involved in the FA dimerization. Furthermore, the study demonstrates the powerful role of polysaccharide-polysaccharide cross-linking trough ester-linked DFAs as a natural resistance mechanism of maize against corn borers (damage by stem tunneling was 29% lower in C3H than in C3L) [124]. On the other hand, Rodríguez and coauthors [31] reported that the maize plants respond to the attack of MCB activating genes involved in the cell-wall organization, and moreover, the inbred line PB130 increase twice its DFAs concentration after MCB attack. Current investigations would confirm these motivating findings.

The maize weevil (MW), *Sitophilus zeamais* Motsch., is a destructive insect feeding on stored maize throughout the world [125]. Grain weight losses could range from 10% to 40% during storage [126]. Grain of MW resistant varieties has been characterized attending to the mechanical fortification of the pericarp cell wall. The principal phytochemicals involved in the mechanical fortification are FA, pCA [33], several isomers of DFAs, total phenolic acids, and structural proteins like a hydroxyproline-rich glycoprotein (HRGPs) [34–36]. Strong negative correlations with susceptibility parameters and a positive correlation with grain hardness were found for simple phenolic acids, DFAs, and HRGPs. In addition, in a more recent study the authors identified nine regions common between QTLs associated with MW susceptibility and cell wall bound compounds [37].

There is some evidence in the literature that FA and DFAs are also involved in plant protection against pathogen invasion [40,47]. Much the research attending to the relationship cell wall components and pathogens focuses in fungi. Maydis leaf blight (MLB), *Helminthosporium maydis* Y. Nisik and C. Miyake is the most serious disease in warm and wet temperate and tropical areas, where yield losses close to 70% have been reported. Lyons *et al.* (1993) [47] investigated the distribution of phenylpropanoids during disease development in resistant and susceptible maize cultivars to MLB and reported greater levels of FA and pCA produced in the resistant cultivar.

Structural hydroxycinnamates have been associated with resistance to *Fusarium* diseases in kernel and stalk pith tissues [38,40,41]. Higher contents of pericarp DFAs might act as a preformed structural barrier restricting fungal infection and mycelial progress from diseased to pericarp-intact neighboring kernels [40], while could function in the pith tissues as preformed resistance barriers prior to infection [38]. Low contents of DFAs in the early stages of infection would leave time for the pathogen to develop abundantly and to produce toxins, breaking a possible resistance response and leading to typical symptoms of disease [38]. The role of hydroxycinnamates in the infection of maize silks by *F. graminearum* was also investigated [42]. Although significant changes in the concentration of hydroxycinnamates were observed after inoculation, no role of these compounds as mechanism of resistance was observed in this tissue. As previously noted differential decrease of FA in resistant and susceptible genotypes suggests a possible role of hemicelluloses in the *F. graminearum* resistance in the silks [42].

Likewise, in warmer regions, Fusarium ear rot is prevalent and is the result of kernel infection by *Fusarium verticillioides* (Saccardo) Nirenberg. It not only reduces yield but also contaminates the infected kernels with mycotoxins, mainly fumonisins (FB1, FB2, and FB3). A recent paper reported on the possible role of DFAs in the genotypic resistance to *Fusarium veticillioides* [43]. The contents of the hydroxycinnamates detected in the pericarp of several maize genotypes suggest their contribution to kernel resistance to *F. verticillioides* [43]. In these experiments, resistant genotypes contained about a 3-fold higher level of total DFAs than susceptible ones. A stepwise regression model suggests that the increase of all individual DFA species contributed to limit fumonisin accumulation and mycelial progress of *F. verticillioides*. Crosslinking cell wall polysaccharides results in pericarp hardness [127]. Cell wall DFAs might also have a direct inhibitory effect on mycotoxin production after being released by fungal esterases and other extracellular enzymes during infection of *F. verticillioides*. The 8-5′-C DFA, the major DFA detected in the pericarp of the genotypes, showed *in vitro* to be as effective as FA to inhibit the biosynthesis of trichothecenes by*F. graminearum*[128].

Less work has been carried out quantifying the dimer content of different cell types, or if these different DFAs have different physiological roles. As better detection methods are developed, the extensive network of cross-links involving these phenolic compounds will be better conceived.

4. LIGNIN

Lignin is the second most abundant polymer found in nature after cellulose. Among the many roles lignin plays in plant growth and development are those providing structural support for land plants and as a resistance mechanism to biotic and abiotic stresses. Lignin is an end product of the phenylpropanoid pathway and a heteropolymer of three hydroxycinnamyl alcohol monomers or monolignols: *p*-coumaryl alcohol, coniferyl alcohol, and sinapyl alcohol [129]. The intermediate products of the monolignol biosynthetic pathway serve as precursors of hydroxycinnamates and other phenolic compounds. Both monolignols and their precursors are synthesized in the cytosol (endoplasmic reticulum) and later transported to the wall where lignin is deposited. Because of monolignols are relatively toxic to the cell, it has been suggested that the exported and storage forms are monolignol-glucosides. Once localized in the wall, monolignols are catalyzed by peroxidase and/or laccase to form three phenylpropanoid units: *p*-hydroxyphenyl (H), guaiacyl (G), and syringyl (S). The deposition of each lignin type in plant tissues is spatially and temporally controlled, first being deposited H units, followed by G units, and finally S units [130]. It has been found that mature maize stalks contain 4% H, 35% G, and 61% S. In general, the S/G ratio increases as the plant matures and the lack of S units creates a more condensed lignin (*i.e.*, the prevalence of C–C linkages and low methoxylation stage). The ratio between S and G units (S/G) are commonly between 1.02 and 1.61 for maize stalks and 0.20 and 0.55 for leaves. The maize the brown midrib mutation results in a reduction of lignin content (≈35%) and a variation on lignin composition with S/G ratios between 0.34 and 1.95 [131,132].

Tissue toughness is one of the key factors that regulate herbivore damage in plants [133,134] and the tougher the tissue, the greater lignin content. Tissues with high lignin concentration and otherwise cross-linked, are less palatable than tissues with low lignin content. Most research carried out to evaluate the effect of lignin on maize resistance to insects based on the analysis of forage fiber fractions (ADF, NDF, and ADL) has been commented on Section 2.

Increased lignin deposition might have additional negative effects on insect fitness because phenoloxidase enzymes are involved in lignin polymerization but also in generation of toxic by-products as reactive oxygen species, quinones and peroxides [135,136]. It is known that several of the defense-related maize metabolites are biosynthetically related.

Maysin, chlorogenic acid and phenolic acids originate from the phenylpropanoid pathway [137].

Cell wall lignification may be also induced when plants are wounded or become infected by pathogens or insects, suggesting that lignin may act as a chemical or physical barrier to protect the rest of the plant cells from further damage [54,138,139]. This lignin induced as a response to pathogen attack is considered "stress lignin", and has higher amount of the simpler (H) lignin subunits [140,141]. Ride [142] pointed out that lignification might hinder fungal growth through plant tissue in several ways. First, increased cell wall lignification and stiffness at the point of fungal attack may make walls more resistant to mechanical penetration and render it resistant to dissolution by fungal enzymes. Second, lignification of walls may restrict diffusion of enzymes and toxins from the fungus to the host and of water and nutrients from the host to the fungus, in essence starving the fungus. Third, phenolic precursors of lignin and free radicals produced during polymerization may inactivate fungal membranes, enzymes, toxins, and elicitors. Accordingly, Lyons and coauthors [47] attributed the increased lignin deposition to the infection of the leaf corn fungus *H. maydis*. In addition, Liu *et al.* [39] suggested that potassium chloride stimulates the rapid expression of PAL, tyrosine ammonia-lyase (TAL), cinnamyl alcohol dehydrogenase (CAD), and phenoloxidase activities in both resistant and susceptible maize genotypes after pathogen inoculation, increasing the induced lignin content, and enhancing the resistance of maize to *Fusarium* stalk rot. The lignification process in maize cobs from resistant and susceptible inbred lines to the pathogenic fungus *A. flavus* was studied by Spangler [45]. Lignin extracted using thioacidolysis revealed that resistant lines contained higher lignin than susceptible ones. Regarding to lignin composition, resistant inbreds had two times more G subunits than susceptible ones. However, real-time quantitative PCR showed no significant differences between genotypes for the transcription levels of the last enzyme of the lignin biosynthetic pathway caffeoyl-coA-Omethyltransferase 1 (CCoAOMT1).

In brief, increased toughness and concentration of toxic by-products of lignin synthesis (phenoloxidase activity) seem to be the most likely mechanisms for the lignin associated protection of plant tissues from pests and diseases.

5. CONCLUSIONS

Several cell wall components show a determinative role in maize resistance to pest and diseases. However, defense mechanisms are very complex and vary among the same plant species, different tissues or even the same tissue at different developmental stages. Probably we do not get a clear relationship with an overall determination of cell wall components in the whole plant of a single genotype, thus, it is important to note that the role of the cell wall components needs to be tested in diverse genotypes and specific tissues where the feeding or attacking by the pathogen takes place. Diverse studies attending to this genotype-tissue-specific content have been noted in the current review.

Understanding the role of cell wall constituents as defense mechanisms may allow modifications of crops to withstand pests and diseases. Nevertheless, we have to take into account that this defense is interconnected with other biological processes, many genes and pathways that work in the insect attack are also involved in disease defense or in responses to other stress agents. Modifications on the cell wall structure and composition should be evaluated on various pest and disease combinations in order to assess the impact on crop agronomic fitness.

ACKNOWLEDGMENTS

The authors acknowledge all workers and collaborators who participated in their own pest and diseases research studies. Studies by the authors' group were supported by the Plan Nacional I+D+I (AGL2003-00961, AGL2006-1314, AGL2009-09611, AGL2012-33415). Nowadays, R. Santiago acknowledges postdoctoral contract "Isidro Parga Pondal" supported by the Autonomous Government of Galicia and the European Social Fund.

REFERENCES

1. Brett, C.T.; Waldron, K. Cell Wall Formation. In *Physiology and Biochemistry of Plant Cell Walls*; Brett, C.T., Waldron, K., Eds.; Chapman & Hall: London, UK, 1996; pp. 75–111.

2. Hazen, S.P.; Hawley, R.M.; Davis, G.L.; Henrissat, B.; Walton, J.D. Quantitative trait loci and comparative genomics of cereal cell wall composition. *Plant Physiol* **2003**, *132*, 263–271.

3. Carpita, N.; McCann, M. The Cell Wall. In *Biochemistry and Molecular Biology of Plants*; Buchanan, B.B., Gruissem, W., Jones, R.L., Eds.; American Society of Plant Physiologists: Rockville, MD, USA, 2000; pp. 52–108.

4. Carpita, N.C.; Gibeaut, D.M. Structural models of primary cell walls in flowering consistency of molecular structure with the physical properties of the walls during growth. *Plant J* **1993**,*3*, 1–30.

5. Cosgrove, D.J. Assembly and enlargement of the primary cell wall in plants. *Annu. Rev. Cell Dev. Biol* **1997**, *13*, 171–201.

6. Whilstler, R.L.; Richards, E.L. Hemicelluloses. In *The Carbohydrates*; Pigman, W., Horton, D., Eds.; Academic Press: New York, NY, USA, 1970; Volume 2a, pp. 447–469.

7. Hartley, R.D. *p*-Coumaric and ferulic components of cell-walls of ryegrass and their relationships with lignin and digestibility. *J. Sci. Food Agric* **1972**, *23*, 1347–1354.

8. Scalbert, A.; Monties, B.; Lallemand, J.Y.; Guittet, E.; Rolando, C. Ether linkage between phenolic-acids and lignin fractions from wheta straw. *Phytochemistry* **1985**, *24*, 1359–1362.

9. Ralph, J.; Grabber, J.H.; Hatfield, R.D. Lignin-ferulate cross-links in grasses. Active incorporation of ferulate polysaccharide esters into ryegrass lignins. *Carbohydr. Res* **1995**,*275*, 167–178.

10. Barros-Ríos, J.; Malvar, R.A.; Santiago, R. Función de la pared celular del maíz (*Zea mays* L.) como mecanismos de defensa frente a la plaga del taladro (*Ostrinia nubilalis* Hüb y *Sesamia nonagrioides* Lef.). *Rev. Educ. Bioquím* **2011**, *30*, 132–142.

11. Rojanaridpiched, C.; Gracen, V.E.; Everett, H.L.; Coors, J.G.; Pugh, B.F.; Bouthyette, P. Multiple factor resistance in maize to European corn borer. *Maydica* **1984**, *29*, 305–315.

12. Ostrander, B.M.; Coors, J.G. Relationship between plant composition and European corn borer resistance in three maize populations. *Crop Sci* **1997**, *37*, 1741–1745.

13. Buendgen, M.R.; Coors, J.G.; Grombacher, A.W.; Russell, W.A. European corn borer resistance and cell wall composition of three maize populations. *Crop Sci* **1990**, *30*, 505–510.

14. Beeghly, H.H.; Coors, J.G.; Lee, M. Plant fiber composition and resistance to European corn borer in four maize populations. *Maydica* **1997**, *42*, 297–303.

15. Martin, S.A.; Darrah, L.L.; Hibbard, B.E. Divergent selection for rind penetrometer resistance and its effects on European corn borer damage and stalk traits in corn. *Crop Sci* **2004**, *44*, 711–717.

16. Cardinal, A.J.; Lee, M. Genetic relationships between resistance to stalk-tunneling by the European corn borer and cell-wall components in maize population B73 × B52. *Theor. Appl. Genet* **2005**, *111*, 1–7.

17. Krakowsky, M.D.; Lee, M.; Holland, J.B. Genotypic correlation and multivariate QTL analyses for cell wall components and resistance to stalk tunneling by the European corn borer in maize. *Crop Sci* **2007**, *47*, 485–490.

18. Krakowsky, M.D.; Lee, M.; Woodman-Clikeman, W.L.; Long, M.J.; Sharopova, N. QTL mapping of resistance to stalk tunneling by the European corn borer in RILs of maize population B73 × De811. *Crop Sci* **2004**, *44*, 274–282.

19. Bergvinson, D.J.; Arnason, J.T.; Hamilton, R.I. Phytochemical changes during recurrent selection for resistance to the European corn borer. *Crop Sci* **1997**, *37*, 1567–1572.

20. Bergvinson, D.J.; Arnason, J.T.; Pietrzak, L.N. Localization and quantification of cell wall phenolics in European corn borer resistant and susceptible maize inbreds. *Can. J. Bot. Rev. Can. Bot* **1994**, *72*, 1243–1249.

21. Bergvinson, D.J.; Arnason, J.T.; Hamilton, R.I.; Mihm, J.A.; Jewell, D.C. Determining leaf toughness and its role in maize resistance to the European corn borer (Lepidoptera, Pyralidae). *J. Econ. Entomol* **1994**, *87*, 1743–1748.

22. Hedin, P.A.; Davis, F.M.; Williams, W.P.; Salin, M.L. Possible factors of leaf feeding resistance in corn to the Southwestern corn borer. *J. Agric. Food Chem* **1984**, *32*, 262–267.

23. Hedin, P.A.; Williams, W.P.; Davis, F.M.; Buckley, P.M. Roles of amino acids, protein, and fiber in leaf feeding resistance of corn to the fall armyworm. *J. Chem. Ecol* **1990**, *16*, 1977–1995.

24. Hedin, P.A.; Davis, F.M.; Williams, W.P.; Hicks, R.P.; Fisher, T.H. Hemicellulose is an important leaf-feeding resistance factor in corn to the fall armyworm. *J. Chem. Ecol* **1996**, *22*, 1655–1668.

25. Williams, W.P.; Davis, F.M.; Buckley, P.M.; Hedin, P.A.; Baker, G.T.; Luthe, D.S. Factors associated with resistance to fall armyworm (Lepidoptera: Noctuidae) and southwestern corn borer (Lepidoptera: Crambidae) in corn at different vegetative stages. *J. Econ. Entomol* **1998**, *91*, 1471–1480.

26. Ramputh, A.I. Soluble and Cell Wall Bound Phenolic Mediated Insect Resistance in Corn and Sorghum. Ph.D. Dissertation, Ottawa-Carleton Institute of Biology, Ottawa, ON, Canada, 2002.

27. Santiago, R.; Butron, A.; Arnason, J.T.; Reid, L.M.; Souto, X.C.; Malvar, R.A. Putative role of pith cell wall phenylpropanoids in *Sesamia nonagrioides* (Lepidoptera: Noctuidae) resistance.*J. Agric. Food Chem* **2006**, *54*, 2274–2279.

28. Santiago, R.; Butron, A.; Revilla, P.; Ana Malvar, R. Is the basal area of maize internodes involved in borer resistance? *BMC Plant Biol* **2011**, *11*, 137.

29. Santiago, R.; Butron, A.; Reid, L.M.; Arnason, J.T.; Sandoya, G.; Souto, X.C.; Malvar, R.A. Diferulate content of maize sheaths is associated with resistance to the Mediterranean corn borer *Sesamia nonagrioides* (Lepidoptera: Noctuidae). *J. Agric. Food Chem* **2006**, *54*, 9140–9144.

30. Santiago, R.; Sandoya, G.; Butron, A.; Barros, J.; Malvar, R.A. Changes in phenolic concentrations during recurrent selection for resistance to the Mediterranean corn borer (*Sesamia nonagrioides* Lef.). *J. Agric. Food Chem* **2008**, *56*, 8017–8022.

31. Rodriguez, V.M.; Santiago, R.; Ana Malvar, R.; Butron, A. Inducible maize defense mechanisms against the corn borer *Sesamia*

nonagrioides: A transcriptome and biochemical approach. *Mol. Plant Microbe Interact* **2012**, *25*, 61–68.

32. Barros-Rios, J.; Malvar, R.A.; Jung, H.J.; Santiago, R. Cell wall composition as a maize defense mechanism against corn borers. *Phytochemistry* **2011**, *72*, 365–371.

33. Classen, D.; Arnason, J.T.; Serratos, J.A.; Lambert, J.D.H.; Nozzolillo, C.; Philogene, B.J.R. Correlation of phenolic acid content of maize to resistance to *Sitophilus zeamais*, the maize weevil, in CIMMYTs collections. *J. Chem. Ecol* **1990**, *16*, 301–315.

34. Arnason, J.T.; Baum, B.; Gale, J.; Lambert, J.D.H.; Bergvinson, D.; Philogene, B.J.R.; Serratos, J.A.; Mihm, J.; Jewell, D.C. Variation in resistance of mexican landraces to maize weevil*Sitophilus zeamais*, in relation to taxonomic and biochemical parameters. *Euphytica* **1994**, *74*, 227–236.

35. Garcia-Lara, S.; Bergvinson, D.J.; Burt, A.J.; Ramputh, A.I.; Diaz-Pontones, D.M.; Arnason, J.T. The role of pericarp cell wall components in maize weevil resistance. *Crop Sci* **2004**, *44*, 1546–1552.

36. Bergvinson, D.; Garcia-Lara, S. Genetic approaches to reducing losses of stored grain to insects and diseases. *Curr. Opin. Plant Biol* **2004**, *7*, 480–485.

37. García-Lara, S.; Burt, A.J.; Arnason, J.T.; Bergvinson, D.J. QTL mapping of tropical maize grain components associated with maize weevil resistance. *Crop Sci* **2010**, *50*, 815.

38. Santiago, R.; Reid, L.M.; Arnason, J.T.; Zhu, X.; Martinez, N.; Malvar, R.A. Phenolics in maize genotypes differing in susceptibility to Gibberella stalk rot (*Fusarium graminearum* schwabe).*J. Agric. Food Chem* **2007**, *55*, 5186–5193.

39. Liu, X.Y.; Jin, J.Y.; He, P.; Gao, W.; Li, W.J. Effect of potassium chloride on lignin metabolism and its relation to resistance of corn to stalk rot. *Sci. Agric. Sin* **2007**, *40*, 2780–2787.

40. Bily, A.C.; Reid, L.M.; Taylor, J.H.; Johnston, D.; Malouin, C.; Burt, A.J.; Bakan, B.; Regnault-Roger, C.; Pauls, K.P.; Arnason, J.T.; Philogene, B.J.R. Dehydrodimers of ferulic acid in maize grain pericarp and

aleurone: Resistance factors to *Fusarium graminearum*. *Phytopathology* **2003**, *93*, 712–719.

41. Assabgui, R.A.; Reid, L.M.; Hamilton, R.I.; Arnason, J.T. Correlation of kernel *E*-ferulic acid content of maize with resitance to *Fusarium graminearum*. *Phytopathology* **1993**, *83*, 949–953.

42. Cao, A.; Reid, L.M.; Butron, A.; Malvar, R.A.; Souto, X.C.; Santiago, R. Role of hydroxycinnamic acids in the infection of maize silks by *Fusarium graminearum* Schwabe.*Mol. Plant Microbe Interact* **2011**, *24*, 1020–1026.

43. Sampietro, D.A.; Fauguel, C.M.; Vattuone, M.A.; Presello, D.A.; Catalán, C.A.N. Phenylpropanoids from maize pericarp: Resistance factors to kernel infection and fumonisin accumulation by *Fusarium verticillioides*. *Eur. J. Plant Pathol* **2012**, *135*, 105–113.

44. Mideros, S.X.; Windham, G.L.; Williams, W.P.; Nelson, R.J. Tissue-specific components of resistance to aspergillus ear rot of maize. *Phytopathology* **2012**, *102*, 787–793.

45. Spangler, L.M. Impact of Lignin and Caffeoyl Coenzyme a *O*-methyltransferase 1 on*Aspergillus flavus*Growth in Maize Cobs. Master's Thesis, The Pennsylvania State University, 2008.

46. Ji, H.C.; Yamakawa, T. Cell wall contents and estimation of genetic parameters for polysora rust resistance in tropical maize (*Zea mays* L.). *J. Fac. Agric. Kyushu Univ* **2011**, *56*, 231–236.

47. Lyons, P.C.; Hipskind, J.; Vincent, J.R.; Nicholson, R.L. Phenylpropanoid dissemination in maize resistant or susceptible to *Helminthosporium maydis*. *Maydica* **1993**, *38*, 175–181.

48. Brodeur-Campbell, S.E.; Vucetich, J.A.; Richter, D.L.; Waite, T.A.; Rosemier, J.N.; Tsai, C.-J. Insect herbivory on low-lignin transgenic aspen. *Environ. Entomol* **2006**, *35*, 1696–1701.

49. Schroeder, F.C.; del Campo, M.L.; Grant, J.B.; Weibel, D.B.; Smedley, S.R.; Bolton, K.L.; Meinwald, J.; Eisner, T. Pinoresinol: A lignol of plant origin serving for defense in a caterpillar. *Proc. Natl. Acad. Sci. USA* **2006**, *103*, 15497–15501.

50. Read, J.; Stokes, A. Plant biomechanics in an ecological context. *Am. J. Bot* **2006**, *93*, 1546–1565.

51. Clissold, F.J. The Biomechanics of Chewing and Plant Fracture: Mechanisms and Implications. In *Advances in Insect Physiology: Insect Mechanics and Control*; Casas, J., Simpson, S.J., Eds.; Elsevier Science Serials: Amsterdam, The Netherlands, 2007; Volume 34, pp. 317–372.

52. Bell, A.A. Biochemical mechanisms of disease resistance. *Annu. Rev. Plant Physiol. Plant Mol. Biol* **1981**, *32*, 21–81.

53. Friend, J. Plant phenolics, lignification and plant disease. *Prog. Phytochem* **1981**, *7*, 197–261.

54. Nicholson, R.L.; Hammerschmidt, R. Phenolic compounds and their role in disease resistance. *Annu. Rev. Phytopathol* **1992**, *30*, 369–389.

55. Matern, U.; Grimmig, B.; Kneusel, R.E. Plant cell wall reinforcement in the disease resistance response. Molecular composition and regulation. *Can. J. Bot. Rev. Can. Bot* **1995**, *73*, S511–S517.

56. Grabber, J.H.; Hatfield, R.D.; Ralph, J. Diferulate cross-links impede the enzymatic degradation of non-lignified maize walls. *J. Sci. Food Agric* **1998**, *77*, 193–200.

57. Taylor, N.G. Cellulose biosynthesis and deposition in higher plants. *New Phytol* **2008**, *178*, 239–252.

58. Somerville, C.; Bauer, S.; Brininstool, G.; Facette, M.; Hamann, T.; Milne, J.; Osborne, E.; Parezdez, A.; Persson, S.; Raab, T.; *et al.* Towards a systems approach to understanding plant cell walls. *Science* **2004**, *306*, 2206–2211.

59. Carpita, N.C. Structure and biogenesis of the cell walls of grasses. *Annu. Rev. Plant Physiol. Plant Mol. Biol* **1996**, *47*, 445–476.

60. Van Soest, P. *Nutritional Ecology of the Ruminant*; Cornell University Press: Ithaca, NY, USA; p. 1994.

61. Grombacher, A.W.; Russell, W.A.; Guthrie, W.D. Resistance to first generation European corn borer (Lepidoptera, Pyralidae) and DIMBOA concentration in midwhorl leaves of the BS9 maize synthetic. *J. Kansas Entomol. Soc* **1989**, *62*, 103–107.

62. Santiago, R.; Souto, X.C.; Sotelo, J.; Butron, A.; Malvar, R.A. Relationship between maize stem structural characteristics and resistance to pink stem borer (Lepidoptera: Noctuidae) attack.*J. Econ. Entomol* **2003**, *96,* 1563–1570.

63. Butron, A.; Malvar, R.A.; Cartea, M.E.; Ordas, A.; Velasco, P. Resistance of maize inbreds to pink stem borer. *Crop Sci* **1999**, *39*, 102–107.

64. Ordas, B.; Butron, A.; Soengas, P.; Ordas, A.; Malvar, R.A. Antibiosis of the pith maize to Sesamia nonagrioides (Lepidoptera: Noctuidae). *J. Econ. Entomol* **2002**, *95*, 1044–1048.

65. Rawlings, J.O. *Applied Regression Analysis: A Research Tool*, 2nd ed; Springer: New York, NY, USA, 1988.

66. Cardinal, A.J.; Lee, M.; Moore, K.J. Genetic mapping and analysis of quantitative trait loci affecting fiber and lignin content in maize. *Theor. Appl. Genet* **2003**, *106*, 866–874.

67. Cordero, A.; Malvar, R.A.; Butron, A.; Revilla, P.; Velasco, P.; Ordas, A. Population dynamics and life-cycle of corn borers in south Atlantic European coast. *Maydica* **1998**, *43*, 5–12.

68. Velasco, P.; Revilla, P.; Monetti, L.; Butron, A.; Ordas, A.; Malvar, R.A. Corn borers (Lepidoptera: Noctuidae; Crambidae) in Northwestern Spain: Population dynamics and distribution. *Maydica* **2007**, *52*, 195–203.

69. Hilal, A. Etude du développement de *Sesamia nonagrioides* et établissement de modèles pour la prévision de ses populations dans la nature. *Eur. Mediterranean Plant Prot. Organ. Bull***1981**, *11*, 107–112.

70. Larue, P. La Sésamie du maïs (*Sesamia nonagrioides* Lef.). Dégâts et actualisation de la lutte.*Def. Veg* **1984**, *227*, 163–179.

71. Ordas, B.; Malvar, R.A.; Santiago, R.; Butron, A. QTL mapping for Mediterranean corn borer resistance in European flint germplasm using recombinant inbred lines. *BMC Genomics* **2010**,*11*, 174.

72. Ordas, B.; Malvar, R.A.; Santiago, R.; Sandoya, G.; Romay, M.C.; Butron, A. Mapping of QTL for resistance to the Mediterranean corn borer attack using the intermated B73 × Mo17 (IBM) population of maize. *Theor. Appl. Genet* **2009**, *119*, 1451–1459.

73. Krakowsky, M.D.; Brinkman, M.J.; Woodman-Clikeman, W.L.; Lee, M. Genetic components of resistance to stalk tunneling by the European corn borer in maize. *Crop Sci* **2002**, *42*, 1309–1315.

74. Wild, C.P.; Turner, P.C. The toxicology of aflatoxins as a basis for public health decisions.*Mutagenesis* **2002**, *17*, 471–481. [

75. Melching, J.S. Corn Rusts: Types, Races, and Destructive Potential. Proceedings of the 30th Annual Corn and Sorghum Research Conference, Washington DC, USA, 1975; pp. 90–115.

76. Logrieco, A.; Bottalico, A.; Mule, G.; Moretti, A.; Perrone, G. Epidemiology of toxigenic fungi and their associated mycotoxins for some Mediterranean crops. *Eur. J. Plant Pathol* **2003**, *109*, 645–667.

77. Munkvold, G.P. Epidemiology of *Fusarium* diseases and their mycotoxins in maize ears. *Eur. J. Plant Pathol* **2003**, *109*, 705–713.

78. Sutton, J.C.; Baliko, W.; Funnell, H.S. Evidence for translocation of zearalenone in corn plants colonized by *Fusarium graminearum*. *Can. J. Plant Sci* **1976**, *56*, 7–12.

79. Grabber, J.H.; Ralph, J.; Lapierre, C.; Barrière, Y. Genetic and molecular basis of grass cell-wall degradability. I. Lignin-cell wall matrix interactions. *C. R. Biol* **2004**, *327*, 455–465.

80. Ralph, J.; Hatfield, R.D.; Quideau, S.; Helm, R.F.; Grabber, J.H.; Jung, H.J.G. Pathway of *p*-coumaric acid incorporation into maize lignin as revealed by NMR. *J. Am. Chem. Soc* **1994**,*116*, 9448–9456.

81. Grabber, J.H.; Quideau, S.; Ralph, J. *p*-coumaroylated syringyl units in maize lignin: Implications for beta-ether cleavage by thioacidolysis. *Phytochemistry* **1996**, *43*, 1189–1194.

82. Bunzel, M.; Funk, C.; Steinhart, H. Semipreparative isolation of dehydrodiferulic and dehydrotriferulic acids as standard substances from maize bran. *J. Sep. Sci* **2004**, *27*, 1080–1086.

83. Bunzel, M.; Heuermann, B.; Kim, H.; Ralph, J. Peroxidase-catalyzed oligomerization of ferulic acid esters. *J. Agric. Food Chem* **2008**, *56*, 10368–10375.

84. Jacquet, G.; Pollet, B.; Lapierre, C. New ether-liked ferulic acid coniferyl alcohol dimers identified in grass straws. *J. Agric. Food Chem* **1995**, *43*, 2746–2751.

85. Iiyama, K.; Lam, T.B.T.; Stone, B.A. Covalent cross-links in the cell wall. *Plant Physiol* **1994**,*104*, 315–320.

86. Bartolome, B.; Faulds, C.B.; Tuohy, M.; Hazlewood, G.P.; Gilbert, H.J.; Williamson, G. Influence of different xylanases on the activity of ferulic acid esterase on wheat bran.*Biotechnol. Appl. Biochem* **1995**, *22*, 65–73.

87. Hatfield, R.; Ralph, J.; Grabber, J.H. A potential role for sinapyl *p*-coumarate as a radical transfer mechanism in grass lignin formation. *Planta* **2008**, *228*, 919–928.

88. Takahama, U.; Oniki, T. Enhancement of peroxidase-dependent oxidation of sinapyl alcohol by an apoplastic component, 4-coumaric acid ester isolated from epicotyls of *Vigna angularis*L. *Plant Cell Physiol* **1997**, *38*, 456–462.

89. Mueller-Harvey, I.; Hartley, R.D.; Harris, P.J.; Curzon, E.H. Linkage of *p*-coumaryl and feruloyl groups to cell wall polysaccharides of barley straw. *Carbohydr. Res* **1986**, *148*, 71–85.

90. Morrison, T.A.; Jung, H.G.; Buxton, D.R.; Hatfield, R.D. Cell-wall composition of maize internodes of varying maturity. *Crop Sci* **1998**, *38*, 455–460.

91. Hatfield, R.D.; Chaptman, A.K. Comparing-corn types for differences in cell wall characteristics and *p*-coumaroylation of lignin. *J. Agric. Food Chem* **2009**, *57*, 4243–4249.

92. Scobbie, L.; Russell, W.; Provan, G.J.; Chesson, A. The newly extended maize internode. A model study of secondary cell wall formation and consequences for digestibility. *J. Sci. Food Agric* **1993**, *61*, 217–225.

93. MacAdam, J.W.; Grabber, J.H. Relationship of growth cessation with the formation of diferulate cross-links and *p*-coumaroylated lignins in tall fescue leaf blades. *Planta* **2002**, *215*, 785–93.

94. Jung, H.-J.G. Maize stem tissues: Ferulate deposition in developing internode cell walls.*Phytochemistry* **2003**, *63*, 543–549.

95. Fontaine, A.S.; Briand, M.; Barriere, Y. Genetic variation and QTL mapping of para-coumaric and ferulic acid contents in maize stover at silage harvest. *Maydica* **2003**, *48*, 75–82.

96. Mechin, V.; Argillier, O.; Menanteau, V.; Barriere, Y.; Mila, I.; Pollet, B.; Lapierre, C. Relationship of cell wall composition to in vitro cell wall digestibility of maize inbred line stems. *J. Sci. Food Agric* **2000**, *80*, 574–580.

97. Harris, P.J.; Hartley, R.D. Detection of bound ferulic acid in cell walls of the Gramineae by ultraviolet fluorescence microscopy. *Nature* **1976**, *259*, 508–510.

98. Waldron, K.W.; Parr, A.J.; Ng, A.; Ralph, J. Cell wall esterified phenolic dimers: Identification and quantification by reverse phase high performance liquid chromatography and diode array detection. *Phytochem. Anal* **1996**, *7*, 305–312.

99. Saulnier, L.; Thibault, J.F. Ferulic acid and diferulic acids as components of sugar-beet pectins and maize bran heteroxylans. *J. Sci. Food Agric* **1999**, *79*, 396–402.

100. Bunzel, M.; Ralph, J.; Funk, C.; Steinhart, H. Isolation and identification of a ferulic acid dehydrotrimer from saponified maize bran insoluble fiber. *Eur. Food Res. Technol* **2003**, *217*, 128–133.

101. Funk, C.; Ralph, J.; Steinhart, H.; Bunzel, M. Isolation and structural characterisation of 8-*O*-4/8-*O*-4- and 8-8/8-*O*-4-coupled dehydrotriferulic acids from maize bran. *Phytochemistry***2005**, *66*, 363–371.

102. Bunzel, M.; Ralph, J.; Bruening, P.; Steinhart, H. Structural identification of dehydrotriferulic and dehydrotetraferulic acids isolated from insoluble maize bran fiber. *J. Agric. Food Chem***2006**, *54*, 6409–6418.

103. Rouau, X.; Cheynier, V.; Surget, A.; Gloux, D.; Barron, C.; Meudec, E.; Louis-Montero, J.; Criton, M. A dehydrotrimer of ferulic acid from maize bran. *Phytochemistry* **2003**, *63*, 899–903.

104. Piber, M.; Koehler, P. Identification of dehydro-ferulic acid-tyrosine in rye and wheat: Evidence for a covalent cross-link between arabinoxylans and proteins. *J. Agric. Food Chem***2005**, *53*, 5276–5284.

105. Zarra, I.; Sanchez, M.; Queijeiro, E.; Pena, M.J.; Revilla, G. The cell wall stiffening mechanism in Pinus pinaster Aiton: Regulation by apoplastic levels of ascorbate and hydrogen peroxide. *J. Sci. Food Agric* **1999**, *79*, 416–420.

106. Fry, S.C. Phenolic components of the primary cell wall and their possible role in the hormonal regulation of growth. *Planta* **1979**, *146*, 343–351.

107. Kamisaka, S.; Takeda, S.; Takahashi, K.; Shibata, K. Diferulic and ferulic acid in the cell wall of avena coleoptiles. Their relationships to mechanical properties of the cell wall. *Physiol. Plant* **1990**, *78*, 1–7.

108. Azuma, T.; Okita, N.; Nanmori, T.; Yasuda, T. Relationship between the deposition of phenolic acids in the cell walls and the cessation of rapid growth in internodes of floating rice. *Plant Prod. Sci* **2005**, *8*, 447–453.

109. Hossain, M.T.; Soga, K.; Wakabayashi, K.; Kamisaka, S.; Fujii, S.; Yamamoto, R.; Hoson, T. Modification of chemical properties of cell walls by silicon and its role in regulation of the cell wall extensibility in oat leaves. *J. Plant Physiol* **2007**, *164*, 385–393.

110. Hatfield, R.D.; Ralph, J.; Grabber, J.H. Cell wall structural foundations: Molecular basis for improving forage digestibilities. *Crop Sci* **1999**, *39*, 27–37.

111. Jung, H.G.; Lamb, J.F.S. Stem morphological and cell wall traits associated with divergent *in vitro* neutral detergent fiber digestibility in alfalfa clones. *Crop Sci* **2006**, *46*, 2054–2061.

112. Jung, H.G.; Casler, M.D. Maize stem tissues: Impact of development on cell wall degradability. *Crop Sci* **2006**, *46*, 1801–1809.

113. Eraso, F.; Hartley, R.D. Monomeric and dimeric phenolic constituents of plant cell walls possible factors influencing wall biodegradability. *J. Sci. Food Agric* **1990**, *51*, 163–170.

114. Iiyama, K.; Lam, T.B.T. Structural characteristics of cell walls of forage grasses—Their nutritional evaluation for ruminants—Review. *Asian Austr. J. Anim. Sci* **2001**, *14*, 862–879.

115. Lam, T.B.T.; Iiyama, K.; Stone, B.A. Hot alkali-labile linkages in the walls of the forage grass Phalaris aquatica and Lolium perenne and their relation to *in vitro* wall digestibility.*Phytochemistry* **2003**, *64*, 603–607.

116. Martin, M.M. The evolution of cellulose digestion in insects. *Philos. Trans. R. Soc. Lond. Ser. B***1991**, *333*, 281–288.

117. McGinnis, A.J.; Kasting, R. Dietary cellulose effect on food consumption and growth of a grasshopper. *Can. J. Zool* **1967**, *45*, 165–167.

118. Raubenheimer, D.; Simpson, S.J. The geometry of compensatory feeding in the locust. *Anim. Behav* **1993**, *45*, 953–964.

119. Scriber, J.M.; Slansky, F. The nutritional ecology of immature insects. *Annu. Rev. Entomol***1981**, *26*, 183–211.

120. Fry, S.C. Cross-linking of matrix polymers in the growing cell walls of angiosperms. *Annu. Rev. Plant Physiol. Plant Mol. Biol* **1986**, *37*, 165–186.

121. Santiago, R.; Malvar, R.A. Role of dehydrodiferulates in maize resistance to pests and diseases. *Int. J. Mol. Sci* **2010**, *11*, 691–703.

122. Malvar, R.A.; Butrón, A.; Ordás, B.; Santiago, R. Causes of Natural Resistance to Stem Borers in Maize. In *Crop Protection Research Advances*; Williams, P.V., Ed.; Nova Science Publishers Inc.: New York, NY, USA, 2008; pp. 57–100.

123. Groh, S.; Gonzalez-de-Leon, D.; Khairallah, M.M.; Jiang, C.; Bergvinson, D.; Bohn, M.; Hoisington, D.A.; Melchinger, A.E. QTL mapping in tropical maize: III. Genomic regions for resistance to Diatraea spp and associated traits in two RIL populations. *Crop Sci* **1998**, *38*, 1062–1072.

124. Barros-Rios, J. Fortificación de la Pared Celular Como Mecanismo de Defensa Frente a la Plaga del Taladro en el Maíz. Ph.D. Dissertation, University of Vigo, Pontevedra, Spain, 2012.

125. Markham, R.H.; Bosque-Pérez, N.A.; Borgemeister, C.; Meikle, W.G. Developing pest management strategies for the maize weevil, *Sitophilus zeamais*, and the large grain borer, Prostephanus truncatus,

in the humid and sub-humid tropics. *FAO Plant Prot. Bull* **1994**, *42*, 125–136.

126. Pingali, P.L.; Pandey, S. Meeting World Maize Needs: Technology Opportunities and Priorities for the Public Sector. In *World Maize Facts and Trends. Meeting World Maize Needs: Technological Opportunities and Priorities for the Public Sector*; Pingali, P.L., Ed.; International Maize and Wheat Improvement Center: Texcoco, Mexico, 2001.

127. Burr, S.J.; Fry, S.C. Feruloylated arabinoxylans are oxidatively cross-linked by extracellular maize peroxidase but not by horseradish peroxidase. *Mol. Plant* **2009**, *2*, 883–892.

128. Boutigny, A.-L.; Barreau, C.; Atanasova-Penichon, V.; Verdal-Bonnin, M.-N.; Pinson-Gadais, L.; Richard-Forget, F. Ferulic acid, an efficient inhibitor of type B trichothecene biosynthesis and Tri gene expression in *Fusarium* liquid cultures. *Mycol. Res* **2009**, *113*, 746–753.

129. Boerjan, W.; Ralph, J.; Baucher, M. Lignin biosynthesis. *Annu. Rev. Plant Biol* **2003**, *54*, 519–46.

130. Bonawitz, N.D.; Chapple, C. The genetics of lignin biosynthesis: Connecting genotype to phenotype. *Annu. Rev. Genet* **2010**, *44*, 337–363

131. Baucher, M.; Monties, B.; van Montagu, M.; Boerjan, W. Biosynthesis and genetic engineering of lignin. *Crit. Rev. Plant Sci* **1998**, *17*, 125–197.

132. Barrière, Y.; Riboulet, C.; Méchin, V.; Maltese, S.; Pignon, M.; Cardinal, A.; Lapierre, C.; Lübberstedt, T.; Martinant, J.P. Genetics and Genomics of Lignification in Grass Cell Walls based on Maize as Model Species. In *Genes, Genomes and Genomics*; Global Science Books: Isleworth, UK, 2007; pp. 133–156.

133. Coley, P.D. Herbivory and defensive characteristics of tree species in a lowland tropical forest. *Ecol. Monogr* **1983**, *53*, 209–233.

134. Raupp, M.J. Effects of leaf toughness on mandibular wear of the leaf beetle, *Plagiodera versicolora*. *Ecol. Entomol* **1985**, *10*, 73–79.

135. Felton, G.W.; Donato, K.; Delvecchio, R.J.; Duffey, S.S. Activation of plant foliar oxidases by insect feeding reduces nutritive quality of foliage for noctuid herbibores. *J. Chem. Ecol* **1989**,*15*, 2667–2694.

136. Stevenson, P.C.; Anderson, J.C.; Blaney, W.M.; Simmonds, M.S.J. Developmental inhibition od *Spodoptera litura* (Fab) larvae by a novel caffeoylquinic acid form the wild groundnut,*Arachis paraguariensis* (Chod et Hassl). *J. Chem. Ecol* **1993**, *19*, 2917–2933.

137. McMullen, M.D.; Frey, M.; Degenhardt, J. Genetics and biochemistry of insect resistance in maize. *Handbook Maize* **2009**.

138. Vance, C.P.; Kirk, T.K.; Sherwood, R.T. Lignification as a mechanism of disease resistance.*Annu. Rev. Phytopathol* **1980**, *18*, 259–288.

139. Stange, R.R.; Ralph, J.; Peng, J.P.; Sims, J.J.; Midland, S.L.; McDonald, R.E. Acidolysis and hot water extraction provide new insights into the composition of the induced "lignin-like" material from squash fruit. *Phytochemistry* **2001**, *57*, 1005–1011.

140. Zhang, S.-H.; Yang, Q.; Ma, R.-C. *Erwinia carotovora* ssp. Carotovora infection induced "defense lignin" accumulation and lignin biosynthetic gene expression in chinese cabbage (*Brassica rapa* L. ssp pekinensis). *J. Integr. Plant Biol* **2007**, *49*, 993–1002.

141. Karkonen, A.; Koutaniemi, S. Lignin biosynthesis studies in plant tissue cultures. *J. Integr. Plant Biol* **2010**, *52*, 176–185.

142. Ride, J.P. The effect of induced lignification on the resistance of wheat cell walls to fungal degradation. *Physiol. Plant Pathol* **1980**, *16*, 187–192.

Chapter 5

Ca^{2+}-Transport through Plasma Membrane as a Test of Auxin Sensitivity

Anastasia A. Kirpichnikova [1], Elena L. Rudashevskaya [1], Vladislav V. Yemelyanov [1,2] and Maria F. Shishova [1,*]

[1]Department of Plant Physiology and Biochemistry, St. Petersburg State University, Universitetskaya emb. 7/9, St. Petersburg 199034, Russia

[2]Department of Genetics and Biotechnology, St. Petersburg State University, Universitetskaya emb. 7/9, St. Petersburg 199034, Russia

ABSTRACT

Auxin is one of the crucial regulators of plant growth and development. The discovered auxin cytosolic receptor (TIR1) is not involved in the perception of the hormone signal at the plasma membrane. Instead, another receptor, related to the ABP1, auxin binding protein1, is supposed to be responsible for the perception at the plasma membrane. One of the fast and sensitive auxin-induced reactions is an increase of Ca^{2+} cytosolic concentration, which is suggested to be dependent on the activation of Ca^{2+} influx through the plasma membrane. This investigation was carried out with a plasmalemma enriched vesicle fraction, obtained from etiolated maize coleoptiles. The magnitude of Ca^{2+} efflux through the membrane vesicles was estimated according to the shift of potential dependent fluorescent dye diS-C_3-(5). The obtained results showed that during coleoptiles ageing (3rd, 4th and 5th days of seedling etiolated growth) the magnitude of Ca^{2+} efflux from inside-out vesicles was decreased. Addition of ABP1 led to a recovery of Ca^{2+} efflux to the level of the youngest and most sensitive cells. Moreover, the efflux was more sensitive, responding from 10^{-8} to 10^{-6} M 1-NAA, in vesicles containing ABP1, whereas native vesicles showed the highest efflux at 10^{-6} M 1-NAA. We suggest that auxin increases plasma membrane permeability to Ca^{2+} and that ABP1 is involved in modulation of this reaction.

KEYWORDS

ABP1; auxin perception; Ca^{2+} signaling; elongation growth

1. INTRODUCTION

One of the most frequent questions, which are faced by investigators of auxin effects is: How can this simple molecule trigger such a huge diversity of physiological reactions? During the last decades, an enormous amount of data on auxin signaling has accumulated. A big step in understanding the auxin mechanism of action was made after elucidation of the soluble hormone-receptor TIR1 and the closely related AFB proteins [1,2]. Nevertheless, not all auxin-induced reactions are mediated by activation of TIR1-signaling. A number of primary reactions, e.g., changes in membrane potential, cytosol acidification, elevation in cytosolic Ca^{2+} concentration, activation of PLA2, protoplast swelling, protein cycling, *etc.*, require another type of receptor, localized in the plasma membrane [3,4,5,6,7]. Nevertheless, the process of auxin sensing at cell surface is still under investigation. The auxin binding protein 1 (ABP1) is considered to be a part of plasma membrane (PM) receptor or closely linked to it [7,8,9,10,11]. The ABP1 appears to be a dimer of 22-kDa subunit, and can easily be solubilized by detergent or acetone from membranes [12,13,14]. A number of *ABP1* genes are known to encode the protein in different plants [15,16,17,18,19]. The ABP1 protein has a single N-glycosylation site, which binds a mannose type glycan [14,20,21]. Two conservative domains (Box A, responsible for auxin binding, and Box B) and an ER targeting marker *C*-terminal KDEL tetrapeptide were determined in the ABP1 structure

[21,22,23]. The main ABP1 pool is localized in ER where it is supposed to be inactive. Only about 2% of the protein is secreted to extracellular space to fulfill its physiological function [24].

It was supposed that ABP1 might be coupled to a transmembrane docking protein [25,26]. Unfortunately, up to now little is known about the nature and function of this protein. Based on the knowledge of different animal hormone-receptor complexes at the PM, several models were developed for the auxin receptor. The first one suggested that a kinase could fulfill the role of a transmembrane domain [27]. Another model implicates G-protein-coupled receptor as the auxin one [28]. The third model assumes that ABP1 binds/interacts with a Ca^{2+}-permeable ion channel, which fulfills the role of a docking protein [29]. It is possible that ABP1 has more than one binding partner, differing in plant tissues and stage of development.

The absence of ABP1 or its reduction leads to significant changes: arrest of embryo development and elongation intensity [30,31]. Even heterozygous *ABP1/abp1* insertion mutants show a number of developmental disturbances confirmed by reduction of sensitivity to auxin and shift in the intensity of early auxin-regulated *Aux/IAAs* genes expression [32,33]. Decrease in ABP1 via antisense transformation leads to significant decrease in elongation intensity [31] and cell enlargement/protoplast swelling [34,35,36].

It was shown earlier that addition of exogenous ABP1 to a model system like protoplasts increased the amplitude of auxin-induced PM hyperpolarization [37]. Recently, a fast ABP1-related auxin-induced shift in the membrane potential (MP) was shown in a similar model system, by use of a sensitive fluorescent dye [38]. The advantage of the latter investigation was the ascertainment that the effect was triggered even by the *C*-terminal peptide of ABP1 and was blocked by antibodies against it. Overexpression of *ABP1* enhances the K^+-transport by activation of K^+-channels and quantity of their expression [39,40]. Thus, it could be concluded that ABP1 is an important modulator of cell sensitivity to the hormone at plasma membrane, but the mechanism of this regulation is still debated.

One of the fast and sensitive reactions triggered by auxin is an elevation of Ca^{2+} concentration in the cytosol. This reaction was estimated for different plant cells, including maize coleoptile parenchyma cells [9,41] Most probably it reflects the activation of plasma membrane channels, permeable for Ca^{2+} [9]. The coleoptile is a juvenile organ, the main function of which is to protect the first leaf at the initial stages of grass seedling development. Coleoptiles are very sensitive to auxin [42]. In maize coleoptiles, the native growth slows down tremendously from the 3rd to 5th day of seedling development [43]. The most intensive growth decrement appears at transition from the 3rd to 4th day of seedling development [44]. This phenomenon coincides with a loss of auxin-induced growth of coleoptile segments [43] and a significant decrease of auxin induced $[Ca^{2+}]_{cyt}$elevation [44]. Thus, a possible reduction in cell sensitivity to the hormone is due to probable changes in auxin signal perception and early transduction. The current investigation focuses on the involvement of a plasma membrane Ca^{2+}-transport system in auxin signal perception under the control of ABP1.

2. RESULTS AND DISCUSSION

The intensity of Ca^{2+} transport through vesicle membranes, obtained from maize coleoptiles of different ages was estimated as ΔMP, determined by a shift in fluorescence of diS-C_3-(5) dye, commonly used to test transmembrane potential not only in purified vesicles, but also at whole cell level, like protoplast or bacterial cell [45,46].

Our model system contained two types of vesicles: right-side-out, which copy the native cell orientation, and inside-out ones. Only Ca^{2+} ions had a gradient across the vesicle membrane (Figure 1a). Addition of IAA into the incubation medium led to a fast shift of dye fluorescence (Figure 1b), similar to our earlier results [47]. The detected shift in MP was due to Ca^{2+} efflux from the vesicles. We assume that right-side-out vesicles do not participate in ΔMP generation because transport of Ca^{2+} out of the cell is carried out by active systems like Ca^{2+}-ATPase and by the Ca^{2+}/proton antiporter systems (for review see [48]). Conditions for activation of these transporters were absent; therefore, the estimated ΔMP was due to flux of Ca^{2+} ions across membranes of inverted vesicles, which correspond to the flow directed into cell *in vivo*. The revealed IAA-induced shift in MP was similar to the effect obtained after addition of 1-NAA, an active synthetic auxin, but not after 2-NAA addition, a non-active synthetic analogue (Figure 1b).

The fluorescence intensity of diS-C_3-(5) is calibrated mainly in a model system with gradients of K^+ and Na^+ ions. The maximal amplitude of dye fluorescence in our model system decreases to −110 mV of the K^+-diffusion potential as calculated by the Nernst equation when we added valinomycin into the Na^+ incubation medium containing K^+-loaded vesicles [47]. According to earlier results the IAA-induced change in dye fluorescence did not exceed −30 to 33 mV in case of membrane vesicles purified from coleoptiles at 4th day of seedling development. In the absence of Na^+, detected changes of the amplitude of the auxin-induced fluorescence signal are presented inFigure 2 in arbitrary units.

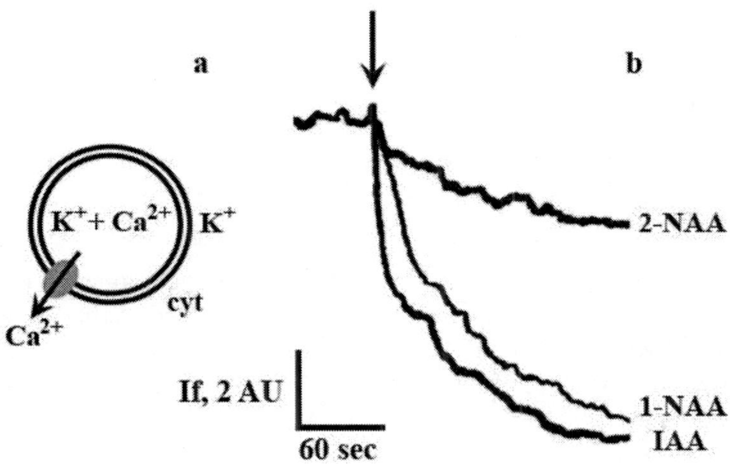

Figure 1. Auxin-induced generation of the membrane potential in a model system represented by plasma membrane vesicles from maize coleoptile cells isolated at the 3rd day of seedling development. (**a**) Scheme of vesicles loading (inside-out orientation of vesicle), cyt—cytosolic side of vesicle, arrow—direction of Ca^{2+} flux; (**b**) Single traces of diS-C_3-(5) fluorescence shift after auxin addition, arrow—addition of auxin.

The distinct auxin concentration dependence of the fluorescence difference was determined. In our experiments, IAA and 1-NAA caused a maximum effect by addition of hormone at 10^{-6} M (Figure 2). Similar results were obtained with vesicles isolated from seedling of all tested ages. However, the amplitude of the fluorescence signal decreased in vesicles from older seedlings. Only in the youngest seedlings the effect of IAA was slightly higher than that of 1-NAA. The 2-NAA-induced shift of diS-C_3-(5) fluorescence increased slightly but almost linearly with concentration. The reaction did not exceed 4 arbitrary units (Figure 2). Moreover, no difference was obtained for 2-NAA during seedling ageing. Thus we conclude that this reaction is auxin-specific.

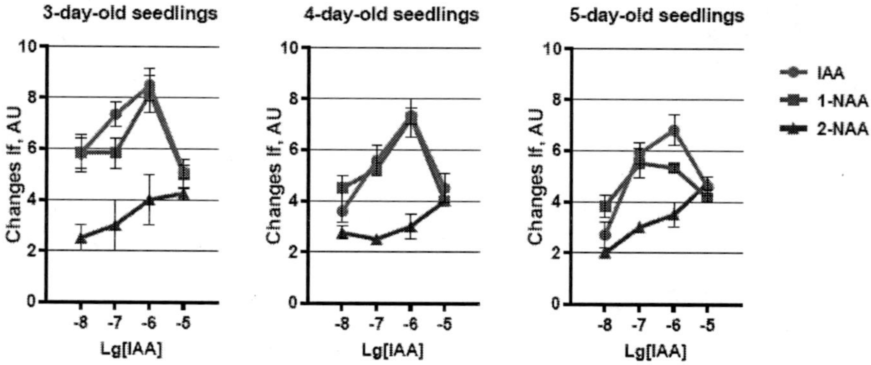

Figure 2. Auxin-induced generation of the membrane potential in a model system. Plasma membrane vesicles from maize coleoptile cells were isolated at the 3rd, 4th and 5th day of seedling development. Histograms represent mean values of the shift in diS-C_3-(5) intensity of fluorescence (If) in arbitrary units (AU) ± SEM.

The obtained results show that physiologically active native and synthetic auxins trigger transport of Ca^{2+} through plasma membrane. The magnitude of this transport decreases within seedlings ageing. That coincides well with earlier results on protoplasts from maize-coleoptile cells, which showed a decrement of $[Ca^{2+}]_{cyt}$ elevation after auxin addition correlated with age of the coleoptile [44]. All this points out a possible decrease of auxin sensitivity at the plasma membrane and early transduction steps, like cytosolic Ca^{2+} elevation, over ageing [44]. If the idea is correct, that ABP1 is an important component of a hormone receptor at the plasma membrane then addition of this protein to the model system might restore the sensitivity for auxin lost within ageing.

Therefore, we modified the model system by loading of ABP1 (10^{-9} M) into the vesicles (Figure 3a). Addition of IAA 10^{-6} M to this system did not change the dynamics of fluorescent response and almost did not affect the amplitude in case of the youngest seedlings (Figure 3b).

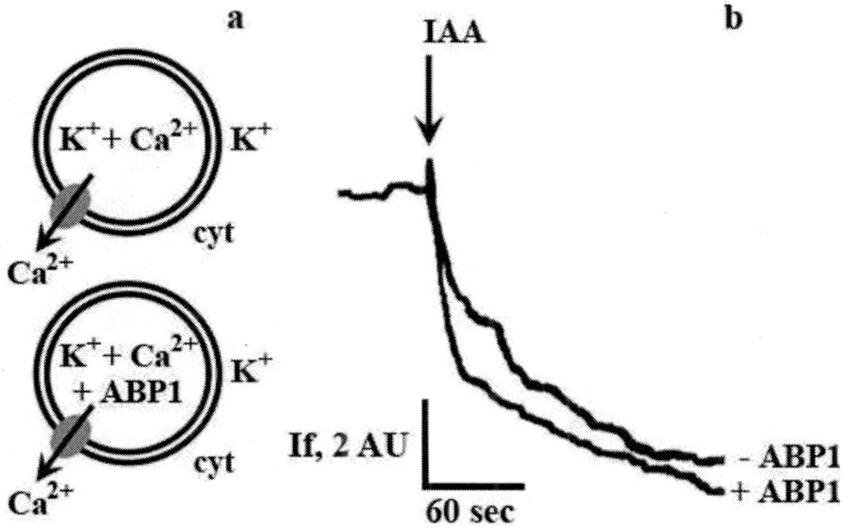

Figure 3. Effect of auxin binding protein 1 (ABP1) on auxin-induced generation of the membrane potential in a model system represented by plasma membrane vesicles from maize coleoptile cells at the 3rd day of seedling development. (**a**) Scheme of vesicles loading (inside-out orientation of vesicle), cyt—cytosolic side of vesicle, arrow—direction of Ca^{2+} flux; (**b**) Single traces of diS-C_3-(5) fluorescence shift after auxin addition.

The presence of ABP1 inside the vesicles led to significant changes in concentration dependence of the response. IAA and 1-NAA-triggered Ca^{2+} efflux at the 3rd day of development did not increase the maximum amplitude but it reached a maximum sensitivity at 10^{-8} M (Figure 4). Thus, we found a significant increase in PM sensitivity to auxins.

A change in auxin sensitivity was shown in older seedlings at 4th and especially at 5th day of seedling development (Figure 4). However, in none of the cases did ABP1 addition increase the maximum value of ΔMP generated in vesicles obtained at the 3rd day of seedling development, when the sensitivity to the hormone had its highest value. Thus, it might be concluded that ABP1 is a limiting factor, which determines the sensitivity of the plant cell at the cell surface. A developmental decrease in ABP1 concentration may occur with seedling ageing toward the end of coleoptile physiological function, and may be coincided with a slowdown of the Ca^{2+} transport through the plasma membrane. Supplementary ABP1 restored the amplitude of Ca^{2+} transport and, as we assume, increased sensitivity to the hormone.

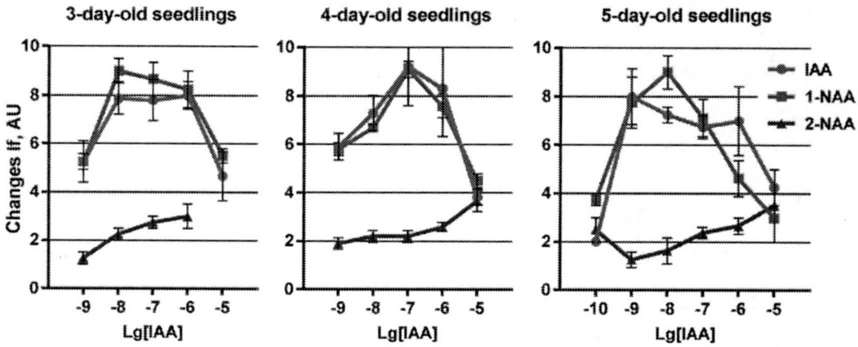

Figure 4. The role of ABP1 in determination of auxin-induced generation of the membrane potential in a model system, represented by plasma membrane vesicles from maize coleoptile cells at the 3rd, 4th and 5th day of seedling development. Histograms represent mean values of the shift in diS-C_3-(5) intensity of fluorescence (If) in arbitrary units (AU) ± SEM.

Special attention was paid to ABP1 because recently more evidences indicate an important role of this protein in the perception of the auxin signal at the plasma membrane. In a number of electrophysiological investigations done on tobacco leaf protoplasts maize ABP1 and antibodies against it strongly affected sensitivity to auxin [26,27,49]. Addition of ABP1 to protoplasts from plants transformed with *rol* genes of *Agrobacterium rhizogenes* raised sensitivity to auxin 100- to 1000-fold. It was supposed that ABP1 might increase the number of active perception units at plasma membrane, which shift protoplast sensitivity [27]. Both stimulatory effects of ABP1 and inhibitory effects of antibodies to this protein were also found in another electrophysiological model system—the whole cell patch clamp [50]. ABP1 mediated an auxin-induced shift in cytosolic pH and a flux of K^+ [51,52]. Recent investigations showed that overexpression of ABP1 enhanced sensitivity of guard cells to auxin [40] and affected enlargement and division of plant cells [31,53,54].

Besides, the whole ABP1 protein, also its binding domain, as represented by the surface of an antibody D16 having auxin activity and the *C*-terminal peptide of ABP1 have physiological activity. D16 was able to trigger MP hyperpolarisation [55] and stimulation of the anion channel [37,56] in the absence of auxin. On the other hand, a synthetic peptide containing 12 residues of the*C*-terminus of ABP1, could mimic the action of high auxin concentrations in regulation of the K^+current, MP value and cytosol alkalinization in guard cells [38,51,52]. Thus, the suggestion was that this peptide played an important role in auxin-ABP1 coupling to intracellular signal cascades [57].

All listed results indicate the importance of ABP1 at the plasma membrane level but does not reveal the mechanism of its action. Recent publications frequently assume that ABP1 participates in regulation of endocytosis [58,59]. Endomembrane trafficking is a process of great importance, which maintains the intracellular re-localization of macromolecules and membranes by secretion and endocytosis. In plant cells traffic of newly synthesized proteins, translocation to the endoplasmic reticulum (ER), and subsequent protein processing and targeting occur via vesicle trafficking through the secretory pathway. Vesicle secretion is a very important process in plant cell growth and signaling [60]. Several publications showed that during auxin-dependent formation of lateral roots, the endomembrane system plays an important role, establishing cellular localization and polarity of the auxin transporters [61,62,63]. Polarity of auxin transporters leads to the formation of auxin gradients determining both initiation of lateral root and the establishment of the primordium cell patterning [64,65]. According to the recently accepted opinion, ABP1 effects are due to ROP GTPase signaling [59]. Nevertheless, the mechanism of transition of auxin signal from extracellular ABP1 to ROP-signaling in the cytosol is still under investigation.

In this manuscript we present data on a fast auxin activation of Ca^{2+} transport through the plasma membrane. This effect is paralleled by the increase of cytosolic Ca^{2+} concentration found before [44]. The latter event might trigger another process, namely vesicle secretion. A fast initiation of exocytosis by elevation of $[Ca^{2+}]_{cyt}$ is well known for neurons and now is shown for plant cells [66,67,68,69]. Even if taking into consideration that exocytosis in animal cells is much more sensitive to cytosolic Ca^{2+} in comparison to cereal coleoptile protoplasts, the phenomenon might indicate common mechanisms of signaling. Analysis of electronic micrographs of oat coleoptile cells reveals a ~12% difference in number of vesicles which are correlated with alterations in growth capacity of the tested cells, preferentially in cells which start elongation [70]. We assume the secretory mechanism is the same in cells from the same tissue, whereas the pool size of vesicles may vary from cell to cell in identical cell types. Elevation of [Ca^{2+}] induced by auxin is estimated to be in the range of approximate 100 nM after auxin treatment [71,72]. This is enough to activate the secretory system within 10–20 s after the increase in Ca^{2+} concentration [66].

The importance of exocytosis in the mechanism of auxin action nicely correlates with an earlier suggestion about hormone-induced increase in a number of H^+-ATPases in plasma membrane via vesicle secretion [73], which will lead along with post-transcriptional phosphorylation [74] to membrane hyperpolarization, shift in ion transport through plasmalemma, increase in acidification of cell wall and cell extension further on. The amplitude of Ca^{2+} rise might be considered as a threshold. Our data show that it depends on the amount of ABP1 and the external concentration of Ca^{2+} ([72], data presented here). Exocytosis is usually accompanied by recycling of vesicles. Electron micrograph data shows that about 60% of delivered vesicles to plasma membrane are recycled [75]. The intensity of recycling will depend on ROP GTPase signaling [76] and concentration of the cytosolic Ca^{2+} concentration [77]. Recently, ROP GTPase signaling was suggested to be linked to ABP1 [78] in support to our model (Figure 5). If cells already pass a step of elongation and have a loss or significant decrease in ABP1 perception facility then this could be another reason for the suggested scheme modulation.

3. EXPERIMENTAL

3.1. Plant Material

Maize seedlings (*Zea mays* L., cv. Moldavsky-215) were grown in darkness at 26 ± 1 °C and 95% relative humidity. Seedlings were illuminated for 4 min every 24 h with a weak light in order to get straight coleoptiles and decrease mesocotyl growth. Seedlings were grown 48 h on wet filter paper and were thereafter transferred on glass tube raft on 1/10 Chesnokov nutrient solution. The ratios between different organs of the seedlings (leaf: coleoptile: mesocotyl: root) in cm were approximately 1.4:1.9:1.6:4.4; 2.3:4.2:2.9:9.1; 7.5:4.9:5.6:10.4, for 3-, 4- and 5- day-old seedlings, respectively.

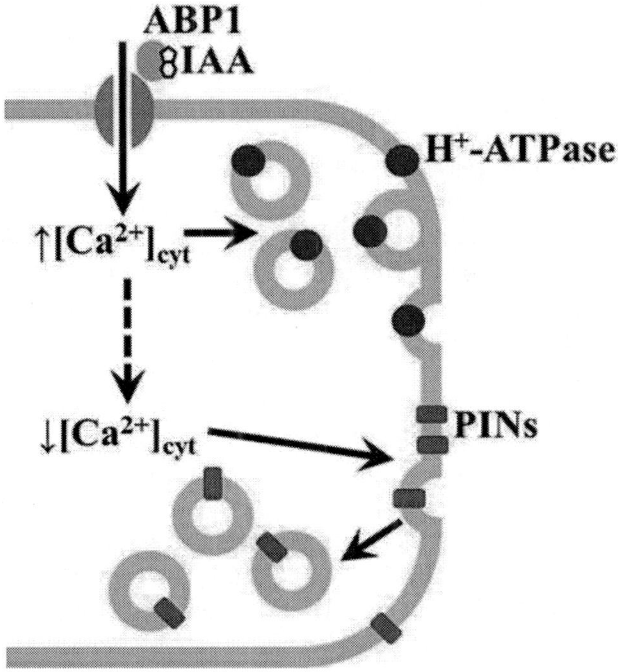

Figure 5. Hypothetical scheme of auxin perception at the plasma membrane and primary signal transduction events in coleoptile cells at early stages of development. In physiologically young coleoptile cell auxin interacts with extracellular ABP1 and triggers Ca^{2+} transport inside the cell ([72], data presented here). Cytosolic Ca^{2+} elevation might initiate exocytosis [66]. This will result in the increase in H^+-ATPases protein in the plasma membrane [73]. Further Rho-like GTPase from plants (ROPs) signaling cascade, which is suggested to be linked to ABP1 [78], leads to endocytosis and redistribution of plasma membrane (PM) proteins, including auxin transporters PINs.

3.2. Measurement of Calcium Transport Rate across Plasma Membrane Vesicles from Maize Coleoptile Cells

The fraction enriched with plasma membranes was isolated from the decapitated coleoptiles of 3, 4 and 5-day-old etiolated maize seedlings. The membrane fraction was prepared by differential centrifugation with subsequent purification in a sucrose density gradient [47]. The purified membranes were collected at the interface between sucrose layers of 1.13

and 1.17 g/cm^3 density. Inhibitor analysis showed no appreciable contamination with vacuolar membranes. The membrane fraction contained vesicles of right-side-out and inside-out orientation in the proportion 1:1, according to the enzymatic test with alamethicin.

Membrane vesicles were loaded by osmotic shock with a medium containing 150 mM K_2SO_4, 1 mM Tris-MES, 150 mM sucrose, pH 6.8 (K^+-medium) or with 1 mM $CaCl_2$ in addition (Ca^{2+} + K^+-medium). ABP1 10^{-9} M, was added to the loading medium depending on the experiment scheme. After loading, the vesicles were concentrated by centrifugation (99,500× *g* for 1 h). The pellet was re-suspended in the K^+-medium.

The change in membrane potential (MP), generated by ion transport through the membrane vesicles due to Ca^{2+} ion gradients, was recorded by the change in fluorescence intensity of 3,3'-dipropylthiodicarbocyanine iodide (diS-C_3-(5)) fluorescent probe (Molecular Probes, Eugene, OR, USA) [79]. This dye is highly sensitive and can be used at a low concentration (0.8 μM). The fluorescence intensity is proportional to MP in a wide range. Fluorescence was measured with a spectrofluorometer constructed similar to MPF44-a Hitachi-Perkin-Elmer model (for details see [47]). Experiments were performed with a standard 1-cm quartz cuvette (volume of 0.3 mL). Excitation and emission wavelengths were 570 and 668 nm, respectively. Fluorescence was measured at an angle of 90 to the direction of excitation light. Aliquots of the dye solution and vesicle preparations were added to the cuvette with K^+-incubation medium. Auxins were added after mixing the membrane fraction with the incubation medium and reaching a steady level in MP.

Experiments were performed in 6–7 replicates in 3–4 independent assays. Figures are made with GraphPad Prism 6 and represent mean values and standard error of the mean (SEM) for n = 4–8.

4. CONCLUSIONS

In summary, we would like to suggest the following hypothetical scheme for auxin perception at the plasma membrane and further primary transduction (Figure 5). The sensitivity of coleoptile cells to auxin is determined by the amount of ABP1, which is postulated to vary during elongation growth. At high external Ca^{2+} concentration auxin induces a fast (10–20 s) elevation of the cytosolic concentration of this ion. In young cells, containing a vesicle pool, this may lead to activation of exocytosis. A further intensification of H^+-ATPase activity, a shift in membrane potential and magnitude of ion transport, cell wall acidification and initiation of elongation growth is supposed to occur. The cytosolic concentration of Ca^{2+} decreases rapidly and, in turn, auxin may induce a shift in endocytosis activity, which is followed by redistribution of PIN proteins and further formation of a new local auxin gradient.

ACKNOWLEDGMENTS

Authors particularly want to thank R. Napier for a kind supply with ABP1 samples. Financial support from Russian Foundation of Basic Research (projects 13-04-00945-a and 12-04-01029-a) and Ministry of Education and Science of Russian Federation (agreement No. 8093 from 23.07.2012, project 2012-1.2.1-12-000-1013-003) is greatly appreciated.

AUTHOR CONTRIBUTIONS

Anastasia A. Kirpichnikova performed the experiments, summarize a state of the art in the field and prepared a manuscript; Elena L. Rudashevskaya conducted the experiments, analyzed the data and prepared the manuscript; Vladislav V. Yemelyanov analyzed and discussed the data, prepared the manuscript, Maria F. Shishova designed the experiments, analyzed the data, prepared the manuscript and suggested the final scheme.

REFERENCES

1. Dharmasiri, N.; Dharmasiri, S.; Estelle, M. The F-box protein TIR1 is an auxin receptor.*Nature* **2005**, *435*, 441–445.
2. Kepinski, S.; Leyser, O. The Arabidopsis F-box protein TIR1 is an auxin receptor. *Nature***2005**, *435*, 446–451.
3. Badescu, G.O.; Napier, R.M. Receptors for auxin: Will it all end in TIRs? *Trends Plant Sci.***2006**, *11*, 217–223.

4. Scherer, G.F.; Zahn, M.; Callis, J.; Jones, A.M. A role for phospholipase A in auxin-regulated gene expression. *FEBS Lett.* **2007**, *581*, 4205–4211.

5. Mockaitis, K.; Estelle, M. Auxin receptors and plant development: A new signaling paradigm. *Annu. Rev. Cell Dev. Biol.* **2008**, *24*, 55–80.

6. Vanneste, S.; Friml, J. Auxin: A trigger for change in plant development. *Cell* **2009**, *136*, 1005–1016.

7. Scherer, G.F. Auxin-binding-protein1, the second auxin receptor: What is the significance of a two-receptor concept in plant signal transduction? *J. Exp. Bot.* **2011**, *62*, 3339–3357.

8. Napier, R.M.V.; Venis, M.A. Tansley review No-79—Auxin action and auxin-binding proteins. *New Phytol.* **1995**, *129*, 167–201.

9. Shishova, M.; Lindberg, S. A new perspective on auxin perception. *J. Plant Physiol.* **2010**, *167*, 417–422.

10. Perrot-Rechenmann, C. Cellular responses to auxin: Division *versus* expansion. *Cold Spring Harb. Perspect. Biol.* **2010**, *2*.

11. Sauer, M.; Kleine-Vehn, J. Auxin binding protein1: The outsider. *Plant Cell* **2011**, *23*, 2033–2043.

12. Venis, M.A. Solubilisation and partial purification of auxin-binding sites of corn membranes.*Nature* **1977**, *266*, 268–269.

13. Shimomura, S.; Sotobayashi, T.; Futai, M.; Fukui, T. Purification and properties of an auxin-binding protein from maize shoot membranes. *J. Biochem.* **1986**, *99*, 1513–1524.

14. Napier, R.M.; Venis, M.A.; Bolton, M.A.; Richardson, L.I.; Butcher, D.W. Preparation and characterisation of monoclonal and polyclonal antibodies to maize membrane auxin-binding protein. *Planta* **1988**, *176*, 519–526.

15. Palme, K. Molecular analysis of plant signaling elements: Relevance of eukaryotic signal transduction models. *Int. Rev. Cytol.* **1992**, *132*, 223–283.

16. Shimomura, S.; Liu, W.; Inohara, N.; Watanabe, S.; Futai, M. Structure of the gene for an auxin-binding protein and a gene for 7SL RNA from *Arabidopsis thaliana*. *Plant Cell Physiol.***1993**, *34*, 633–637.

17. Shimomura, S.; Watanabe, S.; Ichikawa, H. Characterization of auxin-binding protein 1 from tobacco: Content, localization and auxin-binding activity. *Planta* **1999**, *209*, 118–125.

18. Lazarus, C.M.; Macdonald, H. Characterization of a strawberry gene for auxin-binding protein, and its expression in insect cells. *Plant Mol. Biol.* **1996**, *31*, 267–277.

19. Watanabe, S.; Shimomura, S. Cloning and expression of two genes encoding auxin-binding proteins from tobacco. *Plant Mol. Biol.* **1998**, *36*, 63–74.

20. Löbler, M.S.K.; Hesse, T.; Klämbt, D. Auxin receptors in target tissues. In *Molecular Biology of Plant Growth Control*; Fox, J.E.J.M., Alan, R., Eds.; Liss: New York, NY, USA, 1987; pp. 279–288.

21. Hesse, T.; Feldwisch, J.; Balshusemann, D.; Bauw, G.; Puype, M.; Vandekerckhove, J.; Lobler, M.; Klambt, D.; Schell, J.; Palme, K. Molecular cloning and structural analysis of a gene from*Zea mays* (L.) coding for a putative receptor for the plant hormone auxin. *EMBO J.* **1989**, *8*, 2453–2461.

22. Pelham, H.R. Control of protein exit from the endoplasmic reticulum. *Annu. Rev. Cell Biol.***1989**, *5*, 1–23.

23. Jones, A.M. Auxin-binding proteins. *Annu. Rev. Plant Physiol. Plant Mol. Biol.* **1994**, *45*, 393–420.

24. Diekmann, W.; Venis, M.A.; Robinson, D.G. Auxins induce clustering of the auxin-binding protein at the surface of maize coleoptile protoplasts. *Proc. Natl. Acad. Sci. USA* **1995**, *92*, 3425–3429.

25. Klambt, D. A view about the function of auxin-binding proteins at plasma membranes. *Plant Mol. Biol.* **1990**, *14*, 1045–1050.

26. Barbier-Brygoo, H.; Ephritikhine, G.; Klambt, D.; Maurel, C.; Palme, K.; Schell, J.; Guern, J. Perception of the auxin signal at the plasma membrane of tobacco mesophyll protoplasts.*Plant J.* **1991**, *1*, 83–93.

27. Barbier-Brygoo, H. Tracking auxin receptors using functional approaches. *Crit. Rev. Plant Sci.* **1995**, *14*, 1–25.

28. MacDonald, H. Auxin perception and signal transduction. *Physiol. Plant* **1997**, *100*, 423–430.

29. Shishova, M. Membrane Mechanism of Auxin Action on Plant Cell. Ph.D. Thesis, St. Petersburg University, Moscow, Russia, 1999.

30. Chen, J.G.; Shimomura, S.; Sitbon, F.; Sandberg, G.; Jones, A.M. The role of auxin-binding protein 1 in the expansion of tobacco leaf cells. *Plant J.* **2001**, *28*, 607–617.

31. Chen, J.-G.; Ullah, H.; Young, J.C.; Sussman, M.R.; Jones, A.M. ABP1 is required for organized cell elongation and division in *Arabidopsis* embryogenesis. *Genes Dev.* **2001**, *15*, 902–911.

32. Effendi, Y.; Rietz, S.; Fischer, U.; Scherer, G.F. The heterozygous *abp1/ABP1* insertional mutant has defects in functions requiring polar auxin transport and in regulation of early auxin-regulated genes. *Plant J.* **2011**, *65*, 282–294.

33. Effendi, Y.; Scherer, G.F. Auxin binding-protein1 (ABP1), a receptor to regulate auxin transport and early auxin genes in an interlocking system with PIN proteins and the receptor TIR1. *Plant Signal. Behav.* **2011**, *6*, 1101–1103.

34. Steffens, B.; Feckler, C.; Palme, K.; Christian, M.; Bottger, M.; Luthen, H. The auxin signal for protoplast swelling is perceived by extracellular ABP1. *Plant J.* **2001**, *27*, 591–599.

35. Christian, M.S.B.; Schenck, D.; Burmester, S.; Böttger, M.; Lüthen, H. How does auxin enhance cell elongation? Roles of auxin-binding proteins and potassium channels in growth control. *Plant Biol.* **2006**, *8*, 346–352.

36. Yamagami, M.; Haga, K.; Napier, R.M.; Iino, M. Two distinct signaling pathways participate in auxin-induced swelling of pea epidermal protoplasts. *Plant Physiol.* **2004**, *134*, 735–747.

37. Barbier-Brygoo, H.; Zimmermann, S.; Thomine, S.; White, I.R.; Millner, P.; Guern, J. Elementary response chains at the plasma membrane involve external ABP1 and multiple electrogenic ion transport proteins. *Plant Growth Regul.* **1996**, *18*, 23–28.

38. Dahlke, R.I.; Luethen, H.; Steffens, B. ABP1: An auxin receptor for fast responses at the plasma membrane. *Plant Signal. Behav.* **2010**, *5*, 1–3.

39. Philippar, K.; Fuchs, I.; Luthen, H.; Hoth, S.; Bauer, C.S.; Haga, K.; Thiel, G.; Ljung, K.; Sandberg, G.; Bottger, M.; *et al.* Auxin-induced K^+ channel expression represents an essential step in coleoptile growth and gravitropism. *Proc. Natl. Acad. Sci. USA* **1999**, *96*, 12186–12191.

40. Bauly, J.M.; Sealy, I.M.; Macdonald, H.; Brearley, J.; Droge, S.; Hillmer, S.; Robinson, D.G.; Venis, M.A.; Blatt, M.R.; Lazarus, C.M.; *et al.* Overexpression of auxin-binding protein enhances the sensitivity of guard cells to auxin. *Plant Physiol.* **2000**, *124*, 1229–1238.

41. Monshausen, G.D.; Miller, N.D.; Murphy, A.S.; Gilroy, S. Dynamics of auxin-dependent $Ca^{2+}$and pH signaling in root growth revealed by integrating high-resolution imaging with automated computer vision-based analysis. *Plant J.* **2011**, *65*, 309–318.

42. Cleland, R.E. The outer epidermis of *Avena* and maize coleoptiles is not a unique target for auxin in elongation growth. *Planta* **1991**, *186*, 75–80.

43. Rudashevskaya, E.L.; Yemelyanov, V.V.; Kirpichnikova, A.A.; Burova, E.A.; Bobinova, O.A.; Shishova, M.F. The dependence of age changes in maize coleoptile and mesocotyl growth activity on indole-3-acetic acid content. *Bull. St. Petersburg Univ.* **2002**, *3*, 99–106.

44. Shishova, M.; Yemelyanov, V.; Rudashevskaya, E.; Lindberg, S. A shift in sensitivity to auxin within development of maize seedlings. *J. Plant Physiol.* **2007**, *164*, 1323–1330.

45. Oka, K.; Naitou, S.; Yoshida, M.; Ishikawa, H.; Ohta, E.; Sakata, M. Membrane potential measurement of protoplasdts isolated from *Vigna mungo* hypocotil using a fluorescent probe, diS-C_3-(5). *Plant Cell Physiol.* **1987**, *28*, 843–849.

46. Suzuki, H.; Wang, Z.Y.; Yamakoshi, M.; Kobayashi, M.; Nozawa, T. Probing the transmembrane potential of bacterial cells by voltage-sensitive dyes. *Anal. Sci.* **2003**, *19*, 1239–1242.

47. Shishova, M.F.; Inge-Vechtomova, N.I.; Vykhvalov, K.A.; Rudashevskaya, E.L.; Polevoi, V.V. Auxin-dependent transport of K^+ and Ca^{2+} across the membrane of plasmalemma vesicles from coleoptile cells. *Russ. J. Plant Physiol.* **1998**, *45*, 67–73.

48. Kudla, J.; Batistic, O.; Hashimoto, K. Calcium signals: The lead currency of plant information processing. *Plant Cell* **2010**, *22*, 541–563.

49. Barbier-Brygoo, H.; Ephritikhine, G.; Klambt, D.; Ghislain, M.; Guern, J. Functional evidence for an auxin receptor at the plasmalemma of tobacco mesophyll protoplasts. *Proc. Natl. Acad. Sci. USA* **1989**, *86*, 891–895.

50. Rück, A.P.K.; Venis, M.A.; Napier, R.M.; Felle, H. Patch-clamp analysis establishes a role for an auxin-binding protein in the auxin stimulation of plasma membrane current in *Zea mays*protoplasts. *Plant J.* **1993**, *4*, 41–46.

51. Thiel, G.; Blatt, M.R.; Fricker, M.D.; White, I.R.; Millner, P. Modulation of K^+ channels in*Vicia* stomatal guard cells by peptide homologs to the auxin-binding protein *C*-terminus.*Proc. Natl. Acad. Sci. USA* **1993**, *90*, 11493–11497.

52. Gehring, C.A.; McConchie, R.M.; Venis, M.A.; Parish, R.W. Auxin-binding-protein antibodies and peptides influence stomatal opening and alter cytoplasmic pH. *Planta* **1998**, *205*, 581–586.

53. Fellner, M.E.G.; Barbier-Brygoo, H.; Guern, J. An antibody raised to a maize auxin-binding protein has inhibitory effects on cell division of tobacco mesophyll protoplasts. *Plant Physiol. Biochem.* **1996**, *34*, 133–138.

54. Jones, A.M.; Im, K.H.; Savka, M.A.; Wu, M.J.; DeWitt, N.G.; Shillito, R.; Binns, A.N. Auxin-dependent cell expansion mediated by overexpressed auxin-binding protein 1. *Science* **1998**,*282*, 1114–1117.

55. Venis, M.A.; Napier, R.M.; Barbier-Brygoo, H.; Maurel, C.; Perrot-Rechenmann, C.; Guern, J. Antibodies to a peptide from the maize auxin-binding protein have auxin agonist activity.*Proc. Natl. Acad. Sci. USA* **1992**, *89*, 7208–7212.

56. Zimmermann, S.; Thomine, S.; Guern, J.; Barbier-Brygoo, H. An anion current at the plasma membrane of tobacco protoplasts shows ATP-dependent voltage regulation and is modulated by auxin. *Plant J.* **1994**, *6*, 707–716.

57. Leblanc, N.; Perrot-Rechenmann, C.; Barbier-Brygoo, H. The auxin-binding protein Nt-ERabp1 alone activates an auxin-like transduction pathway. *FEBS Lett.* **1999**, *449*, 57–60.

58. Robert, S.; Kleine-Vehn, J.; Barbez, E.; Sauer, M.; Paciorek, T.; Baster, P.; Vanneste, S.; Zhang, J.; Simon, S.; Covanova, M.; *et al.* ABP1 mediates auxin inhibition of clathrin-dependent endocytosis in *Arabidopsis*. *Cell* **2010**, *143*, 111–121.

59. Murphy, A.S.; Peer, W.A. Vesicle trafficking: ROP-RIC roundabout. *Curr. Boil.* **2012**, *22*, R576–R578.

60. Bassham, D.C.; Blatt, M.R. SNAREs: Cogs and coordinators in signaling and development.*Plant Physiol.* **2008**, *147*, 1504–1515.

61. Geldner, N.; Friml, J.; Stierhof, Y.D.; Jurgens, G.; Palme, K. Auxin transport inhibitors block PIN1 cycling and vesicle trafficking. *Nature* **2001**, *413*, 425–428.

62. Dhonukshe, P.; Aniento, F.; Hwang, I.; Robinson, D.G.; Mravec, J.; Stierhof, Y.D.; Friml, J. Clathrin-mediated constitutive endocytosis of PIN auxin efflux carriers in *Arabidopsis*. *Curr. Boil.* **2007**, *17*, 520–527.

63. Friml, J. Subcellular trafficking of PIN auxin efflux carriers in auxin transport. *Eur. J. Cell Boil.* **2010**, *89*, 231–235.

64. Benkova, E.; Michniewicz, M.; Sauer, M.; Teichmann, T.; Seifertova, D.; Jurgens, G.; Friml, J. Local, efflux-dependent auxin gradients as a common module for plant organ formation.*Cell* **2003**, *115*, 591–602.

65. Perez-Henriquez, P.; Raikhel, N.V.; Norambuena, L. Endocytic trafficking towards the vacuole plays a key role in the auxin receptor SCF(TIR)-independent mechanism of lateral root formation in *A. thaliana*. *Mol. Plant* **2012**, *5*, 1195–1209.

66. Sutter, J.U.; Homann, U.; Thiel, G. Ca^{2+}-stimulated exocytosis in maize coleoptile cells. *Plant Cell* **2000**, *12*, 1127–1136.

67. Sutter, J.U.; Denecke, J.; Thiel, G. Synthesis of vesicle cargo determines amplitude of Ca^{2+}-sensitive exocytosis. *Cell Calcium* **2012**, *52*, 283–288.

68. Thiel, G.; Sutter, J.U.; Homann, U. Ca^{2+}-sensitive and Ca^{2+}-insensitive exocytosis in maize coleoptile protoplasts. *Pflug. Archiv. Eur. J. Physiol.* **2000**, *439*, R152–R153.

69. Campanoni, P.; Blatt, M.R. Membrane trafficking and polar growth in root hairs and pollen tubes. *J. Exp. Bot.* **2007**, *58*, 65–74.

70. Quaite, E.; Parker, R.E.; Steer, M.W. Plant cell extension: Structural implications for the origin of the plasma membrane. *Plant Cell Environ.* **1983**, *6*, 429–432.

71. Felle, H. Auxin causes oscillations of cytosolic free calcium and pH in *Zea mays* coleoptiles.*Planta* **1988**, *174*, 495–499.

72. Shishova, M.; Lindberg, S. Auxin induces an increase of Ca^{2+} concentration in the cytosol of wheat leaf protoplasts. *J. Plant Physiol.* **2004**, *161*, 937–945.

73. Hager, A.; Debus, G.; Edel, H.G.; Stransky, H.; Serrano, R. Auxin induces exocytosis and the rapid synthesis of a high-turnover pool of plasma-membrane H_+-ATPase. *Planta* **1991**, *185*, 527–537.

74. Takahashi, K.; Hayashi, K.; Kinoshita, T. Auxin activates the plasma membrane H_+-ATPase by phosphorylation during hypocotyl elongation in *Arabidopsis*. *Plant Physiol.* **2012**, *159*, 632–641.

75. Phillips, G.D.; Preshaw, C.; Steer, M.W. Dictyosome vesicle production and plasma membrane turnover in auxin-stimulated outer epidermal cells of coleoptile segments from*Avena sativa* (L.). *Protoplasma* **1988**, *145*, 59–65.

76. Yalovsky, S.; Bloch, D.; Sorek, N.; Kost, B. Regulation of membrane trafficking, cytoskeleton dynamics, and cell polarity by ROP/RAC GTPases. *Plant Physiol.* **2008**, *147*, 1527–1543.

77. Hwang, J.U.; Gu, Y.; Lee, Y.J.; Yang, Z. Oscillatory ROP GTPase activation leads the oscillatory polarized growth of pollen tubes. *Mol. Biol. Cell* **2005**, *16*, 5385–5399.

78. Xu, T.; When, M.; Nagawa, S.; Fu, Y.; Chen, J.-G.; Wu, M.-J.; Perrot-Rechenmann, K.; Friml, J.; Jones, A.; Yang, Z. Cell surface- and Rho GTPase-based auxin signaling controls cellular interdigitation in *Arabidopsis*. *Cell* **2011**, *143*, 99–110.

79. Ivkova, M.N.; Pechatnikov, V.A.; Ivkov, V.G.; Pletnev, V.V. Mechanism of the fluorescent response of carbocyanine probe diS-C_3-(5) to membrane potential change. *Biofizika* **1983**, *28*, 160–170.

Chapter 6

Homologous NF-YC2 Subunit from *Arabidopsis* and Tobacco Is Activated by Photooxidative Stress and Induces Flowering

Dieter Hackenberg, Ulrich Keetman † and Bernhard Grimm *

Institute of Biology/Plant Physiology, Humboldt-Universität Berlin, Philippstr.13, Building 12, 10115 Berlin, Germany

ABSTRACT

The transcription factor NF-Y consists of the three subunits A, B and C, which are encoded in *Arabidopsis* in large gene families. The multiplicity of the genes implies that NF-Y may act in diverse combinations of each subunit for the transcriptional control. We aimed to assign a function in stress response and plant development to NF-YC subunits by analyzing the expression of *NF-Y* genes and exploitation of *nf-y* mutants. Among the subunit family, *NF-YC2* showed the strongest inducibility towards oxidative stress, e.g. photodynamic, light, oxidative, heat and drought stress. A tobacco *NF-YC* homologous gene was found to be inducible by photooxidative stress generated by an accumulation of the tetrapyrrole metabolite, coproporphyrin. Despite the stress induction, an *Arabidopsis* *nf-yc2* mutant and *NF-YC2* overexpressors did not show phenotypical differences compared to wild-type seedlings in response to photooxidative stress. This can be explained by the compensatory potential of other members of the *NF-YC* family. However, *NF-YC2* overexpression leads to an early flowering phenotype that is correlated with increased *FLOWERING LOCUS T*-transcript levels. It is proposed that *NF-YC2* functions in floral induction and is a candidate gene among the NF-Y family for the transcriptional activation upon oxidative stress.

KEYWORDS

transcription factor; gene family; expression profile; photooxidative stress; flowering time

1. INTRODUCTION

Many promoters of eukaryotes contain the CCAAT–box motif, which is positioned between nucleotide −50 and −120, upstream of the transcription start site [1], and found also for multiple plant genes [1,2]. The CCAAT-box is recognized by the heterotrimeric transcription factor NF-Y (nuclear factor Y), which is also designated CCAAT-box-binding factor (CBF) or heme-activated proteins (HAP) 2, 3 and 5 [3–5]. The NF-Y-complex consists of three subunits A, B and C, which are characterized by conserved protein domains: The C-terminus of NF-YA, the central region of NF-YB, and the N-terminus of NF-YC show sequence identities of more than 70% in a stretch of 56, 90 and 84 amino acids among the NF-Y subunit homologs, respectively [6].

NF-YB and NF-YC contain also motifs of the histone 2A (H2A) and histone 2B (H2B) subfamilies and form a heterodimer at the interface of the histone-fold-motif (HFM) of each subunit [7,8]. Both subunits build the platform for the subsequent association with the DNA-binding subunit NF-YA [9]. The regulatory functions of NF-Y on the CCAAT-box have been mainly investigated for mammalian and yeast genes [1,3,4]. Each subunit is represented by a single gene in these organisms. Specificity and modulation of transcription activation is expected by interaction with other down-stream trans-acting factors [4].

In contrast, the plant NF-Y subunits are encoded by large gene families [10,11]. Initially 29*Arabidopsis thaliana* genes encoding NF-YA, B or C subunits were presented [12], while subsequently a few additional genes were assigned to the NF-Y families comprising in total 10, 13 and 13 members of the NF-YA, NF-YB and NF-YC subfamilies, respectively [11]. Most of these genes are mainly characterized by diverse expression profiles in different tissues and during vegetative and generative development. The multiplicity of gene members in all three NF-Y families implies a complex role of plant NF-Y in transcriptional control. It has previously been proposed that the members of each gene family specifically contribute to activation of specific genes and the heterotrimeric NF-Y complex might act as combinatorial transcription factor. Thus, multiple genes could be regulated in space and time with a specific combination of different NF-Y subunits [12]. The combinatorial diversity of *Arabidopsis* NF-Y subunits was explored by interactome studies using the yeast-two-hybrid approach [13].

In heterogeneous NF-Y heterotrimers single subunits exhibit specific functions during transcriptional control, which could at least not entirely be substituted by another member of the same subunit family. This was demonstrated when a knock-out mutation of an individual *NF-Y*gene caused an obvious phenotype, e.g., during embryogenesis or stress adaptation. The *A. thaliana* mutants *leafy cotyledon 1 (lec1)* and *L1L* (*Lec 1-like*) showed a defective embryonic development [14,15]. The mutant genes belong to the *NF-YB* family and are described as essential regulators during morphogenesis and maturation of developing embryos. The role of LEC1 was defined for desiccation tolerance of seeds. Ectopic expression of *LEC 1* generates embryo-like structures on the leaf surface of transgenic *Arabidopsis thaliana* seedlings. Apart from this impact of NF-YB9 (LEC1) and NF-YB6 (LEC1-LIKE) other subunits were assigned to function in response to environmental stimuli or stress factors. NF-YA5 and NF-YB9 are required to control *LHCB1-3*genes in response to blue light and abscisic acid (ABA) [16]. The photoperiodically induced flowering-time was modulated in knock-out mutants for *NF-YB2* and *NF-YB3* genes and in the *nf-yc3/nf-yc4/nf-yc9* triple mutant under long day conditions [17,18]. It is proposed that a NF-Y complex bound to the promoter *of FLOWERING LOCUS T* (FT) is an important modulator of CONSTANS (CO)-mediated transcriptional activity of the *FT* gene [19].

Response to drought stress and endoplasmic reticulum (ER) stress has been reported to be mediated through NF-YA5 [20] and NF-YB1 [21] as well as NF-YA4, NF-YB3 and NF-YC2 [22]. Together with the transcription factor bZIP28, NF-Y binds to the *endoplasmic reticulum stress responsive element I* (ERSE-I) in combination with the CCAAT-box element [22].

Following the hypothesis that the differential expression pattern of each member of the three NF-Y gene families may indicate specific functions of single subunits, which cannot necessarily be replaced by other representatives of the same family [11], we aimed at evaluating the expression of *NF-YC* genes during adverse growth conditions, such as abiotic stress or herbicide treatments. A tobacco *NF-YC* gene was initially identified among early inducible genes upon accumulating porphyrin intermediates as a result of deregulated tetrapyrrole biosynthesis. Then, we intended to examine *Arabidopsis* genes homologous to this tobacco *NF-YC* gene. *Arabidopsis* T-DNA insertion mutants were selected and analyzed for phenotypic alterations during plant development and different oxidative stress conditions.

2. RESULTS

2.1. Identification of Early Inducible Genes in Response to Photooxidative Stress Triggered by Accumulation of Coproporphyrin in Tobacco

Transgenic tobacco lines expressing ectopically *CPO* (coproporphyrinogen oxidase) antisense RNA suffer from photodynamic cell death in leaf tissue [23]. CPO catalyzes the oxidative decarboxylation of two propionate side chains to vinyl groups of coproporphyrinogen leading to protoporphyrinogen. The necrotic cell death phenotype of *CPO*-antisense lines (Figure 1A) is explained by accumulation of the photoreactive CPO substrate coproporphyrinogen and the oxidized coproporphyrin. In continuation to the initial characterization of CPO-deficient tobacco plants we were interested to identify early inducible genes in response to coproporphyrin-induced cell death. We made use of a light-dose dependent phenotypical change of coproporphyrin-accumulating lines for the search of early inducible genes upon photooxidative stress. The leaf phenotype of wild-type and the *CPO*-antisense line #41 grown under short day condition (6 h light/18 h dark) at 200 μmol photon $m^{-2} \cdot s^{-1}$ differed from that of plants exposed to long day condition (16 h light/8 h dark) at the same light intensity of 200 μmol photon $m^{-2} \cdot s^{-1}$. Although, the plants already accumulate coproporphyrin under short day condition, leaf necrosis is not detectable. The shift from low light to a higher light dosage under the same light intensity led to visible cell death symptoms within 24–48 h (Figure 1A) [23,24].

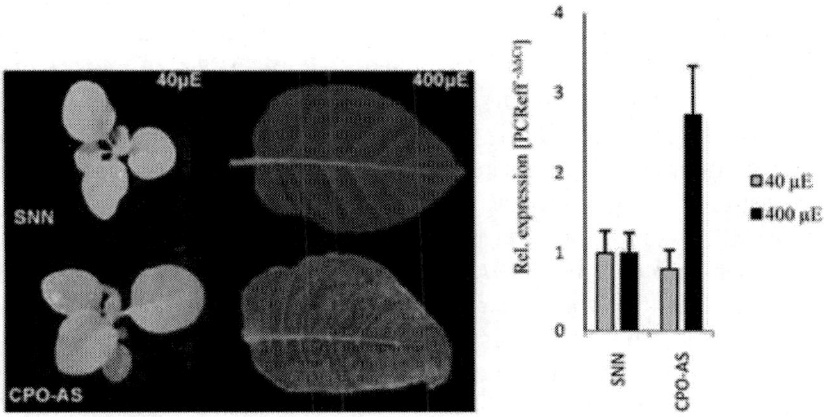

(A) (B)

Figure 1. Tobacco *NF-YC2* is inducible early in *CPO*(coproporphyrinogen oxidase)-antisense plants. (**A**) *CPO*-antisense RNA expressing plants and control plants were grown under low light condition (left panels) before adult plants were transferred to high light. After exposure to high light, *CPO*-antisense plants show leaf necrosis; (**B**) Quantitative RT-PCR analysis of *NtNF-YC2* transcripts of CPO antisense and SNN wild-type plants before and 6 h after transfer from low (40 µM photon $m^{-2}{\cdot}s^{-1}$) to high light intensities (400 µM photon $m^{-2}{\cdot}s^{-1}$). (**A**) (**B**)

To identify genes induced early under photooxidative stress conditions in the porphyrin-accumulating *CPO*-antisense tobacco plants, eight-week-old control and *CPO*-antisense plants (line 1/41) were grown under short day condition before the light exposure was extended for six additional hours and leaf samples were harvested for RNA extractions. Total RNA of control and wild-type plants was subjected to subtractive suppression hybridization. The approach resulted in the identification of, in total, 184 cDNA sequences with differential expression pattern between*CPO* antisense line 1/41 and wild-type plants, which encode 70 different proteins (Table S1). Sixty-one percent of the cDNA sequences were confirmed by reverse northern blotting to be expressed with elevated levels in the *CPO* antisense RNA expressing line.

Among the genes with increased expression in the transgenic line, *NtHAP5b* was identified as encoding the tobacco transcription factor subunit NF-YC. For consistency with the designation of the homologous *Arabidopsis* genes for the NF-Y subunits, we named this gene *NtNF-YC*. The induced *NF-YC* expression upon porphyrin-derived photooxidative stress was confirmed with quantitative reverse transcriptase polymerase chain reaction (qRT-PCR) analysis of tobacco wild-type and *CPO* antisense plants before and after transition to high light doses. The three-week-old plants were grown first under short day and low light condition (6 h light/18 h dark). Then control and transgenic plants were exposed six hours to low and high light before they were harvested for mRNA quantification (Figure 1B). *NtNF-YC* expression was induced during high light and the mRNA accumulated 4-fold compared to low light condition of control and CPO antisense seedlings as well as to high light-exposed wild-type seedlings. In conclusion, this screen enabled identification of a putative candidate gene encoding the C-subunit of the transcription factor NF-Y, which is inducible upon photoreactivity of accumulating tetrapyrrole intermediates.

2.2. Transcript Analysis of 36 Arabidopsis Genes Encoding NF-Y Subunits

Taking advantage of the previous exploitation of the entire *Arabidopsis* genome we searched for*NtNF-YC* homologs among the members of the *Arabidopsis NF-YC* gene family. To benefit from the complete collection of the *NF-Y* gene families for the three subunits we aimed at the identification of the candidate gene among the NF-YC family that is also induced upon porphyric stress. The highest similarity to NtNF-YC exists among a subgroup of the NF-YC family consisting of NF-YC1, NF-YC2, NF-YC3, NF-YC4 and NF-YC9 (Figure 2C*)*. These investigations for the identification of the *NtNF-YC*-homologous gene were embedded in a comprehensive analysis of the transcriptome of the complete set of all *NF-Y* genes in response to different adverse growth conditions including low temperature, heat, low and high light intensities and treatments of photodynamic herbicides, e.g., acifluorfen, norflurazon (inhibitor of carotenoid synthesis), DCMU (inhibitor of the linear photosynthetic electron transport chain), as well as oxidants, e.g., hydrogen peroxide. We quantified the transcript levels of all 36 *NF-Y* genes encoding the NF-YA, NF-

YB and NF-YC subunits, which are currently known in *Arabidopsis* with qRT-PCR.

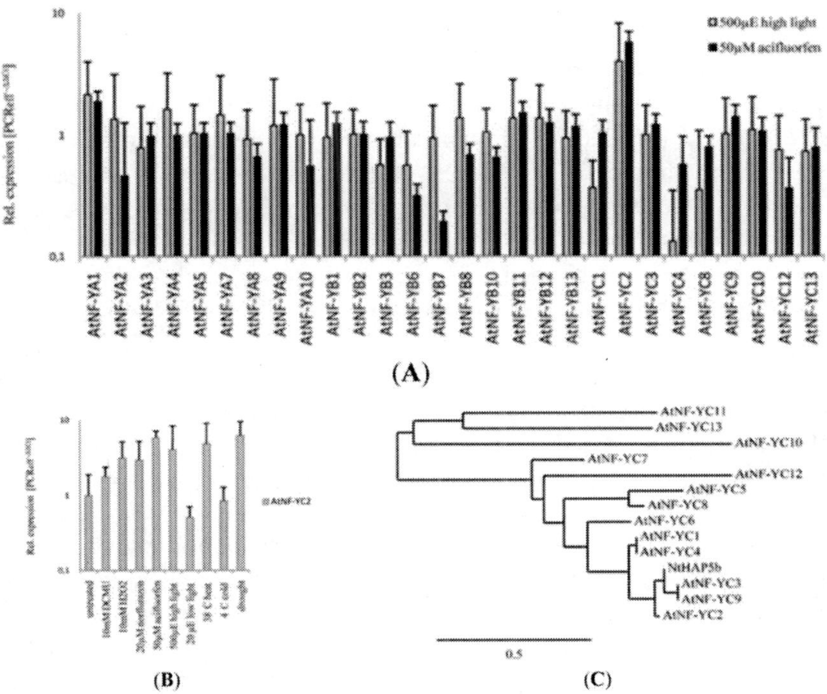

Figure 2. (**A**) Transcript profiles of *AtNF-Y* genes after acifluorfen treatment and high light exposure. Three-week-old*A. thaliana* (Col-0) seedlings were exposed to either 3 h of 500 µmol photon $m^{-2} \cdot s^{-1}$ or treated with 50 µM acifluorfen for 3 h. The transcript levels were normalized to a member of the*SAND* gene family (At2g28390). RNA levels were assayed by qRT-PCR in comparison to those of control seedlings. Representatives of the 36 *Arabidopsis NF-Y* genes not listed in the graph were not detectable in seedlings. Only the *AtNF-YC2*transcript strongly accumulates after high light exposure and acifluorfen treatment; (**B**) *AtNF-YC2* transcript levels after exposure to various stress conditions and after treatment with herbicides. The *NF-YC2* RNA contents were compared in stress-exposed and control seedlings after normalization to transcript levels of the *SAND* gene. The different stress conditions and applications are described in Material and Methods. *AtNF-YC2* shows a strong accumulation of RNA in response to several adverse conditions; (**C**) Phylogenetic tree of selected NF-YC proteins. Full-length amino acid sequences of*Arabidopsis* NF-YC proteins and HAP5b (NtNF-YC) of *Nicotiana tabacum* were used for alignment (using MUSCLE 3.7 and Gblocks 0.91b), phylogenetic analysis (PhyML3.0 aLRT) and

tree rendering (TreeDyn 198.3) performed with the phylogeny resource www.phylogeny.fr. [25].

Table 1 surveys the results of the expression analyses and highlights those *NF-Y* genes which show a rapid and stronger change in expression (<0.5 or >2.0) (or >5.0 for drought stress) after the exposure to different stresses and effectors. The different stress conditions and applications are described in Material and Methods. Among the *NF-Y* genes only a few representatives of the *NF-YA, NF-YB* and *NF-YC* subfamilies reveal pronounced changes of their transcript levels upon application of several inducible stress conditions in comparison to normal growth condition. Drought stress and heat treatments lead to induction of more *NF-Y* genes, most likely because of the intensity of the stress (heat) and the extended stress period (drought). Among those genes with enhanced accumulation of mRNA content in response to stress, *NF-YA1* and *NF-YC2* responded more extensively with increased mRNA levels than other *NF-Y* genes.

In default of a coproporphyrin-accumulating *Arabidopsis cpo* mutant we applied acifluorfen to wild-type seedlings. Acifluorfen is a peroxidising herbicide that inhibits protoporphyrinogen oxidase, an enzyme of tetrapyrrole biosynthesis [27]. Figure 2A displays the transcript profile of all*NF-Y* genes 3 h after application of 50 μM acifluorfen and in a different experiment after 3 h of high light stress. Among these genes ultimately only *NF-YC2* accumulates six-fold higher RNA levels in response to acifluorfen treatment and high light stress (Figures 2A,B). Figure 2B depicts the transcript level of *NF-YC2* after multiple stress treatments indicating the strong oxidative stress inducibility. In summary, elevated *NF-YC2* transcript levels often correlate with oxidative stress (H_2O_2, heat, drought) or photooxidative stress (acifluorfen, norflurazon, high light), whereas *NF-YC2* mRNA amounts are slightly reduced under conditions which predominantly possess a reduced potential for oxidative stress (low temperature, low light). Among the *NF-YC* subfamily (Figure 2C) *NF-YC2* seems to have the strongest responsiveness against oxidative stress conditions. Enhanced *NF-YC2* expression upon porphyrins-induced photooxidative stress resembles the expression of *NtNF-YC* in *CPO*-antisense plants.

Table 1. Overview of AtNF-Y genes with altered expression in response to abiotic stress. AtNF-Y genes revealing relative expression of >2.0 (>5.0 at drought stress) or <0.5 are listed in combination with the determined values. Lower and upper limit of the confidence interval in accordance with [26] appear in parentheses. Growth conditions under stress treatment are described in Material and Methods.

Stress condition	***AtNF-Y* with increased expression**		***AtNF-Y* with lower expression**	
	AtNF-Y	**relative expression**	***AtNF-Y***	**relative expression**
10 mM H_2O_2	*ATNF-YA1*	5.0 (1.6–15.8)	*ATNF-YB7*	0.3 (0.2–0.7)
	ATNF-YC2	3.1 (1.9–5.1)	*ATNF-YC12*	0.5 (0.4–0.7)
20 µM Norflurazon	*ATNF-YA1*	2.3 (1.3–3.9)	*ATNF-YB1*	0.4 (0.2–0.7)
	ATNF-YA10	3.1 (2.0–4.9)	*ATNF-YB6*	0.4 (0.2–0.6)
	ATNF-YC2	2.9 (1.7–5.1)	*ATNF-YC12*	0.5 (0.3–0.8)
50 µM Acifluorfen	*ATNF-YC2*	5.8 (4.7–7.1)	*ATNF-YA2*	0.5 (0.2–1.3)
			ATNF-YB6	0.3 (0.3–0.4)
			ATNF-YB7	0.2 (0.2–0.2)
			ATNF-YC12	0.4 (0.2–0.6)
High light 500 µE	*ATNF-YA1*	2.1 (1.1–4.0)	*ATNF-YC1*	0.4 (0.2–0.6)
	ATNF-YC2	4.0 (1.9–8.3)	*ATNF-YC4*	0.1 (0.1–0.3)
			ATNF-YC8	0.4 (0.1–1.1)
Low light 20 µE	*ATNF-YC4*	2.1 (1.2–3.5)	*ATNF-YC2*	0.5 (0.4–0.7)
			ATNF-YC13	0.4 (0.3–0.5)
Heat stress 38 °C	*ATNF-YA1*	4.2 (2.6–6.8)	*ATNF-YB7*	0.3 (0.2–0.5)
	ATNF-YA5	2.5 (1.7–3.7)	*ATNF-YC13*	0.1 (0.0–0.1)
	ATNF-YA7	2.3 (1.2–4.6)		
	ATNF-YB13	2.5 (1.4–4.4)		
	ATNF-YB6	3.6 (1.9–6.6)		
	ATNF-YB8	4.1 (2.7–6.3)		
	ATNF-YC10	2.1 (1.3–3.3)		
	ATNF-YC2	4.8 (2.5–9.0)		
	ATNF-YC4	9.4 (6.6–134)		
Cold stress 4 °C			*ATNF-YB2*	0.4 (0.3–0.5)
			ATNF-YB3	0.2 (0.2–0.4)
			ATNF-YB8	0.4 (0.2–0.7)
			ATNF-YC12	0.4 (0.2–0.6)
			ATNF-YC4	0.4 (0.2–0.6)
Drought stress	*ATNF-YA1*	21.7 (7.3–64.6)	*ATNF-YC1*	0.4 (0.2–0.8)
	ATNF-YA10	48.0 (6.7–345.6)		
	ATNF-YA3	5.9 (3.1–11.4)		
	ATNF-YA4	5.4 (3.6–8.1)		
	ATNF-YA5	9.7 (4.1–23.1)		
	ATNF-YA7	7.9 (5.4–11.5)		
	ATNF-YB1	9.0 (5.0–16.0)		
	ATNF-YB10	22.6 (8.9–57.7)		
	ATNF-YB6	57.4 (30.7–107.3)		
	ATNF-YB7	5.2 (2.8–9.5)		
	ATNF-YC10	15.2 (9.0–25.8)		
	ATNF-YC2	6.2 (4.0–9.5)		
	ATNF-YC3	7.6 (4.3–13.4)		
	ATNF-YC4	5.7 (3.3–10.2)		

2.3. An Arabidopsis nf-yc2 Mutant and NF-YC2 Overexpressor Plants upon Oxidative Stress

Among the potential T-DNA-insertion mutant lines for *NF-YC2* (SALK_11422; GABI_369E03, GABI_669A05) we identified only one mutant line with a T-DNA-insertion in the coding region (GABI_669A05). This line was designated *nf-yc2*. The exact T-DNA insertion site of *nf-yc2* could be confirmed 90 base pairs upstream of the translation stop codon (Figure 3A,B), while the inserted T-DNA of SALK_11422 and GABI_369E03 interrupted *NF-YC2* in the 5′ and 3′ untranslated region, respectively. The transcript levels of all T-DNA mutant lines were similar to wild-type seedlings (data not shown). But qRT-PCR revealed an increased content of the 3′ part of the cDNA downstream of the T-DNA-insertion site which can be explained with enhanced transcriptional activity due to promoter activity on the T-DNA (Figure 3C). It was expected that homozygous *nf-yc2* synthesizes a truncated protein. But the size, the amount and the stability of NF-YC2 in control plants and the homozygous *nf-yc2* mutant could not be determined. An antibody raised against a small non-conserved peptide sequence of NF-YC2 did not specifically recognize a protein of the expected size in the nuclear extract or total extracts of *Arabidopsis* wild-type seedlings. Efforts to enrich the specific antibody by purification attempts failed.

Additionally, more physiological properties of *nf-yc2* were examined in comparison to wild-type seedlings. Analyses of chlorophyll fluorescence parameters of mutant and wild-type seedlings revealed no significant differences in photosynthetic properties (Figure S1). Additionally, levels of tetrapyrrole intermediates were determined after a 36 h acifluorfen treatment. But the seedlings did not display significant differences in the content of tetrapyrrole intermediates (Figure S2). These results indicate that *nf-yc2* was not more sensitive towards photooxidative stress than control seedlings.

Seedlings of *nf-yc2* and wild type exposed to different abiotic stress conditions did not show differences of the macroscopic phenotype. This holds true, regardless, whether mutant and control plants were either treated with acifluorfen or Rose Bengal as well as exposed to high light or heat stress. When toxic amounts of herbicides were applied to the plants, formation of necrotic tissue upon acifluorfen treatment or bleaching in response to applied norflurazon were quantitatively similar in leaves of mutant and wild-type seedlings. We derived from these experiments that an obvious phenotypic diversification between *nf-yc2* and control seedlings could not be observed.

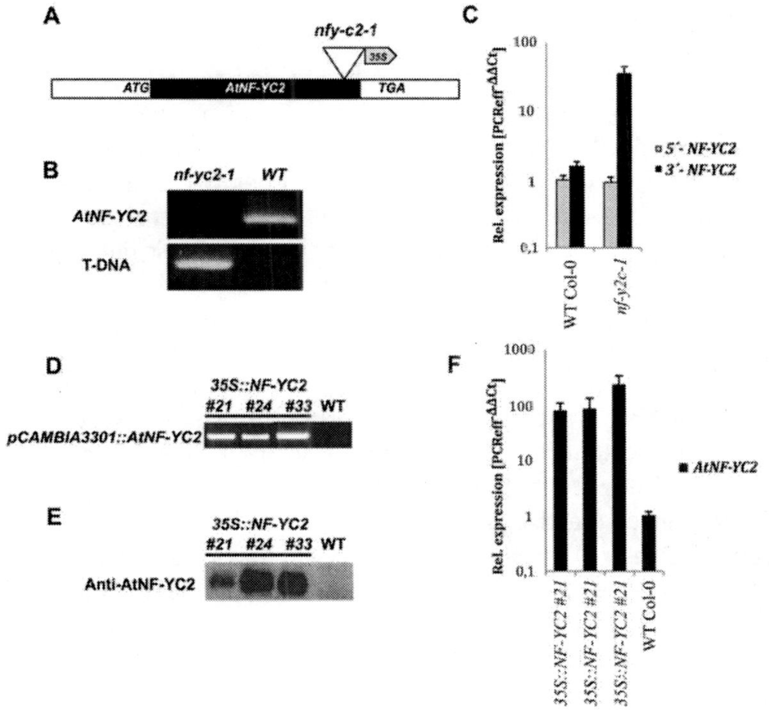

Figure 3. Analysis of the *Arabidopsis nf-yc2* mutant and *NF-YC2*overexpressor lines: (**A**) Schematic T-DNA-insertion site in*AtNF-YC2*; (**B**) Experimental confirmation of homozygosity of the T-DNA-inserted allele in *nf-yc2* in comparison to wild type. A PCR product of the endogenous *NF-YC2* gene could not be generated from DNA of *nf-yc2*; (**C**) The content of *NF-YC2*RNA in *nf-yc2* and control seedlings. The content of the *NF-YC2* RNA was determined with primer pairs upstream and downstream of the T-DNA insertion site in *NF-YC2*. A T-DNA-mediated stimulation leads to accumulation of 3′ transcripts in*nf-yc2* compared to control; (**D**) The *35SCaMV::AtNF-YC2*transgene was detected in the genomic DNA of the three representative transgenic lines #21, #24 and #33 in comparison to the wild-type genome; (**E**) Immunodetection of NF-YC2. Among the primary transformants the lines #21, #24 and #33 were selected to examine the *NF-YC2* expression. By means of anti-NF-YC2 antibodies the overproduction of NF-YC2 could be examined in plant extracts. The wild-type extract contains non-detectable amounts of NF-YC2; (**F**) Relative expression level of *AtNF-YC2* in wild-type and transgenic lines. Progenies of the T2 generation of the three transgenic lines were subjected to semi-quantitative RT PCR.

In order to examine the effect of ectopic *AtNF-YC2* overexpression on photooxidative stress response, the full-length cDNA sequence of *NF*-YC2 was inserted downstream of a CaMV 35S promoter in the pCAMBIA 3301 vector and introduced into the genome of *Arabidopsis thaliana*(Figure 3D). Selection of transformed lines with high expression of the transgene revealed lines #21, #24 and #33 with highly elevated *NF-YC2* mRNA levels (Figure 3F). Using the anti-NF-YC2 antibody, the overproduced protein was immunologically detectable in total cell extracts (Figure 3E), while the content of the endogenous NF-YC2 remained below the detection level in total wild-type extracts. The selected *NF-YC2* overexpressor lines were sprayed with 100 µM acifluorfen. However, no phenotypic differences were detectable on leaves of transgenic lines in comparison to wild-type leaves. Upon application of different concentrations of acifluorfen transgenic and wild-type seedlings showed always a similar formation of leaf necrosis after herbicide treatments. The viability of transgenic lines was also not elevated upon acifluorfen treatments in comparison to control seedlings after herbicide application (data not shown). Also the expression of the stress related *APX* genes (*APX1, APX2* and *APX4*), which are involved in the detoxification of reactive oxygen species (ROS) [28] and are induced by oxidative stress [29], were neither in *nf-yc2* nor in*35SCaMV::NF-YC2* significantly increased in relation to wild-type plants independent if acifluorfen were applied or not (data not shown).

2.4. Characterization of Other nf-yc Mutants Related to nf-yc2

In addition, the impact of photooxidative stress was also tested in other *Arabidopsis* T-DNA insertion mutants of the *Arabidopsis NF-YC* family with high sequence similarity to *NF-YC2* (Figure 3). T-DNA insertion mutants for the family members *AtNF-YC1* (GABI_004H11, SALK_86334),*AtNF-YC3* (GABI_51E10), *AtNF-YC4* (SALK_32163) and *AtNF-YC9* (SALK_058903) were isolated and homozygous lines were generated. With the exception of *nf-yc9,* homozygous *nf-yc1*, *nf-yc3*and *nf-yc4* mutant plants could always be identified with the T-DNA insertion in the coding region. In these T-DNA insertion lines no transcript of the respective *AtNF-YC* gene was detected (data not shown). *nf-yc3* (GABI_51E10) and *nf-yc4* (SALK_32163) as well as *nf-yc1-1* (GABI_004H11) and *nf-yc1-2* (SALK_86334) T-DNA insertion mutants neither showed phenotypically changes in response to treatments with acifluorfen nor to the exposure to high light. The growth of these mutants resembled that of wild-type and *nf-yc2* seedlings.

2.5. Modified Flowering Time of NF-YC1 and NF-YC2 Overexpressor Lines

Apart from the transcript analysis and the assessment of phenotypic differences between mutant and control seedlings upon abiotic stress, *NF-YC2* overexpressor lines and the T-DNA insertion mutant *nf-yc2* were examined for induction of flowering under short and long day conditions. In parallel, *nf-yc1-2* as well as *AtNF-YC1* overexpressor lines were also tested as representatives of the*NF-YC* subfamily with high homology to NF-YC2. Genotyping of *nf-yc1-2* revealed the T-DNA insertion in the exon region (Figure S3A). Homozygous *nf-yc1-2* does not contain detectable amounts of *NF-YC1* transcripts (Figure S3B). *35SCaMV::NF-YC1* expression in transgenic*Arabidopsis* lines yielded a strong accumulation of the specific mRNA. The lines #41, #42 and #57 represent transformants with over-expression of *NF-YC1* (Figure S3C,D).

Thirty-eight-day-old seedlings of the transgenic lines 35SCaMV::NF-YC1 #41, #42, #57 and 35SCaMV::NF-YC2 #21, #24 and #33 showed a significant early flowering phenotype under long day condition compared to wild-type plants and *nf-yc2* and *nf-yc1-2*. Figure 4A shows always one representative plant of each set of overexpressor lines. The overexpressor lines flowered at least 16 days earlier than wild type (Figure 4B). Under short day condition all mutants and overexpressor lines did not significantly modulate the flowering time (Figure 4C). Flowering was set when petals started to be seen during flower development. As a control, *co-1,* a mutant of *CONSTANS* (*co-1*) showed a significantly delayed flowering than wild-type plants under long day condition (Figure 4B). In consistencies with a previous report *co-1* flowered significantly earlier under short day condition (Figure 4C) [30].

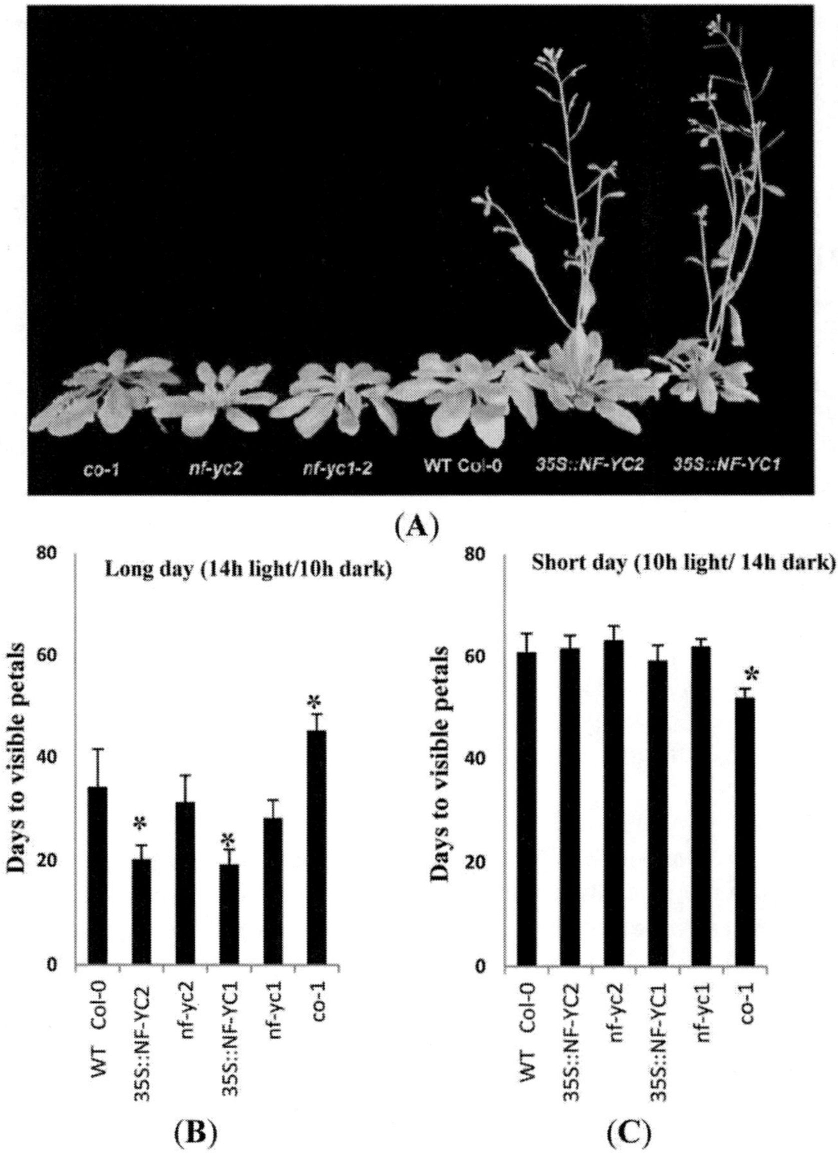

Figure 4. (**A**) Developmental stage and flowering phenotype of*Arabidopsis* wild-type (Col-0), *nf-yc2, nf-yc1-2*, *co-1* and representatives of transgenic lines overexpressing ectopically*NF-YC2* or *NF-YC1* 38 days after germination. Expression of*35S::NF-YC1* and *35S::NF-YC2* leads to induction of early flowering under long day condition in comparison to wild-type plants (**B**). Time period between germination and development of first visible petals is indicated under long day (**B**) or short day condition (**C**).

2.6. Modified Expression of Genes Involved in Flowering Induction in 35SCaMV:NF-YC2 and 35SCaMV:NF-YC1 Overexpressor Lines

In 21-day-old mutant seedlings with modified expression of *NF-YC1* and *NF-YC2* we explored transcript levels of *FLOWERING LOCUS (FT)* and *SUPRESSOR OF OVEREXPRESSION OF CO1 (SOC1)* [31], two genes in the photoperiodic control of flowering by qRT-PCR. Growth under long day condition (14 h light/10 h dark) leads to a 20-fold accumulation of *FT* transcripts in the earlier flowering periods of the *35SCaMV::NF-YC1* and *35SCaMV::NF-YC2* lines (Figure 5A). The two mutants *nf-yc1-2* and *nf-yc2* also showed elevated *FT* RNA levels, which do not correlate with a significant variation in flowering time in comparison to wild-type properties. As control *co-1*contained lower *FT* mRNA levels under long day condition, which correlate with the late flowering phenotype (Figure 5A). Short day exposure (10 h light/14 h dark) of the *NF-YC1* and *NF-YC2* overexpressor lines did not result in modified expression of *FT* (Figure 5B). The *SOC*1 transcript levels were not modulated under long or short day condition in overexpressor plants and T-DNA insertion mutants in comparison to wild-type plants (Figure 5A,B).

3. DISCUSSION

Among the members of the *Arabidopsis NF-YC* family, only *NF-YC2* shows an induced expression under oxidative stress conditions and is a candidate for the transcriptional activation in response to oxidative stress. Besides its inducibility upon several oxidative stress conditions, *AtNF-YC2* expression was drastically enhanced after acifluorfen treatment. The inactivation of protoporphyrinogen oxidase leads to accumulation of the photoreactive protoporphyrinogen, which, upon light exposure, is autooxidized to protoporphyrin and generates reactive oxygen species. The photodynamic stress generated by acifluorfen treatment resembles the photosensitization of tobacco *CPO*-antisense RNA-expressing plants under high light conditions [23,24]. Furthermore, *AtNF-YC2* shares the responsiveness to photooxidative stress with the tobacco homologous *NtNF-YC* gene (Figure 1). The tobacco NF-YC subunit belongs to the same phylogenetic subclade as the *A. thaliana* proteins NF-YC1, 2, 3, 4 and 9 (Figure 2C), but among the*Arabidopsis NF-YC* subfamily NF-YC2 does not show the highest similarity to NtNF-YC.

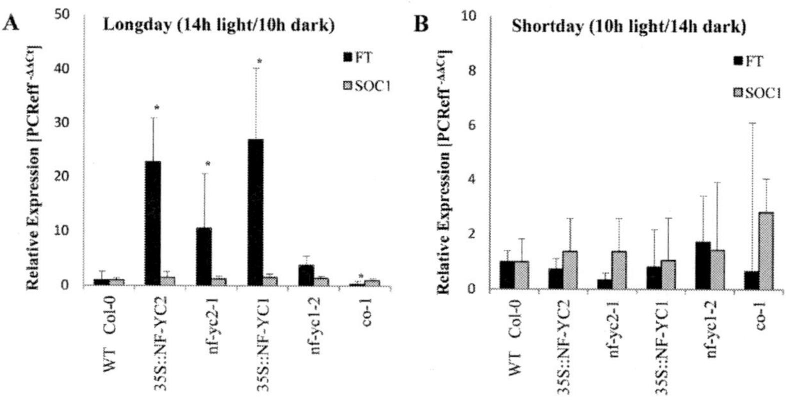

Figure 5. Relative expression profiles of *FLOWERING LOCUS T* (*FT*) and *SUPRESSOR OF CONSTANS1* (*SOC1*) in *Arabidopsis*wild-type (Col-0), *nf-yc2, nf-yc1-2* and a representative line each of *NF-YC2* and *NF-YC1* overexpressor plants. RNA was isolated in plants grown under long day (**A**) and short day condition (**B**). The leaf material was harvested before the expected induction of flowering. Quantitative RT-PCR revealed increased *FT* RNA contents in *NF-YC2* and *NF-YC1*overexpressing lines relative to wild-type RNA. In each RNA sample *FT* and *SOC1* RNA were normalized to constitutive expression of *SAND* (At2g28390).

Porphyrin-accumulating transgenic plants as well as acifluorfen-treated plants trigger a necrotic cell death phenotype indicating a mechanism that integrates accumulation of photooxidative stress during a time period of light exposure. It is suggested that the reactive oxygen species generated during these photodynamic processes might induce the *NF-YC2* expression. The elucidation of additional roles of both homologous representatives of NF-YC subunits in both plant species requires further investigations. Apart from the stress-induced *NtNF-YC* it remains also elusive, how many additional genes belong to the *NF-YC* family in *Nicotiana tabacum*.

Despite of the clear induction pattern of *AtNF-YC2* and its tobacco *NF-YC* homolog, the*Arabidopsis nf-yc2* mutant and the *NF-YC2* overexpressor seedlings do not show phenotypic modulation in their response to (photo)oxidative stress. Derived from the expression data *nf-yc2*mutants can most likely still express a truncated protein which lacks part of the transcription activation domain of NF-YC but still possesses the histone-fold motif. It is not excluded that the truncated NF-YC2 still contributes to the assembly of intact NF-YC heterotrimers. But also NF-YC2 overexpressors did not show a positive or adverse effect upon exposure to oxidative stress. These missing phenotypical alterations in response to oxidative stress are likely explained with the redundancy among NF-YC subunits. When expression of one *NF-YC* gene is depleted, another member of the family might compensate the deficiency. The compensatory capacities among transcription factor families have been proposed in the literature [32]. Neither mutants for *NF-YC1,NF-YC3* and *NF-YC4* nor *nf-yc1/nf-yc2* double mutants (data not shown) or *NF-YC1* overexpressor mutants showed an impaired response upon or increased sensitivity to oxidative stress.

Instead of phenotypical changes during oxidative stress, *NF-YC2* and *NF-YC1* overexpressors displayed early flowering during long day condition (Figure 4). This is correlative with the increased *FT* expression in the overexpressor lines (Figure 5). It is suggested that NF-YC2 and NF-YC1 contribute to the transcriptional control of *FT*. Along these lines *co-1* mutants show an extended flowering time during long day condition which correlates with strong reduction of *FT*expression (Figures 4 and 5) [30].

Day length dependent flowering in *Arabidopsis* is predominantly regulated by two transcription factors CO and FT. Under long day condition CO activates transcription of *FT* in the leaves, whereas the FT protein is transported as phloem mobile flowering signal to the shoot apex. In the apical shoot meristem, FT together and another transcription factor FLOWERING LOCUS D (FD) are responsible for induction of *SOC1* and *APETALA 1* (AP1) encoding two major regulators of the conversion into the inflorescence meristem [33].

The current model of transcriptional control of *FT* includes the function of CO as major regulator via a specific *cis* element inside the *FT* promoter, also described as CONSTANS responsive element (CORE) [19]. The *Arabidopsis* NF-Y complex is assumed to act as modulator of the CO-mediated transcription activation of *FT*. As a potential recruitment site a CCAAT-box inside the *FT* promoter was hypothesized [17], which is not responsible for the transcription activation by CO [19]. Nevertheless diverse NF-Y proteins physically interact with CO and other representatives of the CO-like family [34,35].Therefore, both transcription factors, CO and NF-Y, are suggested to mutually influence each other in their function on the *FT*-promoter.

The exact molecular function of NF-Y proteins during the day length-dependent flowering induction remains still unclear. Down-stream contribution of NF-Ys in other flower-inducing pathways are not excluded. In this context it is worth to be mentioned that gibberellins (GAs) play an important role during the conversion of vegetative to floral tissue at the apical shoot meristem. GAs are embedded in different pathways influencing flower induction in *Arabidopsis* [36]. On the one hand GAs promote flowering independent from the CO/FT-mediated regulatory mechanism under short day conditions in a non-photoperiodic manner [37]. On the other hand GA-mediated effects on flowering time can be observed under short and long day conditions [38]. Moreover, a GA-dependent mechanism regulating the expression of *FT* cannot be excluded [36], especially due to the fact that GAs application can result in increased *FT* transcript levels and early flowering under long day conditions in *Arabidopsis* [39].

The physiological relevance of NF-Y during photoperiodic flower induction became obvious by investigation of several *nf-y* mutants showing modified flowering time. This clear phenotype can be contributed to the mutants *nf-yb2* and *nf-yb3* as well as the *nf-yb2/b3* double and *nf-yc3/c4/c9*triple mutants [17,18]. While the deficit of one or several of the above mentioned NF-Y subunits leads to a delayed flowering phenotype under long day conditions, overexpressing of *NF-YB2* or*NF-YB3* yielded in an early flowering phenotype under identical conditions. On the other hand no effect can be observed in these overexpressor lines in short day light periods [17].

The early flowering phenotypes of both *NF-YC2* and *NF-YC1*-overexpressing mutants underline a similar function of NF-YC2 and NF-YC1 in day length depending floral induction. Similar to *NF-YC3, NF-YC4* and NF-*YC9,* single knock-out mutation of *NF-YC2* or *NF-YC1* did not lead to phenotypical changes in flower induction. While the effect of NF-YC3, C4 and C9 on flower induction could only be observed by simultaneous knock out of all three subunits in the *nf-yc3/nf-yc4/nf-yc9* triple mutant [18], *nf-yc1/nf-yc2* double mutant plants did not show a similar phenotype (data not shown).

4. MATERIAL AND METHODS

4.1. Growth of Plants

Arabidopsis thaliana plants grew for 14, 21 or 34 days under light/dark conditions (120 μM photons $m^{-2} \cdot s^{-1}$, 12 h light/ 12 h dark at 23 °C) before they were exposed to high light (HL—500 μM photons $m^{-2} \cdot s^{-1}$), low light (LL—50 μM photons $m^{-2} \cdot s^{-1}$), increased (38 °C) or low temperature (4 °C) respectively. Alternatively, plants were treated with 50 μM acifluorfen, 10 mM DCMU, 10 mM H_2O_2 or 20 μM norflurazon by spraying leaves with these agents. Leaf material was harvested after exposure time of 3 h. Drought stress was induced when plants were not watered for 8 days.

For RNA quantification *Nicotiana tabacum* var. Samsun NN (SNN) and *CPO*-antisense plants [23] were initially cultured for 14 days under low light conditions (40 μM photons $m^{-2} \cdot s^{-1}$, 12 h light/ 12 h dark at 27 °C) before exposure to high light (400 μM photons $m^{-2} \cdot s^{-1}$, 12 h light/ 12 h dark at 27 °C). Leaf material was harvested before and 6 h after exposure to high light. For *suppression subtractive hybridization* experiments tobacco plants were cultivated for 8 weeks with a low light dose (200 μM photons $m^{-2} \cdot s^{-1}$, 6 h light/ 18 h dark at 25 °C) and were subsequently exposed to a high light dose (400 μM photons $m^{-2} \cdot s^{-1}$) for 12 h before leaf material was harvested.

4.2. RNA—Quantification

For qRT-PCR total RNA of 100 mg leaf material was extracted with RNeasy mini kit (Qiagen, Germany) in accordance to the supplier's manual. The amount of 1.250 ng of total RNA were treated with 2U TURBO DNA-*free* (Ambion, USA) for removing genomic DNA contaminations before first strand cDNA synthesis was performed using SuperScript III reverse transcriptase (Invitrogen, USA) in the presence of 50 ng oligo dT_{18}-primer (Sigma, Germany). As a template for qRT-PCR analysis ten times diluted cDNA were used in 20 μL reactions (1 μL cDNA, 500 nM each primer, 0.4 μL 50 × SYBR Green and 10 μL ImmoMix TM (Bioline, Germany) performed with the Rotor Gene 2000 (Corbert-Research, Australia) or 5 μL PCR reactions (0.5 μL cDNA, 200 nM each primer and 2.5 μL SYBR Green PCR Master Mix (Applied Biosystems, USA) performed in 384 wells plates using the ABI PRISM 7900HT sequence detection system (Applied Biosystems, USA). The amplification program consisted of an initial hot-start activation at 94 °C for 10 min, 45 cycles including 15 s denaturing at 94 °C, 20 s annealing at 60 °C and 20 s elongation phase at 72 °C. Primers of the *Arabidopsis thaliana NF-Y* gene families were kindly provided by the Max-Planck Institute of Molecular Plant Physiology Golm. Oligonucleotide sequences are listed in the Table S2. Transcript levels of each gene were normalized to an *Arabidopsis* member of the *Arabidopsis SAND*gene family (At2g28390) or *Nicotiana tabacum ACTIN* gene respectively. Primer efficiency was determined by linear regression on the Log(fluorescence) of each PCR reaction using the LinRegPCR software [40] and PCR efficiency was defined as mean value for each primer pair. For semi-quantitative RT-PCR, total RNA were isolated with TRIsure (Bioline, Germany) and treated with 1U DNAseI (Fermentas, Germany). First strand cDNA synthesis was performed by RevertAid M-MuLV reverse transcriptase (Fermentas, Germany) in accordance to the manufacturer instructions.

4.3. Plant Mutants

Full length cDNA constructs of *AtNF-YC1* and *AtNF-YC2* were amplified by PCR using specific primers (AtNF-YC1: 5'-ACCATGGATACCAACAACCAGC-3' and 5'-TGACGTCCACCTTGGCCG TCGAGA-3'; AtNF-YC2: 5'-ACCATGGAGCAG TCAGAAGAGG-3' and 5'-TGACGTCCAGACTC ATCAGGGTGTTG-3') and ligated into NcoI and ActII restriction sites of the multiple cloning site of the modified pCAMBIA3301 plasmid. The correct sequence of the *35SCaMV::AtNF-YC1* and*35SCaMV:: AtNF-YC2* gene constructs was confirmed by sequencing and used for *A. tumefaciens*-mediated stable transformation of *Arabidopsis* (ectoype Col-0) plants by means of the floral-dip transformation method described previously [41]. Transformed plants were selected by successive treatments of two, three and four-week-old seedlings with 0.1% (w:v) BASTA (Bayer CropScience AG, Germany). T-DNA insertion mutants of the SALK and GABI-Kat collection were obtained from the European Arabidopsis Stock Centre (Nottingham, UK) or Bielefeld University, Germany. Homozygous T-DNA insertions were validated by PCR using specific oligo primer for either the wild-type allele or the T-DNA left boarder and for indication of different genes (primer sequences in Table S2). Transcript levels of the affected genes in mutant plants were analyzed as described above. AtNF-YC2 protein amounts of *pCAMBIA3301::AtNF-YC2* plants were determined by immunoblot analysis in accordance to standard methods [42].

4.4. Chlorophyll Fluorescence Analyses

In vivo chlorophyll fluorescence was measured with attached leaves of dark-adapted plants as previously described [43]. The following fluorescence parameters were assessed: the maximum photochemical efficiency of PS II in the dark-adapted state $F_v/F_m = (F_m - F_0)/F_m$, effective quantum yield of photochemical energy conversion in PSII ($\Phi P_{SII} = (F_m' - F_t)\, F_m'^{-1}$) and non-photochemical quenching ($NPQ = (F_m - F_m')\, F_m'^{-1}$) in accordance to [44]. Photon flux densities were measured using a quantum sensor (LI-189A, Li-Cor, Lincoln, NE).

4.5. Analysis of Porphyrins

Extraction and analysis of accumulated tetrapyrroles in tobacco leaves were essentially performed as described [23]. Modifications regard the extraction of liquid nitrogen-ground leaf material with 50 mM potassium phosphate buffer (pH 7.8), methano1:0.l M NH_4OH (9:1, *v/v*) and acetone: 0.l M NH_4OH (9:1, *v/v*). The porphyrins were separated by HPLC (Agilent, Germany) on a RP 18 column (Novapak C18, 4 µm particle size, 4.6 × 250 mm). Column eluent was monitored by fluorescence (λ_{ex} 405 nm, λ_{em} 625 nm) and porphyrins were identified and quantified by authentic standards.

4.6. Suppression Subtractive Hybridization

Two µg polyA$^+$ RNA of the sixth tobacco leaf (counting from the top of the plant) from wild-type (driver) and CPO antisense # 41 (tester) plants was extracted and applied to the forward subtracted and the reverse subtracted approach according to the protocol of the supplier (Clontech, Germany). After *E. coli* transformation with pCRII vector (Invitrogen, USA) containing forward subtracted cDNA fragments 1993 colonies were obtained, which were subjected to a colony hybridization using the labeled probes from forward and reverse subtracted cDNA. Finally, 234 bacterial colonies were confirmed and their cDNA sequenced.

4.7. Statistical Analysis

All data shown in graphs are mean values and corresponding standard deviations are displayed as error bars calculated using the Excel 2007 software (Microsoft Corp., USA). For relative expression analysis three biological replicates were used, whereas fluorescence parameter, tetrapyrrole intermediates and days to visible petals were determined by evaluation of 12 biological replicates (different plants). Differences between WT and mutant lines were tested using a two-tailed Student's *t test* and were regarded as significant for $p < 0.05$.

5. CONCLUSION

As a conclusion, it seems to be reasonable that the subunits NF-YC1, NF-YC2, NF-YC3, NF-YC4 and NF-YC9 are involved in the control of day length dependent floral induction. These five members of the *Arabidopsis* NF-YC family can be clustered in one phylogenetic branch possessing a high similarity of their amino acid sequence. The elucidation of specific roles of the individual NF-YC subunits remains a challenging task, due to the obvious overlap of their functional properties.

ACKNOWLEDGEMENTS

The work was supported by the Deutsche Forschungsgemeinschaft (DFG, GR 936/9-1, 9-2), We grateful acknowledge the assistance of Czechowski for the application of the transcriptome platform provided by the Max-Planck-Institute of Molecular Plant Physiology in Potsdam Golm, Germany.

REFERENCES

1. Mantovani, R. A survey of 178 NF-Y binding CCAAT boxes. *Nucleic Acids Res* **1998**, *26*, 1135–1143.

2. Rieping, M.; Schoffl, F. Synergistic effect of upstream sequences, CCAAT box elements, and HSE sequences for enhanced expression of

chimaeric heat shock genes in transgenic tobacco.*Mol. Gen. Genet* **1992**, *231*, 226–232.

3. Mantovani, R. The molecular biology of the CCAAT-binding factor NF-Y. *Gene* **1999**, *239*, 15–27.

4. Maity, S.N.; de Crombrugghe, B. Role of the CCAAT-binding protein CBF/NF-Y in transcription. *Trends Biochem. Sci* **1998**, *23*, 174–178.

5. Matuoka, K.; Chen, K.Y. Nuclear factor Y (NF-Y) and cellular senescence. *Exp. Cell Res* **1999**,*253*, 365–371.

6. Li, X.Y.; Mantovani, R.; van Huijsduijnen, R.H.; Andre, I.; Benoist, C.; Mathis, D. Evolutionary variation of the CCAAT-binding transcription factor NF-Y. *Nucleic Acids Res***1992**, *20*, 1087–1091.

7. Liberati, C.; di Silvio, A.; Ottolenghi, S.; Mantovani, R. NF-Y binding to twin CCAAT boxes: Role of Q-rich domains and histone fold helices. *J. Mol. Biol* **1999**, *285*, 1441–1455.

8. Masiero, S.; Imbriano, C.; Ravasio, F.; Favaro, R.; Pelucchi, N.; Gorla, M.S.; Mantovani, R.; Colombo, L.; Kater, M.M. Ternary complex formation between MADS-box transcription factors and the histone fold protein NF-YB. *J. Biol. Chem* **2002**, *277*, 26429–26435.

9. Sinha, S.; Maity, S.N.; Lu, J.; de Crombrugghe, B. Recombinant rat CBF-C, the third subunit of CBF/NFY, allows formation of a protein-DNA complex with CBF-A and CBF-B and with yeast HAP2 and HAP3. *Proc. Natl. Acad. Sci. USA* **1995**, *92*, 1624–1628.

10. Gusmaroli, G.; Tonelli, C.; Mantovani, R. Regulation of novel members of the Arabidopsis thaliana CCAAT-binding nuclear factor Y subunits. *Gene* **2002**, *283*, 41–48.

11. Siefers, N.; Dang, K.K.; Kumimoto, R.W.; Bynum, W.E.T.; Tayrose, G.; Holt, B.F., III. Tissue-specific expression patterns of Arabidopsis NF-Y transcription factors suggest potential for extensive combinatorial complexity. *Plant Physiol* **2009**, *149*, 625–641.

12. Gusmaroli, G.; Tonelli, C.; Mantovani, R. Regulation of the CCAAT-Binding NF-Y subunits in Arabidopsis thaliana. *Gene* **2001**, *264*, 173–185.

13. Hackenberg, D.; Wu, Y.; Voigt, A.; Adams, R.; Schramm, P.; Grimm, B. Studies on differential nuclear translocation mechanism and assembly of the three subunits of the arabidopsis thaliana transcription factor NF-Y. *Mol. Plant* **2011**.

14. Lotan, T.; Ohto, M.; Yee, K.M.; West, M.A.; Lo, R.; Kwong, R.W.; Yamagishi, K.; Fischer, R.L.; Goldberg, R.B.; Harada, J.J. Arabidopsis LEAFY COTYLEDON1 is sufficient to induce embryo development in vegetative cells. *Cell* **1998**, *93*, 1195–1205.

15. Kwong, R.W.; Bui, A.Q.; Lee, H.; Kwong, L.W.; Fischer, R.L.; Goldberg, R.B.; Harada, J.J. LEAFY COTYLEDON1-LIKE defines a class of regulators essential for embryo development.*Plant Cell* **2003**, *15*, 5–18.

16. Folta, K.M.; Kaufman, L.S. Regions of the pea Lhcb1*4 promoter necessary for blue-light regulation in transgenic Arabidopsis. *Plant Physiol* **1999**, *120*, 747–756.

17. Kumimoto, R.W.; Adam, L.; Hymus, G.J.; Repetti, P.P.; Reuber, T.L.; Marion, C.M.; Hempel, F.D.; Ratcliffe, O.J. The Nuclear Factor Y subunits NF-YB2 and NF-YB3 play additive roles in the promotion of flowering by inductive long-day photoperiods in Arabidopsis. *Planta* **2008**,*228*, 709–723.

18. Kumimoto, R.W.; Zhang, Y.; Siefers, N.; Holt, B.F. NF–YC3, NF–YC4 and NF–YC9 are required for CONSTANS-mediated, photoperiod-dependent flowering in Arabidopsis thaliana. *Plant J* **2010**, *63*, 379–391.

19. Tiwari, S.B.; Shen, Y.; Chang, H.C.; Hou, Y.; Harris, A.; Ma, S.F.; McPartland, M.; Hymus, G.J.; Adam, L.; Marion, C.; Belachew, A.; Repetti, P.P.; Reuber, T.L.; Ratcliffe, O.J. The flowering time regulator CONSTANS is recruited to the FLOWERING LOCUS T promoter via a unique cis-element. *New Phytol* **2010**, *187*, 57–66.

20. Li, W.X.; Oono, Y.; Zhu, J.; He, X.J.; Wu, J.M.; Iida, K.; Lu, X.Y.; Cui, X.; Jin, H.; Zhu, J.K. The Arabidopsis NFYA5 transcription factor is regulated transcriptionally and posttranscriptionally to promote drought resistance. *Plant Cell* **2008**, *20*, 2238–2251.

21. Nelson, D.E.; Repetti, P.P.; Adams, T.R.; Creelman, R.A.; Wu, J.; Warner, D.C.; Anstrom, D.C.; Bensen, R.J.; Castiglioni, P.P.; Donnarummo, M.G.; *et al.* Plant nuclear factor Y (NF-Y) B subunits confer drought

tolerance and lead to improved corn yields on water-limited acres.*Proc. Natl. Acad. Sci. USA* **2007**, *104*, 16450–16455.

22. Liu, J.X.; Howell, S.H. bZIP28 and NF-Y transcription factors are activated by ER stress and assemble into a transcriptional complex to regulate stress response genes in Arabidopsis.*Plant Cell* **2010**, *22*, 782–796.

23. Kruse, E.; Mock, H.P.; Grimm, B. Reduction of coproporphyrinogen oxidase level by antisense RNA synthesis leads to deregulated gene expression of plastid proteins and affects the oxidative defense system. *EMBO J* **1995**, *14*, 3712–3720.

24. Keetman, U.; Mock, H.P.; Grimm, B. Kinetics of antioxidative defence responses to photosensitisation in porphyrin-accumulating tobacco plants. *Plant Physiol. Biochem* **2002**, *40*, 697–707.

25. Dereeper, A.; Guignon, V.; Blanc, G.; Audic, S.; Buffet, S.; Chevenet, F.; Dufayard, J.F.; Guindon, S.; Lefort, V.; Lescot, M.; *et al.* Phylogeny.fr: Robust phylogenetic analysis for the nonspecialist. *Nucleic Acids Res* **2008**, *36*, W465–469.

26. Livak, K.J.; Schmittgen, T.D. Analysis of relative gene expression data using real-time quantitative PCR and the 2(-Delta Delta C(T)) Method. *Methods* **2001**, *25*, 402–408.

27. Witkowski, D.A.; Halling, B.P. Inhibition of plant protoporphyrinogen oxidase by the herbicide acifluorfen-methyl. *Plant Physiol* **1989**, *90*, 1239–1242.

28. Shigeoka, S.; Ishikawa, T.; Tamoi, M.; Miyagawa, Y.; Takeda, T.; Yabuta, Y.; Yoshimura, K. Regulation and function of ascorbate peroxidase isoenzymes. *J. Exp. Bot* **2002**, *53*, 1305–1319.

29. Koussevitzky, S.; Suzuki, N.; Huntington, S.; Armijo, L.; Sha, W.; Cortes, D.; Shulaev, V.; Mittler, R. Ascorbate peroxidase 1 plays a key role in the response of Arabidopsis thaliana to stress combination. *J. Biol. Chem* **2008**, *283*, 34197–34203.

30. Putterill, J.; Robson, F.; Lee, K.; Simon, R.; Coupland, G. The CONSTANS gene of Arabidopsis promotes flowering and encodes a protein showing similarities to zinc finger transcription factors. *Cell* **1995**, *80*, 847–857.

31. Samach, A.; Onouchi, H.; Gold, S.E.; Ditta, G.S.; Schwarz-Sommer, Z.; Yanofsky, M.F.; Coupland, G. Distinct roles of CONSTANS target genes in reproductive development of Arabidopsis. *Science* **2000**, *288*, 1613–1616.

32. Jin, H.; Martin, C. Multifunctionality and diversity within the plant MYB-gene family. *Plant Mol. Biol* **1999**, *41*, 577–585.

33. Turck, F.; Fornara, F.; Coupland, G. Regulation and identity of florigen: FLOWERING LOCUS T moves center stage. *Annu. Rev. Plant Biol* **2008**, *59*, 573–594.

34. Ben-Naim, O.; Eshed, R.; Parnis, A.; Teper-Bamnolker, P.; Shalit, A.; Coupland, G.; Samach, A.; Lifschitz, E. The CCAAT binding factor can mediate interactions between CONSTANS-like proteins and DNA. *Plant J* **2006**, *46*, 462–476.

35. Wenkel, S.; Turck, F.; Singer, K.; Gissot, L.; le Gourrierec, J.; Samach, A.; Coupland, G. CONSTANS and the CCAAT box binding complex share a functionally important domain and interact to regulate flowering of Arabidopsis. *Plant Cell* **2006**, *18*, 2971–2984.

36. Turnbull, C. Long-distance regulation of flowering time. *J. Exp. Bot* **2011**, *62*, 4399–4413.

37. Jang, S.; Torti, S.; Coupland, G. Genetic and spatial interactions between FT, TSF and SVP during the early stages of floral induction in Arabidopsis. *Plant J* **2009**, *60*, 614–625.

38. Rieu, I.; Ruiz-Rivero, O.; Fernandez-Garcia, N.; Griffiths, J.; Powers, S.J.; Gong, F.; Linhartova, T.; Eriksson, S.; Nilsson, O.; Thomas, S.G.; *et al.* The gibberellin biosynthetic genes AtGA20ox1 and AtGA20ox2 act, partially redundantly, to promote growth and development throughout the Arabidopsis life cycle. *Plant J* **2008**, *53*, 488–504.

39. Hisamatsu, T.; King, R.W. The nature of floral signals in Arabidopsis. II. Roles for FLOWERING LOCUS T (FT) and gibberellin. *J. Exp. Bot* **2008**, *59*, 3821–3829.

40. Ramakers, C.; Ruijter, J.M.; Deprez, R.H.; Moorman, A.F. Assumption-free analysis of quantitative real-time polymerase chain reaction (PCR) data. *Neurosci. Lett* **2003**, *339*, 62–66.

41. Clough, S.J.; Bent, A.F. Floral dip: A simplified method for Agrobacterium-mediated transformation of Arabidopsis thaliana. *Plant J* **1998**, *16*, 735–743.

42. Russel, S. *Molecular Cloning: A Laboratory Manual*, 3rd ed; Cold Spring Harbor Laboratory Press: New York, NY, USA, 2001.

43. Lokstein, H.; Tian, L.; Polle, J.E.; DellaPenna, D. Xanthophyll biosynthetic mutants of Arabidopsis thaliana: Altered nonphotochemical quenching of chlorophyll fluorescence is due to changes in Photosystem II antenna size and stability. *Biochim. Biophys. Acta* **2002**,*1553*, 309–319.

44. Rohacek, K. Chlorophyll fluorescence parameters: The definitions, photosynthetic meaning, and mutual relationships. *Photosynthetica* **2002**, *40*, 13–29.

Chapter 7

Physiological Studies on the Marketability of Williams Banana Fruits

M.N. Tourky, **M.E. Tarabih** and **E.E. El-Eryan**

Department of Fruit Handling, Horticulture Research Institute, Agriculture Research Center, Egypt

ABSTRACT

The effect of postharvest treatments with hot water (HW), gibberellic acid (GA_3), salicylic acid (SA) and potassium permanganate ($KMnO_4$) on enhancing or delaying the ripening quality aspect of mature Williams banana fruits were investigated. Untreated and treated fruit had a normal ripening process and similar good freshness at the ripening time (45 days at ambient temperature). This clearly that the used these materials were relatively in delaying ripening and as, can be arrange the appearance of banana fruits in the market with good quality. Such, color development, peeling condition, loss of firmness, increase of pulp/peel ratio, soluble solids content, titratable acidity, pH, total sugar and total starch were used as a good criterion of assessment the banana fruit ripening

1. INTRODUCTION

Bananas (*Musa* spp.) are planted in all moisture tropical area and represent the fourth largest food crop of the world after rice, wheat and maize (Arias *et al.*, 2003). Bananas also today represent the second largest produced fruit after citrus contributing about 16% of the world's total fruit production with an annual production of 129,906,098 metric tons (FAO, 2010).

In Egypt, the total cultivated area of banana reached more than 55000 feddan produced more than 1.100.000 tons (FAO, 2010). Williams (*Musa cavendishii* L.) is one of the most important cultivars newly grown under the local condition, especially at the desert under **drip irrigation**system. Bananas are a usual climacteric fruits wherever previously ripening have been starting by ethylene, they revealed a short life because of rapid declining of peel color and pulp firmness (Seymour, 1993).

Bananas are harvested when reach to full maturity for local market and immature for export. Sometimes, markets do not absorb all the bananas full for harvest. Due to the short life of bananas which reduces their survival in the market a long time for this, prolonging the life of banana to the largest possible period is considered beneficial to the dealer, whether for export or domestic consumption.

Hot water (HW) has been often evaluated as a means for quarantine for banana and extends commercial life of the fruit (Lopez-Cabrera and Marrero-Dominguez, 1998). Also, Lurie (1998) recommended that the decline in fruit hardness affected by hot water treatment was reason for the inhibition of pectin cell wall hydrolysis, reduced level of cell wall degrading enzyme activity or the inhibition of ethylene production due to a reduction in the activity of ethylene-forming enzyme.

Respiration rate as well as chlorophyll degradation were reduced by dipping Kluai Khai banana fruit in hot water at 50°C for 10 min. Also, peel color change was retarded and fruit firmness was maintained, however catalase and peroxidase activities were higher compared to the other treatments (Varit and Songsin, 2011).

The main use, of gibberellins (GA_3) is as a growth regulator to retard ripening, senescence of fruits and extending **shelf-life**, however GA_3delays climacteric peak of respiration peel color development of Dwarf Cavendish banana, total soluble solids accumulation, fruit softening and reduced weight loss during storage (Osman and Abu-Goukh, 2008). The immersion of banana fruits with GA_3 after harvest at rates of 50 to 250 mg L^{-1} for 5 to 30 min resulted in prolonged green and yellow fruit life as a consequence of delayed maturation (Patil and Hulamani, 1998).Friedman (1996) mentioned that GA_3 treatments were effective in reducing the susceptibility of the fruits to be mechanically damaged and delaying senescence of the fruits, prolong storage life and inhibited ethylene production. Also, GA_3 delays chlorophyll degradation and fruit softening (Khader, 1992).

Salicylic acid (SA), a ubiquitous plant phenolic, has been reported to regulate a number of processes in plants. It was shown to interfere with the biosynthesis and/or action of ethylene, abscisic acid and cytokinins in plants. SA treatment has been found to delay the ripening of banana fruits (Srivastava and Dwivedi, 2000). Acetyl salicylic acid (a derivative of SA) application increased the endogenous SA levels, inhibited ethylene production and maintained firmness of kiwi fruit during cold storage (Zhang *et al.*, 2003).

Potassium permanganate ($KMnO_4$) is a stable purple solid that is a strong oxidizing agent and readily oxidizes ethylene (Wills and Warton, 2004). Treatment of $KMnO_4$ as an ethylene absorbent, delayed fruit ripening proved most effective in reducing rot and maintained the physical appearance and quality of fruit in fresh conditions (Bairwa *et al.*, 2002). Postharvest application with $KMnO_4$ had delaying of ripening with good quality of mature Maghrabi banana fruits (Abd El-Naby, 2010).

The purpose of this study was to investigate the response of Williams banana fruits to control ripening with some safe treatments to maintain ideal eating quality and regulate their marketing life.

2. MATERIALS AND METHODS

Mature green bunches of Williams banana were used in the present study during two successive seasons of 2011 and 2012. Bunches were taken from a private orchard at El-Behira governorate, Williams banana plants were in a good physical condition, uniformity and adequate irrigation, fertilizers and other cultural practices recommended for banana plantations in sand soil. Mature banana fruits for both first and second seasons were harvested in first half of January at commercial maturity, i.e., (three quarter full, 120 days after shooting and bunch weighted 23-27 kg with light green color

(Stover and Simmonds, 1987). Bunches were immediately transported to the laboratory of postharvest at Hort. Res. Inst. Mansoura, Egypt. Bunches were left at ambient condition for two days as a sweating period, then bunches were de-handed and individual bananas were washed with tap water and air dried. Physical and **chemical properties** of fruits were measured.

Banana handed were received one of the following treatments as follows:

- Control (Dipping fruits in tap water for 10 min)
- Dipping fruits in hot water at 45°C for 10 min
- Dipping fruits in gibberellic acid at 50 ppm for 10 min
- Dipping fruits in salicylic acid at 50 ppm for 10 min
- Small packet potassium permanganate, each content 20 g

*$KMnO_4$ encased in small polyethylene bags perforated only on one side of banana box to avoid staining the fruits.

For shelf life study, hands of each treatment were placed within polyethylene bag (thickness 0.04 mm) in 9 carton box each box containing 3 hands. The total numbers of boxes were 45 for all treatments. After the treatments, carton boxes were held at room temperature on 20°C±2 and 65-70% relative humidity until the natural ripening (when the fingers reached color between 5-7 degree). Banana fruits in three boxes (3 replicates) for each treatment were taken 15 days intervals to determine the following properties:

Weight loss (%): Was calculated according to the following equation:

$$\text{Weight loss (\%)} = \frac{\text{Initial weight} - \text{Weight at sampling date}}{\text{Initial weight}} \times 100$$

Angulation: Was determined by measuring the equatorial diameter of two different sides and calculated by dividing the difference between the average of total lowest reading and average of total highest reading by the average of total highest reading and multiplying by 100 (Abou-Azizet *al.*, 1993).

Fruit firmness (Ib inch^{-2}): Was measured using PHSH-PULL Dynanomteter Modle DT (01) having plunger of 8 mm diameter probe.

Peel color: Was determined according to the standard color index of the united fruit Co. (7 grades from full green to complete yellow).

Peeling condition: Was scored as follows, (1): Unpeeling, (2): Hard peeling, (3): Peeling and (4): Easy peeling.

Pulp/peel ratio: Pulp and peel are separated, weighed individually and expressed as pulp to peel ratio (i.e., pulp weight divided by the peel weight).

Soluble solids content SSC%: Was determined refractmetrically according to AOAC (2005).

Total acidity (%): Was determined in by titrating by against 0.1 N sodium hydroxide using phenolphthalein as indicator. Results were expressed as percentage of Malic acid in flesh pulp weight according to AOAC (2005).

SSC/acid ratio (%): Was calculated by dividing the value of SSC over the value of titratable acidity of each sample.

Pulp pH: Was measured in the pulp juice with a digital bench top pH meter. pH values give a measure of the alkalinity of a product.

Total sugar: The extract was prepared by taking 0.5 g of fresh pulp and extracting the same with 80% ethanol by centrifuging three times. The supernatant was collected and measured quantity of distilled water was added to it and heated until all ethanol got evaporated. Then volume of sample was made up to 150 mL by adding distilled water. The total sugar was estimated using anthrone's reagent (Ranganna, 1979). The 1 mL of alcoholic extract was taken in a test tube and chilled. After a while 4 mL of anthrone's reagent was carefully run down the walls of the test tube. The test tubes were thereafter immersed in ice water. The tubes were brought to ambient temperatureand boiled in water bath for 10 min. After proper cooling, the absorbance was measured at 625 nm. Total sugars were calculated as gm of glucose 100 g^{-1} weight.

Starch: The banana and solvent mixer was blended and then stirred slowly for 2 h. The suspension was centrifuged at 3500 rpm for 15 min. The supernatant was discarded and the sediment was resuspended in two volumes of distilled water and centrifuged, this step was repeated to wash the sediment. The starch suspension was then neutralized to pH 7 and the sediment starch was dried in an air oven at 50°C (Eggleston *et al.*, 1992).

Peel moisture (%): Was determined by during a pre weighted amount of material in a vacuum oven at 70°C, until it reached a constant weight.

Statistical analysis: Data of both seasons of the study were designed (Completely Randomize Designed) by using analysis of variance (ANOVA), with two factors; time and temperature. Differences among treatment means were statistically analyzed by using the least significant differences test (LSD) at $p = 0.05\%$, means separation using the CoStat program.

3. RESULTS AND DISCUSION

Weight loss percentage: Data presented in Table 1 show clearly that during the two seasons of study, the loss weight gradually increased as the shelf life period prolonged. The data also indicated that all materials treatments significantly reduced the weight loss% than the control after 45 days of shelf life during the two seasons of this study. Since the weight loss%in banana fruits untreated were 2.96 and 4.23% after 45 days held under room temperature in two seasons respectively. While fruits treated with GA_3 had the lowest loss percentage in fruit weight (2.16 and 2.90%) in the two seasons. Although the weight loss% values ranged (2.33-3.23%), (2.40-3.53%) and (2.56-3.80%) for banana fruits treated with HW, SA and $KMnO_4$ after 45 days at shelf life in two seasons, respectively.

With regard to the effect of GA_3, Valero *et al.* (1998) found that gibberellin application is followed by an increase in polyamine levels and in the activities of their biosynthetic enzymes and exogenous polyamines delayed senescence and prolonged fruit storage, physiological processes that are usually accompanied by decreases in plant polyamines. Also, Osman and Abu-Goukh (2008) reported that GA_3 treatment resulted in more reduction of weight loss of banana fruits during shelf life. Recently, salicylic acid has also been shown to inhibit ACC oxidase activity. Ethylene synthesis is normally limited by the supply of the immediate precursor amino cyclopropane-1-carboxylic acid (ACC) (Fan *et al.*, 1996). Thus, the effect of salicylic acid in delaying the ripening of banana fruits may be of due to inhibition of ethylene biosynthesis and action.

Table 1. Effect of post-harvest treatments on weight loss%, angulation% and firmness (Ib inch^{-2}) of Williams banana fruits during ripening at room temperature at 2011and 2012 seasons

	Loss weight (%)				Angulation (%)				Firmness (Ib inch^{-3})			
	Storage period in days											
Treatments	0	15	30	45	0	15	30	45	0	15	30	45
Season 2011												
Control	0.00	0.80	1.00	2.96	14.66	13.46	7.73	5.96	11.16	10.90	9.60	5.16
Hot water at 45°C	0.00	0.46	0.90	2.33	14.66	14.06	8.80	7.23	11.16	10.86	10.23	7.20
GA_3 at 50 ppm	0.00	0.36	0.80	2.16	14.66	14.30	9.33	7.30	11.16	11.10	10.56	7.10
Salicylic acid at 50 ppm	0.00	0.53	1.16	2.40	14.66	13.93	8.96	7.30	11.16	10.93	10.40	7.03
$KMnO_4$ at 20 g	0.00	0.70	1.03	2.56	14.66	13.66	8.16	6.73	11.16	10.80	9.93	6.13
Mean	0.00	0.57	0.98	2.48	14.66	13.88	8.60	6.90	11.16	10.92	10.14	6.52
L.S.D. at 5%												
Treatment	0.098				0.156				0.123			
Time	0.085				0.580				0.353			
Treatment (time)	0.190				0.313				0.246			
Season 2012												
Control	0.00	0.39	1.90	4.23	13.96	12.03	6.90	5.16	10.76	10.23	8.40	4.20
Hot water at 45°C	0.00	0.60	1.53	3.23	13.96	13.13	8.13	6.56	10.76	10.63	9.70	6.20
GA_3 at 50 ppm	0.00	0.46	1.26	2.90	13.96	13.20	8.46	6.66	10.76	10.63	6.96	6.40
Salicylic acid at 50 ppm	0.00	0.76	1.73	3.53	13.96	12.76	8.23	6.50	10.76	10.76	9.73	6.33
$KMnO_4$ at 20 g	0.00	0.86	1.93	3.80	13.96	12.33	7.40	5.93	10.76	10.56	8.66	5.10
Mean	0.00	0.72	1.67	3.54	13.96	12.69	7.82	6.16	10.76	10.56	8.68	5.64
L.S.D. at 5%												
Treatment	0.108				0.262				0.997			
Time	0.090				0.204				0.907			
Treatment (time)	0.203				0.456				2.028			

Moreover, Abd-El-Wahab (2007) reported that hot water dips reduced the percentage of decay, weight loss and phenols of apricot and peach fruits.

In addition, $KMnO_4$ decreases respiration and delays ripening by maintaining ethylene at a low level for a long period (Illeperuma and Jayasuriya, 2002).

Finger angulation percent: From Table 1 the data show clearly angulation percent for Williams banana fruits, gradually decreased with the advanced in shelf life at room temperature during the two season of study. At the end of shelf life period (45 days) angulation, ranged between (7.23-6.56%), (7.30-6.66%), (7.30-6.50%) and (6.73-5.93%) for banana fingers treated with HW, GA_3, SA and $KMnO_4$ in both seasons respectively, while angulation% for control banana finger ranged between (5.96-5.16%) in two seasons. That means that treatment of $KMnO_4$and control gave the best results in this respect in both seasons of study.

Abou-Aziz *et al.* (1993) reported that the angulations percent is one of the principal parameters for using to determine banana fruit maturation.Tourky *et al.* (2006) reported that angulations percent for Williams and Grand Nain banana fruits, gradually decreased with the advance in shelf life at room temperature.

Fruit firmness (Ib inch^{-2}): It is clear from Table 1 that the average fruit firmness had a gradual and continual decrease with the progress of storage period at 20°C±2 and 65-70% R.H. (room temperature) in the two season of investigation.

Results also, revealed that all materials treatments used significantly increased the values of fruit firmness than the control at the first and second seasons. Banana fruits treated with HW, GA_3 and SA recorded the highest values of fruit firmness (7.20-6.20 Ib inch^{-2}), (7.10-6.40 Ib inch^{-2}) and (7.03-6.33 Ib inch^{-2}) after 45 days at ambient condition in the two seasons respectively. On the contrary, untreated banana fruits gave the lowest value of fruit firmness (5.16-4.20 Ib inch^{-2}) at the end of the shelf life period in two seasons, respectively.

These results are in agreement with Martinez-Romero *et al.* (2000) studied the effect of GA_3 at 100 mg L^{-1} on Baby Gold peach then stored at 2°C for 14 days and found that GA_3 maintained higher fruit firmness during storage, the respiration rate and ethylene emission were reduced compared with control.

Table 2. Effect of post-harvest treatments on peel color, peeling condition and pulp/peel ratio of Williams banana fruits through ripening at room temperature during 2011 and 2012 seasons

	Peel color				Peeling condition				Pulp/peel ratio			
	Storage period in days											
Treatments	0	15	30	45	0	15	30	45	0	15	30	45
Season 2011												
Control	2.00	2.66	3.66	6.00	1.00	1.66	3.00	4.00	1.50	1.33	1.66	2.10
Hot water at 45°C	2.00	2.00	2.66	4.66	1.00	1.00	2.00	2.66	1.50	1.53	1.53	1.76
GA_3 at 50 ppm	2.00	2.00	2.33	4.33	1.00	1.00	2.00	2.66	1.50	1.36	1.36	1.96
Salicylic acid at 50 ppm	2.00	2.00	2.33	4.66	1.00	1.33	2.33	3.00	1.50	1.43	1.43	2.13
$KMnO_4$ at 20 g	2.00	2.33	3.33	5.33	1.00	1.66	3.00	4.00	1.50	1.43	1.43	2.03
Mean	2.00	2.20	2.86	5.00	1.00	1.33	2.46	3.26	1.50	1.42	1.48	2.00
L.S.D. at 5%												
Treatment	0.374				0.315				0.126			
Time	0.194				0.222				0.124			
Treatment (time)	0.735				0.497				0.277			
Season 2012												
Control	2.00	3.00	4.00	6.66	1.00	1.66	3.00	4.00	1.33	1.36	1.53	1.90
Hot water at 45°C	2.00	2.00	2.33	4.00	1.00	1.00	2.00	2.66	1.33	1.46	1.60	1.80
GA_3 at 50 ppm	2.00	2.00	2.66	4.33	1.00	1.00	2.33	3.00	1.33	1.33	1.50	1.80
Salicylic acid at 50 ppm	2.00	2.33	2.66	5.00	1.00	1.00	2.66	3.33	1.33	1.40	1.66	2.03
$KMnO_4$ at 20 g	2.00	2.66	4.00	6.33	1.00	1.33	2.66	3.66	1.33	1.50	1.60	2.00
Mean	2.00	2.40	3.13	5.26	1.00	1.20	2.53	3.33	1.33	1.41	1.58	1.90
L.S.D. at 5%												
Treatment	0.243				0.321				0.060			
Time	0.347				0.242				0.097			
Treatment (time)	0.776				0.541				0.217			

Furthermore, Abd El-Naby (2010) reported that postharvest treatments with hot water for Maghrabi banana treatment had the higher firmness at ripening compared to those treated with hot water. Also, treated banana fruits with hot water (50°C for 10 min) were most effective in delaying firmness loss or softening.

Lurie (1998) suggested that the reduction in fruit softening caused by hot water treatment was caused by the inhibition of pectin cell wall hydrolysis, indicative of a reduced level of cell wall degrading enzyme activity of the inhibition of ethylene production due to a reduction in the activity of ethylene-forming enzyme. On the other hand, Srivastava and Dwivedi (2000) observed that SA treatment inhibited the process of banana fruit softening during ripening.

Generally, pulp firmness of most bananas do not change significantly during the early stage of maturation, but as growth progresses changes in pulp firmness may occur. It is therefore important to asce rtain pulp firmness during fruit maturation. Also, the crisp, hard and green fruit turns into a yellow fruit with tender and soft internal pulp at the optimal ripening stage and becomes mushy as it advances towards senescence. The loss of firmness during ripening leads to lower quality and higher incidence of mechanical damage during handling and transportation (Dadzie and Orchard, 1997).

Loss of firmness or softening during ripening has been associated with two or three processes. The first is the breakdown of starch to form sugar. The second is the breakdown of the cell walls or reduction in the middle lamella cohesion due to solubilisation of pectic substances (Smith*et al.*, 1989).

Peel color: The disappearance or loss of peel green color and the corresponding increase in yellowing of the peel during ripening are the obvious manifestations in banana. So, the peel of banana that represents about 40% of the total weight of fresh banana (Tchobanoglous *et al.*, 1993).

Peel color score progressively during storage Williams banana fruits at room temperature Table 2. The untreated fruits reached the full yellow stage (color score 6.0-6.66) after 45 days while, fruits treated by HW, GA_3, SA and $KMnO_4$ reduced score (4.0 to 6.33) at the end of shelf life period in two seasons respectively. That means that treatments with HW, GA_3, SA and $KMnO_4$ resulted delay in peel color development. These results are in agreement with previous reports that GA_3 retarded color development in banana (Ahmed and Tingwa, 1995) also, Osman and Abu-Goukh

(2008) reported that GA_3 and ethylene have opposite effects on fruit ripening and senescent, seeing that GA_3 treatment delayed fruit maturation and retarding ripening process through its effect on retarding the content of chlorophyll degradation in the peel.

During ripening of banana, the flesh color changes from the typical "opaque white" of a product with a high starch content to a "very soft yellow" as the yellowing of the skin intensifies. Also, in fresh banana, the color changes of peel during storage as a result of ripening have been observed as a loss of greenness and increase in reddish and yellowness tones (Salvador *et al.*, 2007). The loss of green color is due to degradation of the chlorophyll structure. External changes in peel color during ripening often reflect changes in pulp color (Wainwright and Hughes, 1990).

Regarding the effect of hot water, Varit and Songsin (2011) showed that treated banana fruit by hot water (45°C for 15 min or 50°C for 10 min) retarded peel color. In this respect, SA treatments have been found to delay the ripening of banana fruits (Srivastava and Dwivedi, 2000).

Bairwa *et al.* (2002) and Abd El-Naby (2010) reported that $KMnO_4$ application as an ethylene absorbent, delayed ripening, proved most effective in reducing rot and maintained the physical appearance and quality of fruit in fresh condition.

Consumers prefer yellow peel banana (score index 6), therefore, application of $KMnO_4$ as ethylene absorber through clay powder as carrier is useful to prolong storage of banana var. Raja Bulu (Santosa *et al.*, 2010).

Peeling condition: Peeling condition is used as a good criterion for evaluation the ripening. Increase in weight loss is usually followed by lack of peel appearance; therefore, control moisture during storage is important. The results in Table 2 showed that all postharvest treatments (HW, GA_3, SA and $KMnO_4$) gave least values of peeling condition at the end of shelf life period (45 days), since HW, GA_3 and SA had values (2.66), (2.66-3.0) and (3.0-3.33) in the two season respectively. While, $KMnO_4$ treatment and control gave the highest values of peeling condition (4.0-3.66) and (4.0) this hold true in the two seasons.

Banana peel represents about 40% of total weight of the fresh fruit (Anhwange *et al.*, 2009). Recent studies demonstrated that banana peel generally include higher **phenolic compounds** than that of banana pulps (Sulaiman *et al.*, 2011).

Abd El-Naby (2010) found that the used hot water and $KMnO_4$ were relatively more effective in delaying ripening of Maghrabi banana fruits, also delay in ripening of banana fruits is reported by SA (Srivastava and Dwivedi, 2000).

In addition, GA_3 can be applied post-harvest to control fruit ripening and maintain quality through packaging, long-distance shipping and **shelf-life** (Osman and Abu-Goukh, 2008).

Pulp/peel ratio: According to Table 2 it is clear that pulp/peel ratio showed a significant progressive and almost uniform trend of increase as shelf life for all treatments during two seasons. At the end of shelf life period (45 days), pulp/peel ratio reach to a maximum values. The ratio were (2.10-1.90), (2.13-2.03) and (2.03-2.00) for control, SA and $KMnO_4$ treatment, at the end of shelf life respectively. While banana fruits treated with HW and GA_3 gave ratio between (1.76-1.80) and (1.96-1.80), respectively.

Pulp to peel ratio is a good and consistent index of ripening of banana. Changes in pulp to peel ratios during ripening of banana indicate differential changes in **moisture content** of the peel and pulp. The increase in pulp to peel ratio during ripening is related to sugar concentration in the two tissues. During ripening, there is a rapid increase in the sugar concentration in the pulp compared to the peel thus contributing to a differential change in osmotic pressure. The peel looses water both by transpiration to the atmosphere and also to the pulp by osmosis, thereby contributing to an increase in the fresh weight of the pulp as the fruit ripens (Stover and Simmonds, 1987). This results in an increase in the pulp to peel ratio during ripening. Abd El-Naby (2010) showed that the pulp/peel ratio was increased of banana fruits by postharvest $KMnO_4$, while hot water treated fruit gave the lowest ratio at fruit ripening. On the other hand, Srivastava and Dwivedi (2000) found that SA treatment decreased the pulp/peel ratio during ripening of banana fruits.

Soulble solids content (SSC): The results in Table 3 show that soluble solids content of Williams banana fruits increased gradually and significantly with increasing shelf life period, reached the maximum values at the end of shelf life periods at the two seasons. Fruits of control and $KMnO_4$ treatments recorded the highest SSC% (14.60-15.40%) and (14.13-15.03%), while treatments of HW, GA_3 and SA gave the least value (12.90-13.36%), (13.16-13.80%) and (13.66-13.46%) compared with other treatments. This is in line with Ahmed and Tingwa (1995)reported that GA_3 decreased sugar accumulation, TSS and sugar/ratio in banana.

Abd El-Naby (2010) indicated that postharvest treatments with HW and KMnO4 to Maghrabi banana fruits gave highest values of TSS in the first season compared to the other treatments.

Fruits including banana, contain many compounds which are soluble in water (sugars, acids, **vitamin C**, **amino acid**s and some pectin). These soluble compounds form the soluble solids content of the fruit. In most ripe fruits including banana, sugar forms the main component of soluble solids, total soluble solids (TSS) are an important postharvest quality attribute in the screening. Since the amount of TSS or sugar in fruits usually increases as they mature and ripen, the soluble solids content of the fruit can be a useful index of maturity or stage of ripeness (Dadzie and Orchard, 1997).

Total acidity: The data in Table 3 noticed that during shelf life periods, increase of total acidity content throughout shelf life period and reached their maximum values at the end of shelf life periods (45 days). Fruit acidity for all materials treatments were lowest values (0.528-0.696%), while fruit acidity of control were about (0.741-0.732) in both seasons respectively at the end of shelf life periods.

Commonly, the content of total acidity in fruit juice were slowly decline as a storage period advanced from harvest till 30 day at cold storage or during marketing at room temperature, which may be attributed to the use of acids as substrate for respiration (Echeverria and Valich, 1989).

Abd El-Naby (2010) found that postharvest $KMnO_4$ treatment for Maghrabi banana fruits gave low values of total acidity at fruit ripening. Al-Qurashi and Awad (2012) show that, acidity concentration was not affected by either SA treatment or storage temperature but, fluctuated during storage with a significant decrease after 40 days of storage and during storage, acidity concentration was much lower than the initial value of harvest.

SSC/acid ratio: It is clear from Table 3 that SSC/Acid ratio% of Williams banana fruits increased gradually and significantly with increasing shelf life period reached the maximum values at the end of shelf life periods at the two seasons. Fruit SSC/Acid ratio% for all materials treatments were highest values (20.23-24.90%), while fruit SSC/Acid ratio% of control were about (19.70-20.96%) in both seasons respectively at the end of shelf life periods.

Table 3. Effect of post-harvest treatments on SSC (%), acidity (%) and SSC/acid ratio of Williams banana fruits through ripening at room temperature during 2011 and 2012 seasons

	SSC (%)				Acidity (%)				SSC/acid ratio			
	Storage period in days											
Treatments	0	15	30	45	0	15	30	45	0	15	30	45
Season 2011												
Control	0.93	4.93	7.76	14.60	0.321	0.458	0.563	0.741	2.86	10.73	13.73	19.70
Hot water at 45°C	0.93	3.96	6.30	12.90	0.321	0.393	0.432	0.540	2.86	10.06	14.50	23.80
GA_3 at 50 ppm	0.93	4.06	6.76	13.16	0.321	0.404	0.476	0.528	2.86	10.03	14.20	24.90
Salicylic acid at 50 ppm	0.93	3.23	6.50	13.66	0.321	0.420	0.516	0.617	2.86	7.63	12.56	22.13
$KMnO_4$ at 20 g	0.93	5.06	6.90	14.13	0.321	0.421	0.524	0.696	2.86	12.00	13.13	20.23
Mean	0.93	4.25	6.84	13.69	0.321	0.419	0.502	0.624	2.86	10.09	13.62	22.15
L.S.D. at 5%												
Treatment	0.184				0.011				0.557			
Time	0.146				0.011				0.475			
Treatment (time)	0.328				0.026				1.062			
Season 2012												
Control	0.96	4.93	6.93	15.40	0.312	0.440	0.620	0.732	3.03	11.20	11.16	20.96
Hot water at 45°C	0.96	4.56	6.23	13.36	0.312	0.386	0.478	0.567	3.03	11.80	13.00	23.50
GA_3 at 50 ppm	0.96	4.26	6.66	13.80	0.312	0.391	0.474	0.558	3.03	10.86	13.50	24.63
Salicylic acid at 50 ppm	0.96	4.10	6.20	13.46	0.312	0.405	0.520	0.621	3.03	10.10	11.90	21.63
$KMnO_4$ at 20 g	0.96	5.20	7.10	15.03	0.312	0.408	0.517	0.690	3.03	12.70	13.66	21.70
Mean	0.96	4.61	6.62	14.21	0.312	0.406	0.522	0.633	3.03	11.33	12.64	22.48
L.S.D. at 5%												
Treatment	0.137				0.016				0.713			
Time	0.162				0.010				0.360			
Treatment (time)	0.363				0.023				0.805			

Table 4. Effect of post-harvest treatments on pH, total sugar and total starch of Williams banana fruits through ripening at room temperature during 2011and 2012 seasons

	pH (%)				Total sugar (%)				Total starch (%)			
	Storage period in days											
Treatments	0	15	30	45	0	15	30	45	0	15	30	45
Season 2011												
Control	5.10	4.90	4.80	4.80	0.78	3.80	7.40	14.06	24.13	20.50	10.23	4.40
Hot water at 45°C	5.10	4.60	4.63	4.40	0.78	3.23	6.10	12.30	24.13	21.33	15.40	8.10
GA_3 at 50 ppm	5.10	4.63	4.60	4.56	0.78	3.40	6.20	12.90	24.13	21.30	15.16	7.96
Salicylic acid at 50 ppm	5.10	4.86	4.70	4.70	0.78	3.16	6.30	13.96	24.13	21.16	14.86	8.36
$KMnO_4$ at 20 g	5.10	4.80	4.70	4.53	0.78	3.76	6.40	13.66	24.13	21.23	10.56	6.96
Mean	5.10	4.76	4.68	4.60	0.78	3.47	6.48	13.38	24.13	21.10	13.24	7.16
L.S.D. at 5%												
Treatment	0.107				0.195				0.087			
Time	0.089				0.149				0.103			
Treatment (time)	0.200				0.333				0.230			
Season 2012												
Control	5.00	4.93	4.90	4.80	0.87	4.10	7.46	15.06	24.86	20.93	10.93	4.76
Hot water at 45°C	5.00	4.50	4.40	4.36	0.87	3.90	6.40	13.23	24.86	21.23	16.26	8.80
GA_3 at 50 ppm	5.00	4.53	4.53	4.46	0.87	4.03	6.50	13.06	24.86	21.03	16.30	8.13
Salicylic acid at 50 ppm	5.00	4.86	4.80	4.60	0.87	3.46	6.03	12.76	24.86	21.33	15.80	8.16
$KMnO_4$ at 20 g	5.00	4.70	4.66	4.60	0.87	4.13	7.10	13.63	24.86	20.80	11.10	7.73
Mean	5.00	4.70	4.66	4.56	0.87	3.92	6.70	13.55	24.86	21.06	14.08	7.52
L.S.D. at 5%												
Treatment	0.119				0.132				0.078			
Time	0.117				0.179				0.087			
Treatment (time)	0.262				0.400				0.196			

During ripening of banana the total soluble solids content increases. However, sugar forms the main component of soluble solids, since the amount of sugar in fruits usually increases as the fruit matures and ripens (Dadzie and Orchard, 1997). Usually **organic acid**s decline during ripening as they are respired or converted to sugar (Wills *et al.*, 1989). Organic acids are important in giving a desired sugar-to-acid balance which results in pleasing fruit taste during ripening. Acidity shows large increases during ripening progresses therefore, total titratable acidity could be used as an index of ripening (Dadzie and Orchard, 1997).

Pulp pH: The pH of banana pulp decreased with the increase in shelf life periods. This is may be to as ripening advances, acidity increases which resulted in decrease of pH. From Table 4 it is clear that the initial values of pH for all treatments ranged from 5.10 to 5.0 which decreased from 4.36 to 4.80 after 45 days of shelf life. Fruit pH of all materials treatments gave decrease of pH at the end of shelf life period while, pH of control banana fruits gained the highest values (4.80%) in the end of shelf life. Yet, all materials application gave the lowest values of fruit pH indicating delayed of fruit ripening.

Pulp pH and total titratable acidity are important post-harvest quality attributes in the assessment of fruit ripening quality. In most bananas, there is a rapid decline in pulp pH in response to increasing ripeness. However the magnitude of decline is cultivar dependent. Generally, when fruits are harvested at matured green stage, the pulp pH is high but as ripening progresses pH drops. Thus, the pulp pH could be used as an index of ripening (Dadzie and Orchard, 1997).

Total sugar (%): According to Table 4, it could observed that the total sugars of Williams banana fruits showed a gradual and highly significantly increase from the beginning of shelf life and it had high sugars values, at the end of shelf life period at both seasons. The data also disclose that, all treatments decreased values of total sugars than the untreated fruits during the two seasons under the study.

Srivastava and Dwivedi (2000) reported that SA treatment resulted in decreased levels of reducing sugar content in banana fruits while an opposite effect on non reducing sugar content, on the other hand SA delays banana fruit ripening. Also, this is in line with earlier reports that GA_3 decreased sugar accumulation, TSS and sugar /ratio in banana fruit Ahmed and Tingwa (1995). GA_3 slows ripening and senescence of bananas fruits by delaying chlorophyll degradation and reducing sugar accumulation (Osman and Abu-Goukh, 2008).

Total starch (%): Table 4 clearly indicated that the amount of starch content in Williams banana fruits decreased significantly as the advance in shelf life period at two season of study. The rate of decreased continued as the shelf life period advanced, reaching to their minimum values at the end of the shelf life period (45 days). The general trend of starch content ended to decrease sharply through 30 and 45 days of shelf life for all treatments at two seasons. Since, starch hydrolyzed enzymatically during ripening process progress. At the beginning of the shelf life of banana fruits, the total starch content ranged between (24.13-24.86%) in banana pulp fruits.

While, at the completion of the shelf life starch was ranged between (4.40-8.80%). Moreover, all materials treatments gave significant decrease of starch content at the end of shelf life period compared with control fruits.

Muthoo and Chetan (1997) who disclosed that the highest GA_3 concentration on Flordasun peach fruits caused a significant increase in SSC compared with control, it may be due to the starch hydrolysis as result of synthesis of alpha-amylase.

The most striking post-harvest chemical change which occurs during the post-harvest ripening of banana is the hydrolysis of starch and the accumulation of sugar which are responsible for the sweetening of the fruit (Dadzie and Orchard, 1997).

Peel moisture (%): Peel and pulp moisture are important post-harvest parameters in the evaluation of the ripening quality of banana. During ripening, the **moisture content** of the peel decreases whereas that of the pulp increases, this is because the peel looses water both to the atmosphere and to the pulp (Dadzie and Orchard, 1997).

Data from Table 5 show that peel moisture% decreased with increase in storage life for untreated or treated fruits at the two seasons of investigation. Postharvest materials had lowest values of peel moisture% compared to control throughout the storage period at the two seasons. Since, banana fruit treated with HW, GA_3, SA and $KMnO_4$ recorded the lowest peel **moisture content** values (90.40-89.80%), (90.13-89.66%), (89.63-88.43%) and (89.73-88.50%), respectively at the end of storage period during the two seasons of study. Data also recorded that the peel **moisture content** of banana fruits seemed to have an opposite trend of pulp **moisture content** during the storage period (45 days). Similar results were obtained by Tourky *et al.* (2006) who reported that peel **moisture content** of banana fruits decreased gradually during storage period.

Pulp moisture (%): According to Table 5, it is clear that the pulp moisture% of treated or untreated banana fruits showed significant increase as ripening process progress, reaching their maximum **moisture content** after 45 days at storage at room ambient during the two successive season of investigation. Banana fruit treated with HW, GA_3, SA and $KMnO_4$ recorded the lowest pulp **moisture content** values (70.76-69.06%), (70.90-69.40%), (71.76-69.20%) and (71.46-69.76%) respectively at the end of storage period during the two seasons of study. While, untreated banana fruits gave highest pulp moisture values (71.86-70.80%) at the end of storage.

These results are confirmed with those reported by Stover and Simmonds (1987) and Tourky *et al.* (2006) who reported that pulp **moisture content** increased gradually during storage periods. Generally, the water content of peel and/or the ripening environment may also play a role in peel de-greening during fruit ripening. It has been shown that there is an increase in weight ratio between pulp and peel during ripening.

The increase in pulp **moisture content** during ripening is due to carbohydrate breakdown and osmotic transfer from the peel to pulp (John and Marchal, 1995). This is through to be the result of the migration of water from the skin (peel) to pulp for the increasing content of sugar in the latter and development of skin color and easily peeling condition. Our data suggested a delay in ripening process of banana fruits treated with HW, GA_3, SA or $KMnO_4$ pre storage at room ambient. Of the entire factor which affects the quality of banana, **moisture content**s probably the most important, since moisture of banana fruit increased with the increase in storage duration in all varieties. This may be due to movement of water from peel to pulp (Sarode and Tayade, 2009).

Table 5. Effect of post-harvest treatments on peel and pulp moisture% of Williams banana fruits through ripening at room temperature during 2011 and 2012 seasons

	Peel moisture (%)				Pulp moisture (%)			
	Storage period in days							
Treatments	0	15	30	45	0	15	30	45
Season 2011								
Control	98.53	97.26	93.53	89.20	61.06	63.33	68.80	71.86
Hot water at 45°C	98.53	98.30	95.20	90.40	61.06	61.76	67.13	70.76
GA_3 at 50 ppm	98.53	98.06	94.50	90.13	61.06	62.03	67.10	70.90
Salicylic acid at 50 ppm	98.53	97.43	94.40	89.63	61.06	62.76	67.40	71.76
$KMnO_4$ at 20 g	98.53	92.06	93.93	89.73	61.06	62.50	67.16	71.46
Mean	98.53	96.62	94.31	89.82	61.06	62.48	67.52	71.35
L.S.D. at 5%								
Treatment	1.152				0.180			
Time	0.856				0.280			
Treatment (time)	1.914				0.628			
Season 2012								
Control	97.90	96.93	92.76	88.13	59.13	61.80	66.80	70.80
Hot water at 45°C	97.90	97.70	94.16	89.80	59.13	59.76	65.20	69.06
GA_3 at 50 ppm	97.90	97.43	93.86	89.66	59.13	59.80	65.53	69.40
Salicylic acid at 50 ppm	97.90	97.20	93.50	88.43	59.13	61.00	66.20	69.20
$KMnO_4$ at 20 g	97.90	96.93	93.10	88.50	59.13	61.06	66.10	69.76
Mean	97.90	97.24	93.48	88.90	59.13	60.68	65.96	69.46
L.S.D. at 5%								
Treatment	0.323				0.209			
Time	0.223				0.143			
Treatment (time)	0.499				0.319			

4. CONCLUSION

From the above mentioned results it can be arrange the appearance of banana fruits in the market by using these materials which cause different ripening rate of banana fruits with gain acceptance by the user. This demonstrate that the used materials were relatively more effective in delaying ripening, since these materials play an integrated role in many of the biochemical changes occur during ripening such as, moisture, starch, sugar, color and texture.

REFERENCES

1. AOAC., 2005. Official Methods of Analysis. 18th Edn., Association of Official Analytical Chemists, Washington, DC., USA

2. Abd El-Naby, S.K.M., 2010. Effect of postharvest treatments on quality aspect of Maghrabi banana fruit. Am.-Eurasian J. Agric. Environ. Sci., 8: 582-587.

3. Abd-El-Wahab, S.M., 2007. Effect of hot water treatments on decay, quality, storage and shelf life of Canino apricot and Florida Prince peach fruits. Ann. Agric. Sci. Moshtohor, 45: 257-276.

4. Abou-Aziz, A.B., A.N.M. Abd El-Kader, M.R. El Sonbaty and M.M.M. Soad, 1993. Effect of different levels of (K-Mg) compouned fertilizer on vegetative growth, yield fruit quality and some leaf-nutrient contents of Maghrabi banana cultivar. Assiut J. Agric. Sci., 24: 345-357.

5. Ahmed, O.K. and P.O. Tingwa, 1995. Effect of gibberellic acid on several parameters of ripening banana fruits (*Musa* sp. cv. *Dwarf cavendish*) [Sudan]. Univ. Khartoum J. Sci., 3: 47-59.

6. Al-Qurashi, A.D. and M.A. Awad, 2012. Postharvest salicylic acid treatment reduces chilling injury of 'Taify' cactus pear fruit during cold storage. J. Food Agric. Environ., 10: 120-124

7. Anhwange, B.A., T.J. Ugye and T.D. Nyiatagher, 2009. Chemical composition of *Musa sepientum* (Banana) peels. Electron. J. Environ. Agric. Food Chem., 8: 437-442.

8. Arias, P., C. Dankers, P. Liu and P. Pilkauskas, 2003. The World Banana Economy 1985-2002. FAO, Rome, ISBN-13: 9789251050576, Pages: 86.

9. Bairwa, L.N., L.K. Dashora and I. Mohan, 2002. Effect of $KMnO_4$ and $AgNO_3$ on postharvest shelf life of banana (*Musa* X *Paradisica*, L.). Ann. Agric. Bio. Res., 7: 79-84.

10. Dadzie, B.K. and J.E. Orchard, 1997. Routine post-harvest screening of banana/plantain hybrids: Criteria and methods. INIBAP Technical Guidelines 2. International Plant Genetic, Resources Institute, Rome, Italy, International Network for the Improvement of Tomato, Montpellier, France, pp: 47.

11. Echeverria, E. and J. Valich, 1989. Enzymes of sugar and acid metabolism in stored Valencia organs. J. Am. Soc. Hort. Sci., 114: 445-449.

12. Eggleston, G., R. Swennen and S. Akoni, 1992. Physicochemical studies on starches isolated from plantain cultivars, plantain hybrids and cooking bananas. Starch, 44: 121-128.

13. FAO, 2010. Food and agriculture organization of the united nations. FAOSTAT-PC, FAO, Rome, Italy.

14. Fan, X., J.P. Mattheis and J.K. Fellman, 1996. Inhibition of apple fruit 1-aminocyclopropane-1-carboxylic acid oxidase activity and respiration by acetylsalicylic acid. J. Plant Physiol., 149: 469-471.

15. Friedman, M., 1996. Food browning and its prevention: An overview. J. Agric. Food Chem., 44: 631-653.

16. Illeperuma, C.K. and P. Jayasuriya, 2002. Prolonged storage of *Karuthacolomban* mango by modified atmosphere packaging at low temperature. J. Horti. Sci. Biotechnol., 77: 153-157.

17. John, P. and J. Marchal, 1995. Ripening and Biochemistry of the Fruit. In: Bananas and Plantains, Gowen, S.R. (Ed.). Chapman and Hall, London, UK pp: 434-466.

18. Khader, S.E.S.A., 1992. Effect of gibberellic acid and Vapor Gard on ripening, amylase and peroxidase activities and quality of mango fruits during storage. J. Hortcult. Sci., 67: 855-860

19. Lopez-Cabrera, J.J. and A. Marrero-Dominguez, 1998. Use of hot water dips to control the incidence of banana crown rot. Acta Hortic., 490: 563-569.

20. Lurie, S., 1998. Postharvest heat treatments. Postharvest Biol. Technol., 14: 257-269.

21. Martinez-Romero, D., D. Valero, M. Serrano, F. Burlo, A. Carbonell, L. Burgos and F. Riquelme, 2000. Exogenous polyamines and gibberellic acid effects on peach (*Prunus persica* L.) storability improvement. J. Food Sci., 65: 288-294.

22. Muthoo, A.K. and T. Chetan, 1997. Effect of different growth regulators and hand thinning on physio-chemical characteristics of peach cv. Flordasun. Adv. Plant Sci., 10: 61-64.

23. Osman, H.E. and A.A. Abu-Goukh, 2008. Effect of polyethylene film lining and gibberellic acid on quality and shelf-life of banana fruits. Univ. Khartoum J. Agric. Sci., 16: 241-260.

24. Patil, S.N. and N.C. Hulamani, 1998. Effect of post-harvest treatments on physical characters and shelf-life of banana fruits. Karnataka J. Agric. Sci., 11: 535-537.

25. Ranganna, S., 1979. Manual of Analysis of Fruit and Vegetables Produts. 2nd Edn., Tata MeGraw-Hill Publ. Co. Ltd., New Delhi, India, Pages: 634.

26. Salvador, A., T. Sanz and S.M. Fiszman, 2007. Changes in colour and texture and their relationship with eating quality during storage of two different dessert bananas. Postharvest Biol. Technol., 43: 319-325.

27. Santosa, E., W.D. Widodo and Kholidi, 2010. The use of clay as potassium permanganate carrier to delay the ripening of Raja Bulu banana. J. Hort. Indonesia, 1: 89-96.

28. Sarode, S.C. and N.H. Tayade, 2009. Physio-chemical changes during ripening in Williams Zeling and Grand nain banana. J. Dairy. Foods Home Sci., 28: 220-224.

29. Seymour, G.B., 1993. Banana. In: Biochemistry of Fruit Ripening, Seymour, G.B., J.E. Taylor and G.A. Tucker (Eds.). Champrnan and Hall, London, pp: 83-106.

30. Smith, N.J.S., G.A. Tucker and J. Jeger, 1989. Softening and cell wall changes in bananas and plantains. Aspects Applied Biol., 20: 57-65.

31. Srivastava, M.K. and U.N. Dwivedi, 2000. Delayed ripening of banana fruit by salicylic acid. Plant Sci., 158: 87-96.

32. Stover, R.H. and N.W. Simmonds, 1987. Banana. 3rd Edn, Longman, London.

33. Sulaiman, S.F., N.A.M. Yusoff, I.M. Eldeen, E.M. Seow, A.A.B. Sajak, Supriatno and K.L. Ooi, 2011. Correlation between total phenolic and mineral contents with antioxidant activity of eight Malaysian bananas (*Musa* sp.). J. Food Compos. Anal., 24: 1-10.

34. Tchobanoglous, G., H. Theisen and S.A. Vigil, 1993. Integrated Solid Waste Management: Engineering Principles and Management Issues. McGraw Hill, New York, USA. pp: 3-22.

35. Tourky, M.N., M.A. El-Shobaky and A.S.H. El-Deen, 2006. Evaluation of two banana cultivars growing in sandy soil under drip irrigation system in relation to growth, fruit quality and successful marketing. J. Agric. Sci. Mansoura Univ., 31: 7859-7870.

36. Valero, D., D. Martinez, F. Riquelme and M. Serrano, 1998. Polyamine response to external mechanical bruising in two mandarin cultivars. HortScience, 33: 1220-1223.

37. Varit, S. and P. Songsin, 2011. Effects of hot water treatments on the physiology and quality of Kluai Khai banana. Int. Food Res. J., 18: 1013-1016.

38. Wainwright, H. and P. Hughes, 1990. Changes in banana pulp colour during ripening. Fruits, 45: 25-28.

39. Wills, R.B.H. and M.A. Warton, 2004. Efficacy of potassium permanganate impregnated into alumina beads to reduce atmospheric ethylene. J. Am. Soc. Hort. Sci., 129: 433-438.

40. Wills, R.H.H., W.B. McGlasson, D. Graham, T.H. Lee and E.G. Hall, 1989. Postharvest: An Introduction to the Physiology and Handling of Fruit and Vegetables. 3rd Edn., Blackwell Scientific Publications, Oxford, UK., ISBN: 978-0851992648, Pages: 280.

41. Zhang, Y., K. Chen, S. Zhang and I. Ferguson, 2003. The role of salicylic acid in postharvest ripening of kiwifruit. Postharvest Biol. Technol., 28: 67-74

Chapter 8

Protocol optimization for enhanced production of pigments in *Spirulina*

Devendra Kumar[1], Neeraj Kumar[2], Sunil Pabbi[1], Suresh Walia[3], Dolly Wattal Dhar[1]

[1]Centre for Conservation and Utilisation of Blue Green Algae, Indian Agricultural Research Institute, New Delhi 110012, India

[2]Department of Microbiology, Kurukshetra University, Kurukshetra, India

[3]Division of Agricultural Chemicals, Indian Agricultural Research Institute, New Delhi, India

ABSTRACT

Spirulina has attracted special attention due to its importance as human foodstuff and natural colours with specific functional properties. These functional properties have been attributed to phycobilins, carotenoids, phenolics and unsaturated fatty acids. Present study was conducted under controlled phytotron conditions to identify the efficient strains of *Spirulina* in terms of pigment synthesis and to optimize their enhanced production. Methodology for enhanced production was standardized by varying specific environmental parameters (light intensity, temperature, carbon dioxide concentration, pH and NaCl level). Different strains of *Spirulina* depicted variability and environmental parameters showed distinct influence on pigments. Growth and pigment production was recorded to be most efficient under optimized conditions of light intensity (70 μmol m^{-2} s^{-1}), temperature (30 °C), CO_2 concentration (550 ppm and 750 ppm), pH (10.5) and NaCl level (2 g L^{-1}).

KEYWORDS

Mass production; Pigments; *Spirulina*

Spirulina, a cyanobacterial genus has been reported to be the complete organic food source and contains high proteins with well balanced amino acids. It is also rich in carbohydrates, vitamins, minerals, phenolics, pigments (chlorophyll, carotenoids and phycobilins) and poly-unsaturated fatty acids (Miranda et al. 1998; Anupama 2000). It is successfully used in aquaculture and poultry industries (Belay et al. 1993; Estrada et al. 2001). This genus is reported to be photosynthetic, filamentous in an open left-hand helix, with predominance from aquatic habitats having high levels of carbonates, bicarbonates and alkaline pH (up to 11). The large scale production of *Spirulina*depends on nutrient availability, temperature, light, pH and CO_2 concentration. In view of this, a study was conducted to optimize the protocol for enhanced production of pigments in *Spirulina*. Ten strains procured from the culture collection of Centre for Conservation and Utilisation of Blue-Green Algae (CCUBGA), IARI, New Delhi were grown in chemically defined Z-medium, which is rich in carbonate and thus, has successively served as a common culture medium (Zarrouk 1966). *Spirulina* strains were grown under diffused light intensity of 55 μmol m^{-2} s^{-1} at 28 °C ± 2 °C temperature with 16:8 h light: dark cycle. Initial biomass of 0.1 g L^{-1} (Costa et al. 2000) was taken as inoculum in 100 mL medium in conical flasks. The influence of different pH (9.5, 10.5, 11.5, 12.5, 13.5) and NaCl concentrations (0, 0.5, 1.0, 1.5, 2.0 g L^{-1}) was studied on the pigment production under culture room facility. Studies were also conducted under varying temperature (25, 30 and 35 °C), light intensity (55, 70 and 85 μmol m^{-2} s^{-1}) and CO_2 concentration (350, 550 and

750 ppm) in growth chambers of National Phytotron Facility, IARI. Homogenized suspension was drawn at 15th day of incubation for the estimation of chlorophyll (Lichtenthaler 1987), carotenoids (Jensen1978) and phycobiliproteins (Bennett and Bogorad 1993).

Most of the strains exhibited uniform suspension in the medium and some showed planktonic growth behaviour. Difference in the thallus colour observed is reported to be under the control of chlorophyll to phycobilin ratio. Pigments can also be subjected to oxidation by molecular oxygen, which is known to be a potent inhibitor of microbial processes including photosynthesis (Paerl and Pinckney 1996). The environmental variables can influence growth and pigment production which in turn may change the composition of the biomass by affecting the metabolism. Varied light intensity exhibited differential influence on the accumulation of chlorophyll, carotenoids and phycobilins. Results indicated that a light intensity of 55 $\mu mol\ m^{-2}\ s^{-1}$ was optimum for maximum chlorophyll and that of 85 $\mu mol\ m^{-2}\ s^{-1}$ was inhibitory in most cases, except in case of CCC538. Variation in the temperature influenced the pigment profile with 35 °C being the optimum for enhanced chlorophyll. Danesi et al. (2001) reported highest biomass production at 30 °C temperature. High temperatures are the key factors for large scale*Spirulina* cultures outdoors with the optimal temperature reported in range of 35–38 °C (Zarrouk 1966; Hu 2004). The temperature below 20 °C and above 40 °C retarded the growth considerably (Borowitzka and Borowitzka 1988). The highest biomass at 30 °C than at 35 °C may be due to higher partial pressure of CO_2 in the medium leading to enhanced concentration of bicarbonates with consequent increase in the rate of photosynthesis (Colla et al. 2004). Further, a longer dark cycle respiratory activity in which cells use reserve material for respiration accordingly can decrease the cell weight (Vonshak et al. 1982). In general, chlorophyll content enhanced with increase in carbon dioxide concentration and a concentration of 750 ppm was optimum. Effect of pH was distinct with 10.5 pH being the optimum for enhanced chlorophyll production as has also been reported by Richmond and Grobbelaar (1986). *Spirulina* requires relatively high pH between 9.5 and 9.8 (Hu 2004), which inhibit the contamination by most algae in the culture. Therefore, high amounts of sodium bicarbonate is essential in the culture medium to sustain high pH. High alkalinity is mandatory for the growth and bicarbonate is used to maintain the high pH (Vonshak and Tomaselli 2000). The significant pigment production may be the result of an increase in pH due to the formation of a $CO_2/H_2CO_3/HCO^{3-}/CO_3^{-2}$ system, which may function as useful buffer system for maintaining the alkaline pH. This is important for the optimum growth and helps to prevent the carbon depletion (Richmond 1990; Vonshak 1997). Sodium chloride concentration of 2 g L^{-1} depicted highest chlorophyll, however, several species of *Spirulina* exhibit inhibition of growth in hostile environments of saline lakes (Henrikson 1994). Exposure of *S. platensis* to an enhanced sodium concentration resulted in a 30 % increase in intracellular accumulation of Na^+ accompanied by small changes in dry mass and chlorophyll content (Kavita and Mohanty 2000) (Table 1). Carotenoids and phycobiliproteins were highest at 70 µmol m^{-2} s^{-1} light intensity and 10.5 pH was optimum for higher carotenoids and phycobiliproteins. In conventional cultivation, light absorption is proportional to the phycocyanin and chlorophyll content of the cells and a slight reduction in phycocyanin may be accompanied by an increase in chlorophyll (Cornet et al. 1992). Carotenoids were maximum at 35 °C temperature and enhanced CO_2 resulted in their increase with 750 ppm to be an optimum concentration. Sodium chloride concentration of 1.5 g L^{-1} showed highest carotenoids. Phycobiliproteins enhanced significantly at 35 °C in some cultures, whereas decline was observed in others. Enhancement in CO_2 concentration increased the phycobiliproteins, and sodium chloride concentration of 2 g L^{-1} was most suitable for phycobiliprotein production (Tables 2, 3). Study clearly showed that the pigments from*Spirulina*, which have the potential applications in biotechnology can be enhanced by manipulating cultural conditions and environmental variables. Standardized protocol for enhanced production of pigments involved an optimized light intensity: 70 µmol m^{-2} $s^{-1}$for chlorophyll and carotenoids, and 85 µmol m^{-2} s^{-1} for phycobiliproteins, temperature: 35 °C for chlorophyll and phycobiliproteins, and 25 °C for carotenoids, CO_2concentration: 500 ppm and 750 ppm, pH: 10.5 and NaCl concentration: 2 g L^{-1}.

Table 1. Influence of environmental variables on chlorophyll (mg g^{-1} dry weight) content of *Spirulina* strains

Environmental variables	*Spirulina* strains[a]										SE (m)±	CD (P = 0.05)
	CCC477	CCC478	CCC479	CCC480	CCC481	CCC482	CCC483	CCC538	CCC539	CCC540		
Light (µmol m^{-2} s^{-1})												
55	1.550	1.934	1.795	2.611	1.445	1.781	3.188	1.725	2.148	2.553	0.20	0 .41
70	1.192	0.924	0.611	1.359	1.245	2.341	2.783	2.156	1.808	2.358	0.23	0 .47
85	1.014	1.250	1.143	1.304	1.227	1.620	1.298	2.134	1.548	1.991	0.14	0.29
Temperature (°C)												
25	0.408	0.587	0.907	1.513	1.147	1.626	1.293	1.650	0.919	1.693	0.16	0.33
30	0.640	0.859	1.103	1.133	1.294	0.932	1.416	1.388	1.157	3.252	1.29	2.63
35	1.416	0.992	1.542	1.848	1.444	2.238	1.916	2.992	1.875	1.804	0.24	0.50
CO_2 concentration (ppm)												
350	2.547	2.648	1.257	2.008	0.973	0.665	2.324	1.314	1.235	0.858	0.27	0.56
550	0.745	0.979	1.225	1.052	1.339	2.176	2.576	2.150	2.074	2.160	0.24	0.50
750	3.163	4.555	4.035	4.320	2.952	5.961	1.140	3.861	3.056	4.466	0.40	0.82
pH												
9.5	1.632	1.207	1.134	2.062	1.365	0.762	1.019	1.332	1.617	0.953	0.27	0.56
10.5	1.752	2.783	1.579	4.412	2.641	2.466	2.206	2.593	1.527	2.141	0.32	0.66
11.5	1.648	1.787	0.959	1.334	1.552	0.731	0.530	0.459	0.486	0.611	0.07	0.15
12.5	1.369	1.902	0.447	0.965	0.771	0.268	0.579	0.359	0.529	0.193	0.11	0.24
13.5	0.318	1.859	0.993	0.421	1.285	0.442	0.114	0.153	0.187	0.198	0.22	0.45
NaCl concentration (g L^{-1})												
0	0.620	0.872	1.243	0.563	0.661	0.531	0.234	0.638	0.566	0.877	0.02	0.05
0.5	0.981	0.457	0.936	0.782	1.835	0.657	0.373	0.668	0.764	2.065	0.04	0.08
1	1.371	1.749	0.357	0.702	0.349	1.860	0.468	0.749	0.463	1.662	0.01	0.02
1.5	0.830	1.348	0.574	0.878	0.654	1.842	0.665	0.645	0.848	1.325	0.02	0.05
2	2.252	1.247	1.562	1.137	1.374	0.481	0.648	0.647	0.704	1.638	0.02	0.04

[a]CCC477, CCC478, CCC479, CCC480-*Spirulina platensis*, CCC481-*Spirulina maxima*, CCC482-*Spirulina lonar*, CCC483-*Spirulina platensis* (mutant), CCC538, CCC539-*Arthrospira* sp., CCC540-*Spirulina* sp.

Table 2. Influence of environmental variables on carotenoids (mg g^{-1} dry weight) content of *Spirulina* strains

Environmental variables	*Spirulina* strains[a]										SE (m)±	CD (P = 0.05)
	CCC477	CCC478	CCC479	CCC480	CCC481	CCC482	CCC483	CCC538	CCC539	CCC540		
Light (μmol m^{-2} s^{-1})												
55	0.243	0.306	0.257	0.287	0.141	0.193	0.268	0.133	0.222	0.266	0.01	0.03
70	0.147	0.144	0.094	0.228	0.151	0.307	0.307	0.233	0.204	0.271	0.02	0.05
84	0.250	0.246	0.202	0.179	0.106	0.231	0.193	0.170	0.202	0.169	0.02	0.04
Temperature (°C)												
25	0.123	0.073	0.157	0.301	0.190	0.235	0.231	0.184	0.252	0.228	0.02	0.05
30	0.149	0.285	0.184	0.186	0.143	0.176	0.146	0.173	0.179	0.208	0.02	0.05
35	0.206	0.224	0.280	0.148	0.166	0.184	0.129	0.181	0.149	0.158	0.02	0.05
CO_2 concentration (ppm)												
350	0.133	0.244	0.180	0.290	0.188	0.216	0.235	0.158	0.181	0.208	0.01	0.03
550	0.159	0.154	0.241	0.393	0.244	0.222	0.274	0.244	0.214	0.203	0.02	0.04
750	0.269	0.208	0.179	0.229	0.263	0.203	0.360	0.312	0.224	0.304	0.03	0.06
pH												
9.5	0.144	0.200	0.358	0.140	0.109	0.114	0.117	0.180	0.183	0.121	0.02	0.04
10.5	0.233	0.292	0.331	0.283	0.203	0.185	0.306	0.269	0.217	0.182	0.05	0.11
11.5	0.180	0.259	0.261	0.073	0.069	0.111	0.107	0.069	0.122	0.112	0.02	0.04
12.5	0.169	0.285	0.166	0.048	0.090	0.030	0.080	0.045	0.124	0.063	0.02	0.05
13.5	0.055	0.152	0.157	0.042	0.013	0.019	0.005	0.003	0.037	0.013	0.01	0.03
NaCl concentration (g L^{-1})												
0	0.079	0.074	0.045	0.094	0.034	0.141	0.044	0.040	0.100	0.103	0.08	0.02
0.5	0.137	0.136	0.057	0.081	0.057	0.230	0.093	0.045	0.096	0.151	0.09	0.03
1	0.034	0.073	0.083	0.048	0.080	0.105	0.036	0.074	0.077	0.056	0.07	0.01
1.5	0.030	0.133	0.114	0.024	0.123	0.105	0.029	0.125	0.120	0.017	0.05	0.01
2	0.042	0.010	0.020	0.034	0.021	0.014	0.019	0.018	0.035	0.030	0.03	0.08

[a]CCC477, CCC478, CCC479, CCC480- *Spirulina platensis*, CCC481-*Spirulina maxima*, CCC482-*Spirulina lonar*, CCC483-*Spirulina platensis* (mutant), CCC538, CCC539-*Arthrospira* sp., CCC540-*Spirulina* sp.

Table 3. Influence of environmental variables on phycobiliproteins (mg g^{-1} dry weight) content of *Spirulina* strains

Environmental variables	*Spirulina* strains[a]										SE(m)±	CD (P = 0.05)
	CCC477	CCC478	CCC479	CCC480	CCC481	CCC482	CCC483	CCC538	CCC539	CCC540		
Light (μmol m^{-2} s^{-1})												
55	1.341	1.160	4.059	3.723	1.558	2.091	1.817	1.111	2.277	1.603	0.09	0.19
70	0.354	0.569	2.325	2.194	1.749	4.650	6.235	4.091	3.574	4.922	0.36	0.75
84	0.785	0.614	2.798	3.030	1.703	4.035	3.300	5.134	5.540	3.622	0.15	0.30
Temperature (°C)												
25	1.153	1.051	5.175	3.151	2.199	1.663	1.278	1.431	0.599	0.731	0.13	0.28
30	0.881	2.333	3.063	3.713	2.660	4.803	4.497	3.283	3.247	2.708	0.38	0.78
35	6.651	3.321	11.746	1.766	8.199	2.346	4.489	3.722	0.955	2.627	0.38	0.78
CO_2 concentration (ppm)												
350	7.266	7.524	4.731	8.557	2.239	1.367	5.178	3.431	2.938	1.848	0.10	0.20
550	3.036	4.255	10.505	2.930	9.699	4.954	4.639	4.866	7.979	4.483	0.09	0.03
750	4.562	6.918	6.805	8.089	7.582	8.455	1.848	6.275	2.477	8.400	0.09	0.02
pH												
9.5	5.023	2.908	2.244	7.183	11.243	0.995	1.462	4.418	9.796	2.485	0.14	0.24
10.5	8.603	3.922	6.318	20.266	5.985	8.995	12.197	15.536	5.126	9.219	0.33	0.66
11.5	2.789	1.818	2.675	2.265	1.760	2.278	4.156	2.705	2.169	3.200	0.23	0.45
12.5	3.635	1.118	1.932	1.575	1.752	1.150	2.915	1.520	1.538	0.739	0.22	0.56
13.5	1.274	2.113	0.754	0.619	1.043	0.872	0.822	0.331	0.806	0.664	0.19	0.34
NaCl concentration (g L^{-1})												
0	3.192	3.796	2.889	2.614	5.782	4.390	3.640	2.946	2.592	2.836	0.07	0.01
0.5	3.697	5.951	6.206	4.583	5.976	6.425	6.306	5.206	4.545	5.937	0.13	0.29
1	5.122	4.622	5.625	4.236	5.713	6.765	4.913	6.931	5.929	5.390	0.15	0.34
1.5	6.349	5.613	5.277	6.512	8.514	6.636	5.475	6.310	6.600	6.798	0.19	0.34
2	6.492	6.611	11.486	5.739	6.764	4.259	5.512	10.323	11.792	4.629	0.09	0.20

[a]CCC477, CCC478, CCC479, CCC480- *Spirulina platensis*, CCC481-*Spirulina maxima*, CCC482-*Spirulina lonar*, CCC483-*Spirulina platensis* (mutant), CCC538, CCC539-*Arthrospira* sp., CCC540-*Spirulina* sp.

ACKNOWLEDGMENTS

The authors are thankful to Director IARI, New Delhi for allowing the present study at the Institute and to NAIP, ICAR for financial support.

REFERENCES

1. Anupama, P. R. (2000). Value-added food: Single cell protein. *Biotechnology Advances, 18,* 459–479.
2. Belay, A., Ota, Y., Miyakawa, K., & Shimamatsu, H. (1993). Current knowledge on potential health benefits of*Spirulina*. *Journal of Applied Phycology, 5,* 235–241.
3. Bennett, A., & Bogorad, L. (1993). Complementary chromatic adaptation in a filamentous blue green alga.*Journal of Cell Biology, 58,* 419–435.
4. Borowitzka, M. A., & Borowitzka, L. J. (1988). *Microalgal biotechnology*. Cambridge: Cambridge University Press.
5. Colla, L. M., Reinehr, C. O., Reichert, C., & Costa, J. A. V. (2004). Production of biomass and nutraceutical compounds by *Spirulina platensis* under different temperature and nitrogen regimes. *Bioresource Technology,98,* 1489–1493.
6. Cornet, J. F., Dussa, C. G., Cluzel, P., & Dubertret, G. A. (1992). Structured model for simulation of cyanobacterium*Spirulina platensis* in photobioreactors: Identification of kinetic parameters under light and mineral limitations.*Biotechnology and Bioengineering, 40,* 826–834.
7. Costa, J. A. V., Linde, G. A., Atala, D. I. P., Mibieli, G. M., & Kruger, R. T. (2000). Modelling of growth conditions for cyanobacterium *Spirulina platensis* in microcosms. *World Journal of Microbiology & Biotechnology, 16,* 15–18.
8. Danesi, E. D. G., Rangel, C. O., Pelzer, L. H., Carvalho, J. C. M., Sat, S. & Moraes, I. O. (2001). Production of*Spirulina platensis* under different temperatures and urea feeding regimes for chlorophyll

attainment.*Proceedings: Eighth International Congress in Engineering and Food* (Vol. 2 In: 1978–1982).

9. Estrada, J. E., Bescos, P., & Villar Del Fresno, A. M. (2001). Antioxidant activity of different fractions of *Spirulina platensis* protein extract. *Il Farmaco, 56*, 497–500.

10. Henrikson, R. (1994). *Microalga Spirulina-Superalimento del future*. Bercelona: Ediciones Urano S A. ISBN 84-7953-047-2.

11. Hu, Q. (2004). Industrial production of microalgal cell mass and secondary products-major industrial species:*Arthrospira (Spirulina) platensis*. In A. Richmond (Ed.), *Handbook of microalgal culture: Biotechnology and applied phycology* (pp. 264–272). Oxford: Blackwell Science Ltd.

12. Jensen, A. (1978). Chlorophylls and carotenoids. In J. A. Hellebust & J. S. Caraig (Eds.), *Phycological methods: Physiological and biochemical methods* (pp. 59–70). Cambridge: Cambridge University Press.

13. Kavita, V., & Mohanty, P.. (2000). Alterations in the structure of phycobilisomes of the cyanobacterium *Spirulina platensis* in response to enhanced Na^+ level. *World Journal of Microbiology & Biotechnology, 16*, 795–798.

14. Lichtenthaler, H. K. (1987). Chlorophyll and carotenoids: pigments of photosynthetic biomembranes. In L. Paker & R. Douce (Eds.), *Methods in enzymology*, (Vol. 148, pp. 350–382). Boca Raton: CRC Press.

15. Miranda, M. S., Cintra, R. G., Barros, S. B. M., & Filho, J. M. (1998). Antioxidant activity of the microalga *Spirulina maxima*. *Brazilian Journal of Medical and Biological Research, 31*, 1075–1079.

16. Paerl, H. W., & Pinckney, J. L. (1996). A minireview of microbial consortia: Their role in aquatic production and biogeochemical cycling. *Microbial Ecology, 31*, 225–247.

17. Richmond, A. (1990). *Handbook of microalgal mass culture*. Boca Raton: CRC Press. ISBN 0-8493-3240-0.

18. Richmond, A., & Grobbelaar, J. U. (1986). Factors affecting the output rate of *Spirulina platensis* with reference to mass cultivation. *Biomass, 10*, 253–264.

19. Vonshak, A. (1997). *Spirulina*: Growth, physiology and biochemistry. In A. Vonshak (Ed.), *Spirulina platensis (Arthrospira)*: *Physiology, cell-biology and biotechnology* (pp. 43–65). London: Taylor & Francis.

20. Vonshak, A., Abeliovich, A., Boussiba, S., Arad, S., & Richmond, A. (1982). Production of *Spirulina* biomass: Effects of environmental factors and population density. *Biomass, 2,* 175–185.

21. Vonshak, A., & Tomaselli, L. (2000). *Arthrospira* (*Spirulina)*: Systematics and ecophysiology. In A. Whitton & E. Potts (Eds.), *The ecology of cyanobacteria* (pp. 505–522). The Netherlands: Kluwer Academic Publishers.

22. Zarrouk, C. (1966). Contribution a l' etude d' une cyanobacterie: Influence de divers facteurs physiques et chimiques et la photosynthese de *Spirulina maxima* (Setchell et Gardener) Geitler. Ph.D. Thesis, University of paris, France.

Chapter 9

Transgenic Plants for Enhanced Phytoremediation – Physiological Studies

Paulo Celso de Mello- Farias[1], Ana Lúcia Soares Chaves[2] and Claiton Leoneti Lencina[1]

[1] Federal University of Rio Grande, Brazil

[2] Federal University of Pelotas, Brazil

1. INTRODUCTION

Natural processes such as volcanic eruptions, continental dusts and anthropogenic activities like mining, combustion of fossil fuel phosphate fertilizers, military activities and metal working industries lead to emission of heavy metals and accumulation of these chemicals in ecosystem. So, the metals found in our environment come from natural weathering process of earth's crust, soil erosion, mining, industrial discharge, urban runoff, sewage effluents, air pollution fall out, pest or disease control agents. The concentrations of the contaminants can vary from highly toxic concentrations from an accidental spill to barely detectable concentrations that, after long-term exposure, can be detrimental to human health (Alexander, 1999; Doty, 2008). Particularly, heavy metals are toxic because they cause DNA damage and their carcinogenic effects in animals and humans are probably caused by their mutagenic ability (Knasmuller et al., 1998; Baudouin et al., 2002; Hooda, 2007). Potential threat is that heavy metals are not degradable and without intervention stay in soil for centuries. The cleanup of most of the contaminated sites is mandatory in order to reclaim the area and to minimize the entry of toxic elements into the food chain. Various engineering – based methods such as soil excavation, soil washing or burning or pump and treat systems are already being used to remediate metal contaminated soils ((Hooda, 2007). As a result over recent decades an annual worldwide release of heavy metals reached 22,000 t (metric ton) for cadmium, 939,000 t for copper, 783,000 t for lead and 1,350,000 t for zinc (Singh et al., 2003).

The cost of cleaning up contaminated sites is extremely high. In the USA alone, U$ 6–8 billion is annually spent in remediation efforts, with global costs in the range of U$ 25–50 billion (Glass, 1999;Tsao, 2003). Engineering methods for the remediation of contaminated sites include excavation, transport, soil washing, extraction, pumping and treating of contaminated water, addition of reactants such as hydrogen peroxide or potassium permanganate, and incineration. A serious consequence of the high cost of remediation technologies is that polluted commercial properties are often abandoned rather than cleaned up. There are over 500,000 of these so-called brownfields in the USA.

Elemental pollutants are particularly difficult to remediate from soil, water, and air because, unlike organic pollutants that can be degraded to harmless small molecules, toxic elements such as mercury, arsenic, cadmium, lead, copper, and zinc, are immutable by all biochemical reactions and hence remain in the ecosystem (Kramer & Chardonnens, 2001). The heavy metals remainings in various ecosystems would seep into surface water, groundwater or even channel into the food chain by crops growing on such a soil (Lin et al., 1998). These heavy metals may adversely affect the soil ecosystem safety, not only agricultural product and water quality, but also the human health (Zhou et al., 2004).

Another popular clean-up method involves augmented bioremediation with the addition of specific microbial strains known to degrade the pollutant. Bacteria and fungi collectively can utilize a vast range of organic molecules. But for bioremediation using microbes at a particular site to be successful, many conditions must be met. These include the ability of the microbes with the desired metabolic activity to survive in that environment, the accessibility or bioavailability of the chemical, and the presence of inducers to activate expression of the necessary enzymes. Many organic pollutants are recalcitrant to degradation and cannot be used as sole carbon source (Doty, 2008). The pollutants are sometimes metabolized by enzymes with other natural substrates; therefore, these substrates sometimes need to be present in order for the genes to be expressed. This requirement is problematic if the inducing chemical is itself a harmful pollutant, such as phenol. Bioremediation also depends on the presence of sufficient carbon and energy sources. Often, thousands of gallons of a food source such as molasses must be pumped down into the site to allow bacterial growth (Doty, 2008).

The use of microorganisms in engineered bioremediation systems has had mixed success. A review of this broad and active field is beyond the scope of this review; a recent book provides an excellent overview of bioremediation of xenobiotics, petroleum, BTEX (benzene, toluene, ethylbenzene, and xylene), explosives, and heavy metals (Fingerman & Nagabhushanum, 2005; Doty, 2008).

Plants are autotrophic organisms capable of using sunlight and carbon dioxide as sources of energy and carbon. However, plants rely on the root system to take up water and other nutrients, such as nitrogen and minerals, from soil and groundwater. As a side effect, plants also absorb a diversity of natural and man-made toxic compounds for which they have developed diverse detoxification mechanisms (Eapen et al., 2007; Van Aken, 2008).

Pollutant-degrading enzymes in plants probably originate from natural defense systems against the variety of allelochemicals released by competing organisms, including microbes, insects and other plants (Singer, 2006; Van Aken, 2008). From this viewpoint, plants can be seen as natural, solar-powered pump-and-treat systems for cleaning up contaminated environments, leading to the concept of phytoremediation (Pilon-Smits, 2005 ; Van Aken, 2008).

First developed for the removal of heavy metals from soil, the technology has since proven to be efficient for the treatment of organic compounds, including chlorinated solvents, polyaromatic hydrocarbons and explosives (Pilon-Smits, 2005 ; Salt et al., 1998; Van Aken, 2008). Beyond the removal of contaminants from soil, phytoremediation involves different processes, such as enzymatic degradation, that potentially lead to contaminant detoxification (Pilon-Smits, 2005 ; Dietz & Schnoor, 2001; Van Aken, 2008). However, despite great promise, rather slow removal rates and potential accumulation of toxic compounds within plants might have limited the application of phytoremediation (Eapen et al., 2007; Van Aken, 2008).

Unlike organic contaminants, metals cannot be degraded. Instead, phytoremediation strategies for metals are based on stabilization, accumulation, and in some cases on volatilization. The phytostabilization of metals may simply involve the prevention of leaching through the upward water flow created by plant transpiration, reduced runoff due to above-ground vegetation, and reduced soil erosion via the stabilization of soil by plant roots (Vassilev et al., 2004; Martínez et al., 2006).

Phytoremediation, as a cost-effective and environmentally friendly method, is an emerging technology based on the use of plants to remove, transform, clean up or stabilize contaminants including organic pollutants located in water, sediments, or soils (Cunningham et al., 1997; Cherian & Oliveira, 2005;Mello-Farias & Chaves, 2008). This method has attracted growing attention because of its distinctive potential and advantages compared with conventional technologies, such as soil replacement, solidification, and washing strategies (Yang et al., 2005a, 2005b ; Mello-Farias & Chaves, 2008). The advantages of phytoremediation over usual bioremediation by microorganisms are that plants, as autotrophic systems with large biomass, require only modest nutrient input and they prevent the spreading of contaminants through water and wind erosion (Pulford & Watson, 2003; Cherian & Oliveira, 2005; Mello-Farias & Chaves, 2008). Plants also supply nutrients for rhizosphere bacteria, allowing the growth and maintenance of a microbial community for further contaminant detoxification (Cherian & Oliveira, 2005; Mello-Farias & Chaves, 2008). Phytoremediation takes advantage of the unique, selective and naturally occurring uptake capabilities of plant root systems, together with the translocation, bioaccumulation and pollutant storage/degradation abilities of the entire plant body. Besides being aesthetically pleasing, phytoremediation is on average tenfold cheaper than other physical, chemical or thermal remediation methods since it is performed *in situ*, is solar driven and can function with minimal maintenance once established (Hooda, 2007).

According to Sarma (2011), there are different strategies of phytoremediation, each having a different mechanism of action for remediating metal-polluted soil, sediment or water, like: 1) Phytoextraction: plants absorb metals from soil through the root system and translocate them to harvestable shoots where they accumulate. Hyperaccumulators mostly used this process to extract metals from contaminated site. The recoveries of extracted metals are also possible through harvesting the plants appropriately; 2) Phytovolatilization: plants used to extract certain metals from soil and then release them into the atmosphere by volatilization; 3) Phytostabilization: plant root microbial interaction can immobilize organic and some inorganic contaminants by binding them to soil particles and as a result reduce migration contaminants to grown water; 4) Phytofiltration: plant roots (rhizofiltration) or seedlings (blastofiltration) absorb or adsorb pollutants, mainly metals from water and aqueous waste streams (Prasad & Freitas, 2003)

Many genes are involved in metal uptake, translocation and sequestration; the transfer of any of these genes into candidate plants is a possible strategy for genetic engineering of plants to improve phytoremediation traits. Depending on the strategy, transgenic plants can be engineered to accumulate high concentrations of metals in harvestable parts. Transfer or overexpression of genes will lead to enhanced metal uptake, translocation, sequestration or intracellular targeting. Genetic engineering of plants for synthesis of metal chelators will improve the capability of plant for metal uptake (Karenlampi et al., 2000; Pilon-Smits & Pilon, 2002; Clemens et al., 2002; Eapen & D´Souza, 2005).

The application of powerful genetic and molecular techniques may surely identify a range of gene families that are likely to be involved in transition metal transport. Considerable progress has been made recently in identifying plant genes encoding metal ion transporters and their homologous in hyperaccumulator plants. Therefore, it is hoped that genetic engineering may offer a powerful new means by which to improve the capacity of plants to remediate environmental pollutants (Yang et al., 2005a, 2005b , Mello-Farias & Chaves, 2008).

2. TYPES OF POLLUTANTS

There is a variety of different pollutants, originated, in most cases, by human action. To facilitate the development of studies on decontamination techniques and also according the different physical and chemical characteristics they present, the different types of contaminants were divided into two major classes: organic and inorganic. These two groups are further subdivided. Organic pollutants include various compounds such as polychlorinatedbiphenyls (PCB’s), polycyclicaromatichydrocarbons (PAH’s), nitroaromatic (explosives), halogenated hydrocarbons, chlorinated solvents. When compared to inorganic, the organic pollutants are relatively less toxic to plants because they are less reactive and do not accumulate readly. Many of these compounds are not only toxic or teratogenic, but also carcinogenic.

The inorganic contaminants include heavy metals, such as mercury, lead, cadmium, among others; and non-metallic compounds like arsenic and radionuclides like uranium, cesium, chromium, strontium, technetium, tritium, etc. Many metals are essential to growth and development of living forms. However, when in high concentrations, they become extremely toxic, leading the organism to oxidative stress with great production of harmful free radicals, highly dangerous to cells and tissues. Some particularly reactive metals interfere in the structure and function of proteins, and also cause the substitution of other essential nutrients (Garbisu et al. 2002; Pulfort; Watson, 2003; Taiz & Zeiger, 2002).

Many elemental pollutants penetrate the plant through regular systems of nutrient absorption. The plants protect themselves from these xenobiotics through degradation of endogenous toxic organic or sequestering them in the vacuoles (Meagher, 2000).

Different technologies of phytoremediation are compatible with a great number of pollutants. Constructed wetlands have been applied for many inorganics, including metals, nitrates, phosphates, cyanides, as well as organics such as explosives and herbicides (Horne 2000; Schnoor et al., 1995 ;Jacobson et al, 2003).

There is a special category of plants called hyperaccumulators (described later), for they accumulate a considerable amount of toxic metals and radionucleides in their tissues (phytoextraction), keeping these compounds above the ground surface. This is the main goal of phytoremediation.

2.1. Inorganics

The absorption of any metal by plants depends on the metal relative bioavailability in the contaminated array. Changes in the soil chemistry, such as decreased pH, may increase the availability of many metals for the absorption by the roots. Many plants can absorb significant levels of metals in some soil conditions. Changes in rhizosphere microbial status (e.g.: presence of mycorrhizae) can also have profound effects (positive or negative) on the uptake of metals by the roots (Smith, 1996). The general consensus of researchers in this area, however, is that phytoremediation, especially for heavy metals, will only be economically viable through the use of hyperaccumulators. The research in the past two decades has shown that certain specialized plants have the ability to accumulate more than 3% (dry weight) of heavy metals and over 25% (dry weight) in sap / latex with no apparent damage to the plant (Baker & Brooks, 1989, Baker et al. 1994; Huang & Cunningham, 1996). The mechanisms that govern this tolerance and absorption of excessive concentrations of metals in leaves were the subject of active research and vary according to the element (Cunningham & Lee, 1995; Huang & Cunningham, 1996). The mechanisms of tolerance include the accumulation of Zn in cell walls; Ni associated with the pectin in large cells; Ni, Co and Zn being chelated by malic acid; phytochelatin associated to Zn, Ni chelation by citrate; and Co associated with oxalate crystals calcium in plant tissues. Knowledge of the mechanisms of tolerance will aid in identifying the genetic characteristics necessary for the transfer of metal tolerance of plants capable of producing greater biomass with deeper rooting. It was suggested that in some cases, the resulting biomass rich in metals (biominery) could be incinerated and have metals economically recycled. This "biomineration" of metals can also be applied as a mining technique for metals with significant economic value (Robinson et al., 1998).

Another type of inorganic compounds that may be susceptible to phytoremediation are radionucleides. The presence of radionucleides in soil and water poses serious risk to human health. These contaminants come from the explosion of atomic bombs or nuclear power plant accidents such as Chernobyl, Ukraine and, more recently in Fukushima, Japan. The selection of an appropriate cleaning technology of these contaminated areas is based on the environmental chemistry of each element, character of deposition and rate of radioactive decay. A variety of physicochemical methods are available, like soil washing, ion exchange, leaching with chelating agents, flocculation and osmosis-ultrafiltration. Recently there has been increasing interest in the use of biological methods to remove radionucleides (Duschenkov, 2003). Negri and Hinchman (2000) reported data in the use of plants for the treatment of ^{3}H, U, Pu, ^{137}Cs and ^{90}Sr.

2.2. Organics

More recently, with the development of the pesticide industry, the metabolic capacity of the plant system began to be assessed. The most modern herbicides are based on the selectivity of crops due to metabolic differences between species of plants. This capability, often created by man, is the cornerstone of the highly profitable market of herbicides. "Desirable" plants rapidly metabolize the herbicide compound in a nontoxic one, while "undesirable" herbs do not, and are therefore dead. This mechanism developed by the natural selection of plants, proves to be potentially exploitable in the remediation of contaminated soils.

This ability of plants to detoxify xenobiotics is widely recognized and with current utility. Besides, plants generally have a metabolic system with differences in the efficiency of degradation of toxic compounds when compared to microorganisms, what makes the union of these two distinct systems in the rhizosphere, an ideal situation for a more efficient phytoremediation. Recent research includes plant selection, alternative patterns of rooting, the composition of exudates produced by the plant and its effect on microbial communities, exudation of specific compounds inducing specific metabolic pathways and, inoculation with rhizosphere microorganisms capable of degrading xenobiotics efficiently (Langenbach, 1994). The plants and their roots can create an environment in the soil which is rich in microbial activity, able to change the availability of organic contaminants or increase the degradation of certain organic compounds, such as hydrocarbons derived from Petroleum.

Siciliano et al. (2003) evaluated the impact of microbial remediation on soil mass and the capacity of microbial community to degrade hydrocarbons in order to determine whether phytoremediation treatments increase the metabolic potential of microbial community by altering its taxonomic structure. It was found that the best remediation system to reduce hydrocarbons in the soil was obtained by increasing the population of bacteria containing genes for the catabolism of hydrocarbons in the rhizosphere community, thus demonstrating the importance of using microorganisms in phytoremediation. However, it is necessary to identify the species of suitable plants that can beneficially alter microbial diversity for soil remediation. According to Pires et al. (2003), the absorption of herbicides by plants is affected by the compound's chemical properties, environmental conditions and the characteristics of plant species. Actually, the probability of a plant being phytoremediator depends on the type of pollutant; plants should be tested to detect that one with the greater resistance to a specific pollutant. Estevez-Nunez et al. (2001) evaluated trinitrotoluene (TNT), and found that its chemical structure influences its biodegradability. According to these authors, the oxygenated metabolism for aromatic compounds by bacteria does not occur in TNT because of its chemical properties generating compounds not metabolized by microorganisms. However, anaerobic processes have advantages because of the absence of oxygen. Therefore the use of fungi for the bioremediation of TNT has generated considerable interest. Estevez-Nunez et al. (2001) concluded that the remediation of TNT by these organisms is a very valid process; and the rhizoremediation by microbes able to colonize the rhizosphere of plants, will provide a fast and efficient mechanism for the removal of this pollutant. Figure 1 shows some types of organic pollutants.

There are other types of contaminants called Persistent Organic Pollutants (POPs) that resist long in the soil. Some examples are Dichlorodiphenyltrichloroethane (DDT), Polychlorinated biphenyls (PCB), Dioxins, etc. Research has shown that a variety of plants can remove persistent compounds, transporting them to aerial plant tissues (Coutinho & Barbosa, 2007). It is important to highlight that, due to variety of contaminants, the study of pollutant is very important to generate an effective phytoremediation.

Most current methods of cleaning metals and volatile organic compounds not on the soil surface are coarse, expensive and physically destructive (Baker et al., 1994). The remediation by conventional methods of engineering often costs 50 to 500 dollars per ton of soil, and certain specialized techniques can cost up to US$ 1000 (Cunningham & Ow, 1996). Phytoremediation associated with biotechnology is an emerging technology that promises a viable remediation when pollutants: a) are close to the surface, b) are relatively non-leachable, and c) have little immediate risk to the environment (Cunningham & Lee, 1995). The results are more effective in slightly or moderately polluted areas. For heavy contamination, the decontamination time is too long (Robinson et al., 1998). The combination of metal hyperaccumulation and degradation or, increased sequestration of organic compounds with greater biomass and deeper rooting systems can result in a powerful technology of phytoremediation that will provide cheaper, permanent and intrusive remediation. Table 1 shows a summary of the techniques applied to the different types of phytoremediated compounds.

3. HYPERACCUMULATORS

Over recent years a special interest has emerged in the phenomenon of heavy-metal hyperaccumulation since this property may be exploited in the remediation of heavy-metal-polluted soils through phytoextraction and phytomining (Robinson et al., 1997; Martínez et al., 2006).

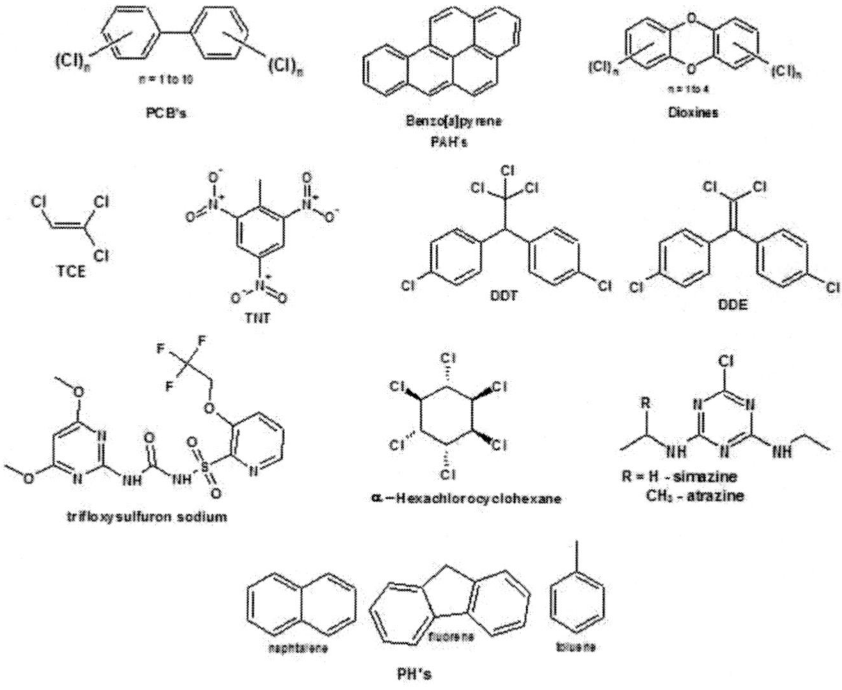

Figure 1. Different types of organic pollutants

The concept of hyperaccumulation was originally introduced to plants containing more than 0.1% (1000 $mg.kg^{-1}$) of Ni in dried plant tissues (Jaffré et al., 1976). At present, the criteria used for hyperaccumulation vary per metal, ranging from 100 $mg.kg^{-1}$ dry mass for Cd, to 1000 $mg.kg^{-1}$ for Cu, Co, Cr, and Pb, to 10000 $mg.kg^{-1}$ for Zn and Mn. These values exhibit a shoot-to-soil metal concentration ratio, the so-called bioaccumulation factor that is higher than 1 (Baker et al., 1994).

An ideal plant for environmental cleanup should have a high biomass production, combined with superior capacity for pollutant tolerance, accumulation, and degradation, depending on the type of pollutant and the phytoremediation technology of choice. Hyperaccumulators are good candidates in phytoremediation, particularly for the removal of heavy metals. Phytoremediation efficiency of plants can be substantially improved using genetic engineering technologies (Cherian & Oliveira, 2005;Mello-Farias & Chaves, 2008).

Table 1. Outline of phytoremediated chemicals (**Nwoko, 2010**). *BTEX = benzene, toluene, ethyl benzene, xylenes; #PCB = Polychlorinated biphenyl

Type of phytoremediation	Chemicals Treated
Phytoaccumulation/extraction	Cd, Cr, Pb, Ni, Zn, radionuclides, BTEX*, penachlorophenol, short chained aliphatic compounds
Phytodegradation/transformation	Nitrobenzene, nitroethane, nitrotoluene, atrazine, chlorinated solvents (chloroform, carbon tetrachloride, etc)
Phytostabilization	Heavy metals in ponds, phenols and chlorinated solvents
Phytostimulation	Polycyclicaromatic hydrocarbon, BTEX, PCB#, tetrachloroethane
Phytovolatilization	Chlorinated solvents, Hg, Se.
Phytofiltration	Heavy metals, organics and radionucleides.

*BTEX = benzene, toluene, ethyl benzene, xylenes;
#PCB = Polychlorinated biphenyl

Transferring the genes responsible for the hyperaccumulating phenotype to higher shoot-biomass-producing plants has been suggested as a potential avenue for enhancing phytoremediation as a viable commercial technology (Pilon-Smits & Pilon, 2002; Martínez et al., 2006).

Some of the plants belonging to *Brassicaceae* such as *Alyssum* species, *Thlaspi* species and *Brassica juncea*, *Violaceae* such as *Viola calaminaria*, *Leguminosae* such as *Astragalus racemosus* are known to take up high concentrations of heavy metals and radionucleides (Reeves & Baker, 2000; Negri & Hinchman, 2000, Eapen & D´Souza, 2005).

To date, there are approximately 400 known hyperaccumulators metal in the world (Reeves & Baker, 2000; Eapen & D´Souza, 2005) and the number is increasing. However, the remediation potential of many of these plants is limited because of their slow growth and low biomass (Chaney et al., 2000;Lasat, 2002; Mc Grath et al., 2002; Eapen & D´Souza, 2005).

4. CELLULAR AND MOLECULAR MECHANISMS INVOLVED IN PHYTOREMEDIATION

Exposure to pollutants may cause a series of symptoms in plants. Pollutant action can result in inhibition of cellular activity or rupture of cell structure, due to possible damages of essential components (Coutinho & Barbosa, 2007). Plants show some potential cellular and molecular mechanisms and strategies, which can be involved in detoxification of organic and inorganic pollutants such as herbicides, explosives and heavy metals. These mechanisms can be related to the cell wall composition and root environment, plasma membrane properties and integrity, enzymatic transformation, complexation with ligands and vacuolar compartmentalization (Hall, 2002; Cherian & Oliveira, 2005). Depending on the nature of pollutant (organic or inorganic) plant cells can use one or some of these systems of remediation (Coutinho & Barbosa, 2007; Cherian & Oliveira, 2005; Mello-Farias & Chaves, 2008).

4.1. Cell Wall Composition and Rhizosphere

Organic contaminants, when in contact with roots, may be sorbed to the root structure. The hydrophobic or hydrophilic nature of the organic compounds also determines their possible uptake. Hemicellulose in the cell wall and the lipid bilayer of plant membranes can bind hydrophobic organic pollutants effectively (Pilon-Smits, 2005). In addition, the root uptake of chemicals depends on many factors as plant's uptake efficiency, the transpiration rate, and the chemical concentration in soil water. Further, organic pollutants can be degraded or mineralized by plants, either independently or in association with microorganisms. For example, organics like polycyclic aromatic hydrocarbons (PAHs), polychlorinated biphenyls, and petroleum hydrocarbons are sufficiently degraded by rhizospheric microbial activity (Olson et al., 2003). Plants have significant metabolic activity in both roots and shoots, and some of the enzymes involved in these metabolic processes (e. g. nitroreductases, dehalogenases, laccases, peroxidases, etc.) are useful in the remediation process (Schnoor et al. 1995 ;Wolfe & Hoehamer, 2003).

Even though mycorrhizas and ectomycorrhizas are not considered in general reviews of plant metal tolerance mechanisms, they can ameliorate the effects of metal toxicity on the host plant. However, the mechanism involved in conferring this increase of tolerance is not yet well explained; they may be quite diverse and show considerable species and metal specificity since large differences in response to metals have been observed, both between fungal species and to different metals (specially Zn, Cu and Cd) within species (Hall, 2002).

Finally, cell walls may play an important role in detoxifying metals in plant cells of the Ni and Zn/Cd hyperaccumulating plant species. About 60–70% of Ni and/or Zn accumulated is distributed in the apoplast cell walls (Krämer et al., 2000; Li et al., 2005; Yang et al., 2005b). However, molecular bases of metal detoxification by cell walls are not well understood (Yang et al., 2005b).

4.2. Plasma Membrane Properties and Integrity

Although there is no direct evidence for a role for plasma membrane efflux transporters in heavy metal tolerance in plants, recent research has revealed that plants possess several classes of metal transporters that must be involved in metal uptake and homeostasis in general and, thus, could play a key role in tolerance (Yang et al., 2005a). Transport proteins and intracellular high-affinity binding sites mediate the uptake of metals across the plasma membrane. A comprehensive understanding of the metal transport processes in plants is essential for formulating effective strategies to develop genetically engineered plants that can accumulate specific metals (Yang et al., 2005b).

Several classes of proteins have been implicated in heavy metal transport in plants. These include the heavy metal (or CPx-type) ATPases that are involved in the overall metal-ion homeostasis and tolerance in plants, the natural resistance-associated macrophage protein (Nramp) family of proteins, and the cation diffusion facilitator (CDF) family proteins (Williams et al., 2000), zinc–iron permease (ZIP) family proteins, etc. (Yang et al., 2005a; 2005b).

CPx-type heavy metal ATPases have been identified in a wide range oforganisms and have been implicated in the transport of essential as well aspotentially toxic metals like Cu, Zn, Cd, Pb across cell membranes (Williams et al.,2000; Yang et al., 2005a; 2005b). These transporters use ATP to pump a variety of charged substrates across cell membranes and are distinguished by the formation of a charged intermediate during the reaction cycle (Yang et al., 2005a; 2005b).

Transport of metals and alkali cations across plant plasma membrane and organellar membranes is essential for plant growth, development, signal transduction, and toxic metal phytoremediation (Cherian and Oliveira, 2005).

Another factor concerning plasma membrane seems to be the maintenance of its physical integrity in presence of heavy metals, in order to prevent or reduce their entry in the cell, besides the efflux mechanisms described above (Hall, 2002; Coutinho & Barbosa, 2007).

4.3. Enzimatic Transformation

Enzimatic transformation in plants concerns mainly organic pollutants and it can be considered a case of phytotransformation. In this process plants uptake organic pollutants and, subsequently, metabolize or transform them into less toxic metabolites. Once taken up and translocated the organic chemicals generally undergo three transformation stages: (a) chemical modification (oxidations, reductions, hydrolysis); (b) conjugation (with glutathione, sugars, amino acids); and (c) sequestration or compartmentalization (conjugants are converted to other conjugates and deposited in plant vacuoles or bound to the cell wall and lignin) (Ohkawa et al., 1999; Cherian and Oliveira, 2005).

Plant enzymes that typically catalyze the first phase of the reactions are P450 monoxygenases and carboxylesterases (Coleman et al., 1997; Burken, 2003). The second phase involves conjugation to glutathione (GSH), glucose, or amino acids, resulting in soluble, polar compounds (Marrs, 1996). For instance, detoxification of herbicides in plants is attributed to conjugation with glutathione catalyzed by glutathione S-transferase (GST) (Lamoureux et al., 1991). It was also reported that a group of GSTs mediate conjugation of organics to GSH in the cytosol (Kreuz et al., 1996; Neuefeind et al., 1997). Sometimes organic pollutants, such as atrazine and TNT, are partially degraded and stored in vacuoles as bound residues (Burken & Schnoor, 1997). The third phase of plant metabolism is compartmentalization and storage of soluble metabolites either in vacuoles or in the cell wall matrix. The glutathione S-conjugates are actively transported to the vacuole or apoplast by ATP-dependent membrane pumps (Martinoia et al., 1993). Also, an alternate conjugation-sequestration mechanism for organics exists in plants and involves coupling of a glucose or malonyl group to the organic compound, followed by the transport of the conjugate to the vacuole or the apoplast (Coleman et al., 1997).

Mechanisms as complexation whit ligands and vacuolar compartmentalization are described below.

4.4. Complexation With Ligands

Complexation with ligands is a process associated to heavy metal pollutants, and it can be an extracellular or an intracellular molecular event. These ligands can be chelators as organic acids or peptides such phytochelatins (PCs), methallothioneins (MTs) or glutathione (GSH) (Mello-Farias & Chaves, 2008).

Plant tolerance to heavy metals depends largely on plant efficiency in the uptake, translocation, and further sequestration of heavy metals in specialized tissues or in trichomes and organelles such as vacuoles. The uptake of metals depends on their bioavailability, and plants have evolved mechanisms to make micronutrients bioavailable (Cherian and Oliveira, 2005). Chelators such as siderophores, organic acids, and phenolics can help release metal cations from soil particles, increasing their bioavailability. For example, organic acids (malate, citrate) excreted by plants act as metal chelators. By lowering the pH around the root, organic acids increase the bioavailability of metal cations (Ross, 1994 *)*. However, organic acids may also inhibit metal uptake by forming a complex with the metal outside the root. Citrate inhibition of Al uptake resulting in aluminum tolerance in several plant species is an example of this mechanism (De la Fuente et al., 1997; Pineros & Kochian, 2001; Papernik et al. 2001). Copper tolerance in Arabidopsis is also the result of a similar mechanism (Murphy et al., 1999).

Intracellular complexation involves peptide ligands, such as metallothioneins (MTs) and phytochelatins (PCs) (Yang et al., 2005b). Chelation of metals in the cytosol by high-affinity ligands is potentially a very important mechanism of heavy-metal detoxification and tolerance (Hall, 2002).

Metallothioneins (MTs) are cysteine-rich proteins that have high affinity to cations such as Cd, Cu, and Zn (Cobbett & Goldsbrough, 2002; Singh et al., 2003; Cherian & Oliveira, 2005). They confer heavy-metal tolerance and accumulation in yeast. Overexpression of genes involved in the synthesis of metal chelators may lead to enhanced or reduced metal uptake and enhanced metal translocation or sequestration, depending on the type of chelator and on its role and location (Cherian & Oliveira, 2005;Pilon-Smits, 2005). MT proteins were originally isolated as Cu, Cd and Zn binding proteins in mammals. There is now good evidence that four categories of these proteins occur in plants, which are encoded by at least seven genes in *Arabidopsis thaliana* (Cobbett & Goldsbrough, 2002; Hall, 2002;Gratão et al., 2005).

The biosynthesis of MTs is regulated at the transcriptional level and is induced by several factors, such as hormones, cytotoxic agents, and metals, including Cd, Zn, Hg, Cu, Au, Ag, Co, Ni, and Bi (*Yang et al.,* 2005a).

Phytochelatins are a class of post-translationally synthesized (cysteine-rich metal-chelating) peptides that play a pivotal role in heavy-metal tolerance in plants and fungi by chelating these substances and decreasing their free concentrations (Vatamaniuk et al., 1999). PCs have been most widely studied in plants, particularly in relation to Cd tolerance (Cobbett, 2000; Goldsbrough, 2000). PCs consist of only three amino acids, glutamine (Glu), cysteine (Cys), and glycine (Gly). They are structurally related to the tripeptide glutathione (GSH), and are enzymatically synthesized from GSH. PCs form a family of structures with increasing repetitions of the -Glu-Cys dipeptide followed by a terminal Gly, (-Glu-Cys)*n*-Gly, where *n* is generally in the range of 2–5, but can be as high as 11 (Cobbett, 2000; Yang et al., 2005b).

Many plants cope with the higher levels of heavy metals by binding them to PCs and sequestering the complexes inside their cells (Yang et al., 2005a). As mentioned above, PCs are synthesized non-translationally, using glutathione as a substrate by PC synthase, an enzyme that is activated in the presence of metal ions (Cobbett, 2000). So, PCs are structurally related to glutathione (GSH; γ-GluCysGly), and numerous physiological, biochemical, and genetic studies have confirmed that GSH (or, in some cases, related compounds) is the substrate for PC biosynthesis (Cobbett, 2000; Cobbett and Goldsbrough, 2002).

Although PCs clearly can have an important role in metal detoxification, alternative primary roles of PCs in plant physiology have also been proposed. These have included roles in essential metal ion homeostasis and in Fe or sulphur metabolism (Sanita di Toppi & Gabbrielli, 1999; Cobbett and Goldsbrough, 2002). However, there is currently no direct evidence that PCs have functions outside of metal detoxification.

Because of MTs and PCs peptidic nature and because they bind metals in thiolate complexes, these peptide molecules demand a greater input of amino acids (especially cysteine), sulfur and nitrogen from the plant as the level of accumulated metals rise. Their synthesis is energy expensive and requires significant amounts of the growth limiting elements sulfur and nitrogen. Increased synthesis might thus at some point affect plant growth and therefore limit their use as phytoremediators (Tong et al., 2004).

4.5. Vacuolar Compartmentalization

The vacuole is generally considered to be the main storage site for metals in yeast and plant cells and there is evidence that phytochelatin–metal complexes are pumped into the vacuole in fission yeast (*Schizosaccharomyces pombe*) and in plants (Tong et al., 2004; Yang et al., 2005b). Compartmentalization of metals in the vacuole is also part of the tolerance mechanism of some metal hyperaccumulators. The Ni hyperaccumulator *Thlaspi goesingense* enhances its Ni tolerance by compartmentalizing most of the intracellular leaf Ni into the vacuole (Krämer et al., 2000; Tong et al., 2004). High-level expression of a vacuolar metal ion transporter TgMTP1 in *T. goesingense* was proposed to account for the enhanced ability to accumulate metal ions within shoot vacuoles (Persans et al., 2001; Tong et al., 2004; Yang et al., 2005b).

5. GENETICALLY ENGINEERED PLANTS FOR PHYTOREMEDIATION

The genetic and biochemical basis is becoming an interesting target for genetic engineering, because the knowledge of molecular genetics model organisms can enhance the understanding of the essencial metal metabolism components in plants.A fundamental understanding of both uptake and translocation processes in normal plants and metal hyperaccumulators, the regulatory control of these activities, and the use of tissue specific promoters offer great promise that the use of molecular biology tools can give scientists the ability to develop effective and economic phytoremediation plants for soil metals (Chaney et al., 1997; Fulekar et al., 2008). Plants such as *Populus angustifolia, Nicotiana tabacum* or*Silene cucubalis* have been genetically engineered to overexpress glutamylcysteine syntlietase, and thereby provide enhanced heavy metal accumulation as compared with a corresponding wild type plant (Fulekar et al., 2008).

Candidate plants for genetic engineering for phytoremediation should be a high biomass plant with either short or long duration (trees), which should have inherent capability for phytoremediation. The candidate plants should be amicable for genetic transformation. Some of high biomass hyperaccumulators for which regeneration protocols are already developed include Indian mustard (*Brassica juncea*), sunflower (*Helianthus annuus*), tomato (*Lycopersicon esculentum*) and yellow poplar (*Liriodendron tulipifera*) (Eapen & D'Souza, 2005; Mello-Farias & Chaves, 2008).

The application of powerful genetic and molecular techniques may surely identify a range of gene families that are likely to be involved in transition metal transport. Considerable progress has been made recently in identifying plant genes encoding metal ion transporters and their homologous in hyperaccumulator plants. Therefore, it is hoped that genetic engineering may offer a powerful new means by which to improve the capacity of plants to remediate environmental pollutants (Yang et al., 2005a; Mello-Farias & Chaves, 2008).

Brassica juncea was genetically engineered to investigate rate-limiting factors for glutathione and phytochelatin production. To achieve this, *Escherichia coli* gshl gene was introduced. The γ-ECS transgenic seedlings showed increased tolerance to cadmium and had higher concentrations of phytochelatins, γ-GluCys, glutathione, and total nonprotein thiols compared to wild type seedlings (Ow, 1996; Fulekar et al., 2008). Study showed that c-glutamylcysteine synthetase inhibitor, L-buthionine-[S,R]-sulphoximine (BSO), dramatically increases As sensitivity, both in non-adapted and As-hypertolerant plants, showing that phytochelatin-based sequestration is essential for both normal constitutive tolerance and adaptative hypertolerance to this metalloid (Schat et al., 2002; Fulekar et al., 2008).

Some genes have been isolated and introduced into plants with increased heavy metal (Cd) resistance and uptake, like *AtNramps* (Thomine *et al.*, 2000), *AtPcrs* (Song et al., 2004), and *CAD1* (Ha et al., 1999) from *Arabidopsis thaliana*, library enriched in Cd-induced cDNAs from *Datura innoxia* (Louie et al., 2003), *gshI*, *gshII* (Zhu et al., 1999a) and PCS cDNA clone (Heiss et al., 2003) from *Brassica juncea*.

There are some examples of transgenic plants for metal tolerance/phytoremediation, as tobacco with accumulation of Cd, Ca and Mn transformed with gene *CAX-2* (vacuolar transporters) from *A. thaliana*(Hirschi et al., 2000); *A. thaliana* tolerant to Al, Cu, and Na with gene *Glutathione-S-transferase* from tobacco (Ezaki et al., 2000); tobacco with Ni tolerance and Pb accumulation with gene *Nt CBP4* from tobacco (Arazi et al., 1999); tobacco (Goto et al., 1998) and rice (Goto et al., 1998; 1999) with increased iron accumulation with gene *Ferretin* from soybean; *A. thaliana* and tobacco resistant to Hg with gene *merA* from bacteria (Rugh et al., 2000; Bizily et al., 2000; Eapen & D'Souza, 2005); indian mustard tolerant to Se transformed with a bacterial glutathione reductase in the cytoplasm and also in the chloroplast (D´Souza et al., 2000); transgenic *A. thaliana* plants expressing SRSIp/ArsC and ACT 2p/γ-ECS together showed high tolerance to As, these plants accumulated 4- to 17-fold greater fresh shoot weight and accumulated 2- to 3-fold more arsenic per gram of tissue than wild plants or transgenic plants expressing γ-ECS or ArsC alone (Dhankher et al., 2002; Mello-Farias & Chaves, 2008).

Even though there is a variety of different metal tolerance mechanisms, and there are many reports of transgenic plants with increased metal tolerance and accumulation, most, if not all, transgenic plants created to date rely on overexpressing genes involved in the biosynthesis pathways of metal-binding proteins and peptides (Zhu et al., 1999b; Mejäre & Bülow, 2001; Bennett et al., 2003; Gisbert et al., 2003), genes that can convert a toxic ion into a less toxic or easier to handle form, or a combination of both (Dhankher et al., 2002; Yang et al., 2005b; Mello-Farias & Chaves, 2008).

At least three different engineering approaches to enhanced metal uptake can be envisioned (Clemens et al., 2002), which include enhancing the number of uptake sites, alteration of specificity of uptake system to reduce competition by unwanted cations and increasing intracellular binding sites. Each metal has specific molecular mechanism for uptake, transport and sequestration (Eapen & D'Souza, 2005; Mello-Farias & Chaves, 2008).

New metabolic pathways can be introduced into plants for hyperaccumulation or phytovolatilization as in case of *MerA* and *MerB* genes which were introduced into plants which resulted in plants being several fold tolerant to Hg and volatilized elemental mercury

(Bizily et al., 2000; Dhankher et al., 2002; Eapen & D'Souza, 2005) developed transgenic *Arabidopsis* plants which could transport oxyanion arsenate to aboveground, reduce to arsenite and sequester it to thiol peptide complexes by transfer of *Escherichia coli ars C* and *γ-ECS* genes (Eapen & D'Souza, 2005).

Alteration of oxidative stress related enzymes may also result in altered metal tolerance as in the case of enhanced Al tolerance by overexpression of glutathione-S-transferase and peroxidase (Ezaki et al., 2000; Eapen & D'Souza, 2005). Overexpression of 1-aminocyclopropane-1-carboxylic acid (ACC) deaminase led to an enhanced accumulation of a variety of metals (Grichko et al., 2000; Eapen & D'Souza, 2005).

According to Eapen & D'Souza (2005), it is essential to have plants with highly branched root systems with large surface area for efficient uptake of toxic metals. Experiments had shown that *Agrobacterium rhizogenes* could enhance the root biomass in some hyperaccumulator plants (Eapen, unpublished work). The hairy roots induced in some of the hyperaccumulators were shown to have high efficiency for rhizofiltration of radionuclide (Eapen et al., 2003) and heavy metals (Nedelkoska and Doran, 2000).

Nowadays there are many different examples of genes that have been used for the development of transgenic plants for metal tolerance and/or phytoremediation, as shown on Table 2.

6. ADVANTAGES AND DISADVANTAGES OF PHYTOREMEDIATION

Admittedly, phytoremediation has benefits to restore balance to a stressed environment, but it is important to proceed with caution. Plants enjoy enormous reduction in energy cost and utilization by virtue of deriving energy from solar radiation. The plant tolerates a wide range of environmental conditions.

The molecular composition of plants, mainly related to their enzyme and protein profiles, is of great interest to phytoremediation, because this technology can exploit plant molecular and cellular mechanisms of detoxification, through the use of genetic engineering tools.

Table 2. Selected examples of transgenic plants for metal tolerance/phytoremediation (from Eapen & D'Souza, 2005)

Gene transferred	Origin	Target plant species	Effect
MT2 gene	Human	Tobacco, oil seed rape	Cd tolerance
MT1 gene	Mouse	Tobacco	Cd tolerance
MTA gene	Pea	*Arabidopsis*	Cu accumulation
CUP-1 gene	Yeast	Cauliflower	Cd accumulation
CUP-1 gene	Yeast	Tobacco	Cu accumulation
γ-Glutamylcysteine synthetase	*E. coli*	Indian mustard	Cd tolerance
Glutathione synthetase	Rice	Indian mustard	Cd tolerance
Cysteine synthetase	Rice	Tobacco	Cd tolerance
CAX-2 (vacuolar transporters)	*Arabidopsis*	Tobacco	Accumulation of Cd, Ca and Mn
At MHX	*Arabidopsis*	Tobacco	Mg and Zn tolerance
Nt CBP4	Tobacco	Tobacco	Ni tolerance and Pb accumulation
FRE-1 and FRE-2	Yeast	Tobacco	More Fe content
Glutathione-s-Transferase	Tobacco	*Arabidopsis*	Al, Cu, Na tolerance
Citrate synthase	Bacteria	*Arabidopsis*	Al tolerance
Nicotinamine amino transferase (NAAT)	Barley	Rice	Grew in iron deficient soils
Ferretin	Soybean	Tobacco	Increased iron accumulation
Ferretin	Soybean	Rice	Increased iron accumulation
Zn transporters ZAT (At MTPI)	*Arabidopsis*	*Arabidopsis*	Zn accumulation
Arsenate reductase γ- glutamylcysteine synthetase	Bacteria	Indian mustard	As tolerance
Znt A-heavy metal transporters	*E. coli*	*Arabidopsis*	Cd and Pb resistance
Selenocysteine methyl transferase	*A. bisculatus*	*A. thaliana*	Resistance to selenite
ATP sulfurylase CAPS		Indian mustard	Se tolerance
Cystathione-gamma synthase (CGS)		Indian mustard	Se volatilization
Glutathione-S-transferase, peroxidase		*Arabidopsis*	Al tolerance
Glutathione reductase		*B. juncea*	Cd accumulator
ACC-deaminase	Bacteria		Many metal tolerance
YCF1	Yeast	*Arabidopsis*	Cd and Pb tolerance
Se-cys lyase	Mouse	*Arabidopsis*	Se tolerance and accumulation
Phytochelatin synthase (Ta PCS)	Wheat	*Nicotiana glauca*	Pb accumulation

The nature of plants is still an advantage because they are able to develop, over time, complex mechanisms to absorb nutrients, detoxify pollutants and control the local geochemical conditions. The plants play an important role in regulation water contant in soil avoiding the penetration of liquids by infiltration, which is the main mechanism of entry of contaminants. Plant roots supplement microbial nutrients and provide aeration to the soil, increasing consequently microbial population compared to non-vegetated area. Above all, phytoremediation gives better aesthetic appeal than other physical means of remediation.

On the other hand, phytoremediation has several limitations that require further intensive research on plants and soil conditions. A major disadvantage is that this method of detoxification is too slow or only seasonally effective. Regulatory agencies often require significant progress in remediation to be made in only a few years, making most phytoremediation unsuitable. In many cases, like trichloroethylene and carbon tetrachloride, the concentration of pollutant is not reduced satisfactorily. Besides, in some contaminated sites, the pollutants can reach phytotoxic concentration, making the plant ineffective. For this reason, recent studies have been conducted with the aim of increasing the phytoremediation potential of plants using genetic engineering (Danh et al. 2009). In phytoremediation technology, multiple metal contaminated soil and water require specific metal hyperaccumulator species and therefore, a wide range of research prior to the application. Other factors are also tied to the success of phytoremediation such as the existence of a pollutant in a bio-available form. If the metal is strongly linked to the organic soil it will not be available to the plant. Moreover, the plants are quite specific to certain pollutants. Hyperaccumulators of Cd and Zn (*Thlaspi caerulescens*) can be sensitive to other metals, such as Cu, not allowing the detoxification of polluted areas with different pollutants (Mijovilovich et al., 2009). Despite the current limitations, present day phytoremediation technology is used worldwide and several researchers are working to overcome these limitations. Table 3 resumes advantages and limitations of some of the sub-process of phytoremediation.

7. PERSPECTIVES ON BIOTECHNOLOGY - BASED PHYTOREMEDIATION

The environmental contamination by pollutants, organic or inorganic, has great importance due to its impacts on human and animal health. Thus, the most effective and inexpensive technologies to promote detoxification are necessary in the recovery of affected biomes. Great efforts have been made in identifying plant species and their detoxification mechanisms more efficient on those places. The mechanisms of pollutant uptake, accumulation, exclusion, among others, vary according to each plant species and are very important, for they will determine its specific role in phytoremediation.

Plants can have their detoxification capabilities significantly enhanced through the identification of specific genes in certain promising species and the transmission of these to other species, using genetic engineering tools. This can play a significant role in the more effective detoxification of contaminated sites by improving the cost-benefit.

Table 3. Advantages and limitations of some of the phytoremediation sub-processes (Prasad, 2004; Gratão et al. 2005)

Advantage	Limitation
Phytoextraction	
The plant must be able to produce abundant biomass in short time. e. g.: in a greenhouse experiment, gold was harvested from plants.	Metal hyperaccumulators are generally slow-growing and bioproductivity is rather small and shallow root systems. Phytomass after process must be disposed off properly
Phytostabilization	
It circumvents the removal of soil, low cost and is less disruptive and enhances ecosystem restoration/re-vegetation	Often requires extensive fertilization or soil modification using amendments; long-term maintenance is needed to prevent leaching.
Phytovolatilization	
Contaminant/Pollutant will be transformed into less-toxic forms. e. g.: elemental mercury and dimethyl selenite gas. Atmospheric processes such as photochemical degradation for rapid decontamination/transformation.	The contaminant or a hazardous metabolite might accumulate in plants and be passed on in later products such as fruit or lumber. Low levels of metabolites have been found in plant tissue.
Phytofiltration/rhizofiltration	
It can be either *in situ* (floating rafts on ponds) or *ex situ* (an engineered tanks system); terrestrial or aquatic.	pH of the medium to be monitored continually for optimizing uptake of metals; chemical speciation and interactions of all species in the influent need be understood; functions like a bioreactor and intensive maintenance is needed.

Studies on phytoremediation are developed in order to benefit the environment. Several pollutants are bringing some kind of harm to all habitats. Thus, the use of specific techniques already represents hope. The necessary mechanisms are different, however, the organisms, especially plants, have specific ways for the removal, detention or conversion of specific pollutants. The study and subsequent evaluation of the interaction between the soil and its microorganisms, plant and pollutant is very necessary and guiding.

All things considered, more studies must be carried out in this area to better know the phytoremediation capacity of living organisms and their possible use in combating pollution through plant transformation technology.

REFERENCES

1. M. Alexander, 1999 Biodegradation and bioremediation. Ithaca, NY, USA: Academic Press.

2. T. Arazi, R. Sunkar, B. Kaplan, H. A. Fromm, 1999Tobacco plasma membrane calmodulin binding transporter confers Ni+ tolerance and Pb2+ hypersensitivity in transgenic plants. Plant Journal, 20 171 182

3. J. M. Baker, R. R. Brooks, 1989Terrestrial higher plants which hyperaccumulate metallic elements- a review of their distribution, ecology and phytochemistry. Biorecovery, 1 81 126

4. J. M. Baker, S. P. Mc Grath, C. M. D. Sidoli, R. D. Reeves, 1994The possibility of in situ heavy metal decontamination of polluted soils using crops of metal accumulating plants. Resources, Conservation and Recycling, 11 42 49

5. Baudouin, M. Charveron, R. Tarrouse, Y. Gall, 2002Environmental pollutants and skin cancer. Cell Biology and Toxicology, 18 341 348

6. L. E. Bennett, J. L. Burkhead, K. L. Hale, N. Terry, M. Pilon, E. A. H. Pilon-Smits, 2003Analysis of transgenic Indian mustard plants for phytoremediation of metalcontaminated mine tailings. Journal of Environmental Quality, 32 432 440

7. S. P. Bizily, C. L. Rugh, R. B. Meagher, 2000Phytodetoxification of hazardous organomercurials by genetically engineered plants. Nature Biotechnology, 18 213 217

8. J. G. Burken, J. L. Schnoor, 1997Uptake and metabolism of atrazine by poplar trees. Enviromental Science and Technology, 31 1399 1406

9. J. G. Burken, 2003Uptake and metabolism of organic compounds: green-liver models. In: Phytoremediation: Transformation and Control of Contaminants, McCutcheon, S.C. & Schnoor, J. L. (Ed.), 59 84Wiley: New York.

10. R. L. Chaney, M. Malik, Y. M. Li, S. L. Brown, J. S. Angle, A. J. M. Baker, 1997Phytoremediation of soil metals. Current Opinion in Biotechnology, 8 279 284

11. R. L. Chaney, Y. M. Li, S. L. Brown, F. A. Homer, M. Malik, J. S. Angle, A. J. M. Baker, R. D. Reeves, M. Chin, 2000Improving metal hyperaccumulator wild plants to develop phytoextraction systems: approaches and progress. In: Phytoremediation of contaminated soil and water, Terry, N. & Banuelos, G. (Ed.) 129 158Boca: Lewis Publishers.

12. S. Cherian, M. M. Oliveira, 2005Transgenic Plants in Phytoremediation: Recent Advances and New Possibilities. Environmental Science & Technology, 39 24 9377 9390

13. S. Clemens, M. G. Palmgren, U. Krämer, 2002A long way ahead: understanding and engineering plant metal accumulation. Trends in Plant Science, 7 309 314

14. S. Cobbett, 2000Phytochelatin biosynthesis and function in heavy-metal detoxification. Current Opinion in Plant Biology, 3 211 216

15. Cobbett, P. Goldsbrough, 2002Phytochelatins and Metallothioneins: Roles in Heavy Metal Detoxification and Homeosteasis. Annual Review of Plant Biology, 53 159 182

16. J. O. D. Coleman, M. M. A. Blake-Kalff, T. G. E. Davies, 1997Detoxification of xenobiotics by plants: chemical modification and vacuolar compartmentation. Trends in Plant Science, 2 144 151

17. H. Coutinho, A. R. Barbosa, 2007Fitorremediação: Considerações Gerais e Características de Utilização. Silva Lusitana, 15 1 103 117Portugal.

18. S. D. Cunningham, C. R. Lee, 1995Phytoremediation: Plant-based remediation of contaminated soils and sediments. In: Bioremediation: Science and Applications, Skipper, H.D. & Turco, R.F. (Ed.), 145 156SSSA Special Publication 43, Soil Science Society of America, Madison, WI.

19. S. D. Cunningham, D. W. Ow, 1996Promises and prospects of phytoremediation. Plant Physiology, 110 3 715 719

20. S. D. Cunningham, J. R. Shann, D. E. Crowley, T. A. Anderson, 1997Phytoremediation of contaminated water and soil: Phytoremediation of soil and water contaminations. ACS Symposium Series, 664 2 17

21. L. T. Danh, P. Truong, R. Mammucari, T. Tran, N. Foster, 2009Vetiver grass, Vetiveria zizanioides: A choice plant for phytoremediation of heavy metals and organic wastes. International Journal of Phytoremediation, 11 664 691

22. J. M. De la Fuente, V. Ramirez-Rodriguez, J. L. Cabrera-Ponce, L. Herrera-Estrella, 1997Aluminium tolerance in transgenic plants by alteration of citrate synthesis. Science, 276 1566 1568

23. O. P. Dhankher, Y. Li, B. P. Rosen, J. Shi, D. Salt, J. F. Senecoff, N. A. Sashti, R. B. Meagher, 2002Engineering tolerance and hyperaccumulation of arsenic in plants by combining arsenate reductase and g-glutamylcysteine synthetase expression. Nature Biotechnology, 20 1140 1145

24. C. Dietz, J. L. Schnoor, 2001Advances in phytoremediation. Environmental Health Perspectives, 109 163 168

25. S. L. Doty, Enhancing phytoremediation through the use of transgenics and endophytes. (2008 New Phytologist, 179 318 333

26. Souza. M. P. D´, E. A. H. Pilon-Smits, N. Terry, 2000The physiology and biochemistry of selenium volatilization by plants. In: Phytoremediation of toxic metals: using plants to clean up the environment, Raskin, I. & Ensley, B.D. (Ed.) 171 188

27. S. Dushenkov, 2003Trend in phytoremediation of radionuclides. Plant and Soil, 249 167 175

28. S. Eapen, K. Suseelan, S. Tivarekar, S. Kotwal, R. Mitra, 2003Potential for rhizofiltration of uranium using hairy root cultures of Brassica juncea and Chenopodium amaranticolor. Environmental Research, 91 127 133

29. S. Eapen, S. F. D'Souza, 2005Prospects of genetic engineering of plants for phytoremediation of toxic metals. Biotechnology Advances, 23 97 114

30. S. Eapen, S. Singh, Souza. S. F. D´, 2007Advances in development of transgenic plants for remediation of xenobiotic pollutants. Biotechnology Advances, 25 442 451

31. Estevez-Núnez, A. Caballero, J. L. Ramos, 2001Biological degradation of 2,4,6-trinitrotoluene. Microbiology and Molecular Biology Reviews, 65 3 335 352

32. Ezaki, R. C. Gardner, Y. Ezaki, H. Matsumoto, 2000Expression of aluminium induced genes in transgenic Arabidopsis plants can ameliorate aluminium stress and/or oxidative stress. Plant Physiology, 122 657 665

33. M. Fingerman, R. Nagabhushanum, 2005Bioremediation of aquatic and terrestrial ecosystems. Enfield, NH, USA: Science Publishers.

34. M. H. Fulekar, A. Singh, A. M. Bhaduri, 2008Genetic engineering strategies for enhancing phytoremediation of heavy metals. African Journal of Biotechnology, 8 4 529 535

35. C. Garbisu, A. J. Hernandez, O. Barrutia, J. Alkorta, J. M. Bacerril, 2002Phytoremediation: a technology using green plants to remove contaminants from pollutes areas. Reviews of Environmental Health, 17 3 173 188

36. C. Gisbert, R. Ros, A. De Haro, D. J. Walker, M. P. Bernal, R. Serrano, J. Navarro-Avino, 2003A plant genetically modified that accumulates Pb is especially promising for phytoremediation. Biochemical and Biophysiscal Research Communities, 303 440 445

37. D. Glass, 1999U.S. and International Markets for Phytoremediation, 1999 2000Needham, MA, USA: D. Glass Associates, Inc.

38. P. Goldsbrough, 2000Metal tolerance in plants: the role of phytochelatins and metallothioneins. In: Phytoremediation of contaminated soil and water, Terry N. & Bañuelos G. (Ed.), 221 233CRC Press LLC.

39. Goto, T. Yoshihara, H. Saiki, 1998Iron accumulation in tobacco plants expressing soybean ferritin gene. Transgenic Research, 7 173 180

40. Goto, T. Yoshihara, N. Shigemoto, S. Toki, F. Takaiwa, 1999Iron accumulation in rice seed by soybean ferritin gene. Nature Biotechnology, 17 282 286

41. P. L. Gratão, M. N. V. Prassad, P. F. Cardoso, P. J. Lea, R. A. Azevedo, 2005Phytoremediation: green technology for the clean up of toxic metals in the environment. Brazilian Journal of Plant Physiology, 17 1 53 64

42. V. P. Grichko, B. Filby, B. R. Glick, 2000Increased ability of transgenic plants expressing the bacterial enzyme ACC deaminase to accumulate Cd, Co, Cu, Ni, Pb and zinc. Journal of Biotechnology, 81 45 53

43. S. B. Ha, A. P. Smith, R. Howden, W. M. Dietrich, S. Bugg, M. J. O'Connell, P. B. Goldsbrough, C. S. Cobbett, 1999Phytochelatin synthase genes from Arabidopsis and the yeast Schizosaccharomyces pombe. Plant Cell, 11 1153 1163

44. J. L. Hall, 2002Cellular mechanisms for heavy metal detoxification and tolerance. Journal of Experimental Botany, 53 1 11

45. S. Heiss, A. Wachter, J. Bogs, C. Cobbett, T. Rausch, 2003Phytochelatin synthase (PCS) protein is induced in Brassica juncea leaves after prolonged Cd exposure. Journal of Experimental Botany, 54 1833 1839

46. K. D. Hirschi, V. D. Korenkov, N. L. Wilganowski, G. J. Wagner, 2000Expression of Arabidopsis CAX2 in tobacco altered metal accumulation and increased manganese tolerance. Plant Physiology, 124 125 133

47. V. Hooda, 2007Phytoremediation of toxic metals from soil and waste water. Journal of Environmental Biology, 28 2 367 376

48. J. Horne, 2000Phytoremediation by constructed wetlands: In: Phytoremediation of contaminated soil and water, Terry, N. & Bañuelos, G. (Ed.), 13 40Boca Raton: Lewis.

49. J. W. Huang, S. D. Cunningham, 1996Lead phytoextraction: species variation in lead uptake and translocation. New Phytologist, 134 1 75 84

50. T. Jaffré, R. R. Brooks, J. Lee, R. D. Reeves, 1976 Sebertia acuminata: a hyperaccumulator of nickel from New Caledonia. Science, 193 579 580

51. M. E. Jacobson, S. Y. Chiang, L. Gueriguian, L. R. Weshtholm, J. Pierson, 2003Transformation kinetics of trinitrotoluene conversion in aquatic plants. In: Phytoremediation: Transformation and control of contaminants, McCutcheon, S.C. & Schnoor, J.L. (Ed.), 409 427Wiley, NY.

52. S. Karenlampi, H. Schat, J. Vangronsveld, J. A. C. Verkleij, D. Van der Lelie, M. Mergeay, A. L. Tervahauta, 2000Genetic engineering in the improvement of plants for phytoremediation of metal polluted soils. Environmental Pollution, 107 225 231

53. S. Knasmuller, E. Gottmann, H. Steinkellner, A. Fomin, C. Pickl, A. Paschke, R. God, M. Kundi, 1998Detection of genotoxic effects of heavy metal contaminated soils with plant bioassays. Mutation Research, 420 37 48

54. U. Krämer, I. J. Pickering, R. C. Prince, I. Raskin, D. E. Salt, 2000Subcellular localization and speculation of nickel in hyperaccumulator and non-accumulator Thlaspi species. Plant Physiology, 122 1343 1353

55. U. Krämer, A. N. Chardonnens, 2001The use of transgenic plants in the bioremediation of soils contaminated with trace elements. Applied Microbiology and Biotechnology, 55 661 672

56. K. Kreuz, R. Tommasini, E. Martinoia, 1996Old enzymes for a new job: herbicide detoxification in plants. Plant Physiology, 111 349 353

57. L. Lamoureux, D. G. Rusness, P. Schroder, H. Rennenberg, 1991Diphenyl ether herbicide metabolism in a spruce cell suspension culture: the identification of two novel metabolites derived from

glutathione conjugation. Pesticide Biochemistry and Physiology, 39 291 301

58. T. Langenbach, 1994Degradação de xenobiontes: potencialidades e limites. In: Microrganismos de importância agrícola, Araújo, R.S. & Hungia, M. (Ed.), 218 236Brasília: EMBRAPA-SPI.

59. M. M. Lasat, 2002Phytoextraction of toxic metals: a review of biological mechanisms. Journal of Environmental Quality, 31 109 120

60. T. Q. Li, X. E. Yang, Z. L. He, J. Y. Yang, X. X. Long, 2005Zn accumulation and subcellular distribution in leaves of Zn hyperaccumulator species of Sedum alfredii Hance. Pedosphere, 16 5 616 623

61. C. F. Lin, S. S. Lo, H. Y. Lin, Y. C. Lee, 1998Stabilization of cadmium contaminated soils using synthesized zeolite. Journal of Hazardous Material, 60 217 226

62. M. Louie, N. Kondor, J. G. Witt, 2003Gene expression in cadmium-tolerant Datura innoxia: Detection and characterization of cDNAs induced in response to Cd2+. Plant Molecular Biology, 52 81 89

63. K. A. Marrs, 1996The functions and regulation of glutathione S-transferase in plants. Annual Review of Plant Physioogy and Plant Molecular Biology, 47 127 158

64. M. Martínez, P. Bernal, C. Almela, D. Vélez, P. García-Agustín, R. Serrano, J. Navarro-Aviñó, 2006An engineered plant that accumulates higher levels of heavy metals than Thlaspi caerulescens, with yields of 100 times more biomass in mine soils. Chemosphere, 64 478 485

65. E. Martinoia, E. Grill, R. Tomasini, K. Kreuz, N. Amrehin, 1993ATP-dependent glutathione S-conjugate 'export' pump in the vacuolar membrane of plants. Nature, 364 247 249

66. S. P. Mc Grath, F. J. Zhao, E. Lombi, 2002Phytoremediation of metals, metalloids and radionuclides. Advances in Agronomy, 75 1 56

67. R. B. Meagher, 2000Phytoremediation of toxic elemental and organic pollutants. Current Opinion of Plant Biology, 3 153 162

68. M. Mejäre, L. Bülow, 2001Metal-binding proteins and peptides in bioremediation and phytoremediation of heavy metals. Trends in Biotechnology, 19 67 73

69. P. C. Mello-Farias, A. L. S. Chaves, 2008Biochemical and molecular aspects of toxic metals phytoremediation using transgenic plants. In: Transgenic Approach in Plant Biochemistry and Physiology, Tiznado-Hernandez, M.E.; Troncoso-Rojas, R. & Rivera-Domínguez, M. A. (Ed.) 253 266Research Signpost, Kerala, India.

70. B. Mijoviloch, W. Leitenmaier, P. M. Meyer-Klauche, B. Kroneck, B. Götz, H. Kupper, 2009Complexation and toxicity of copper in higher plants. II. Differents mechanisms for copper versus cadmium detoxification in the cooper-sensitive cadmium/zinc hyperaccumulator Thlaspi caerulescens (Ganges Ecotype). Plant Physiolology, 151 715 731

71. S. Murphy, W. R. Eisenger, J. E. Shaff, L. V. Kochian, L. Taiz, 1999Early copper-induced leakage of K+ from Arabidopsis seedlings is mediated by ion channels and coupled to citrate efflux. Plant Physiology, 121 1375 1382

72. T. J. Nedelkoska, P. M. Doran, 2000Hyperaccumulation of cadmium by hairy roots of Thlaspi caerulescens. Biotechnology Bioengineering, 67 5 607 615

73. M. C. Negri, R. R. Hinchman, 2000The use of plants for the treatment of radionuclides. In: Phytoremediation Of Toxic Metals: Using Plants to Clean Up the Environment, Raskin, I. & Ensley, E.D. (Ed.), 107 132New York: Wiley.

74. T. Neuefeind, P. Reinemer, B. Bieseler, 1997Plant glutathione S-transferases and herbicide detoxification. Biological Chemistry, 378 199 205

75. O. Nwoko, 2010Trends in phytoremediation of toxic elemental and organic pollutants. African Journal of Biotechnology, 9 37 6010 6016

76. Ohkawa, H. Imaishi, N. Shiota, 1999In: Pesticide Chemistry and Biosciences: The Food- Environment Challenge, Brooks, J.T. & Roberts, T.R. (Ed.), 259 264Royal Society of Chemistry, Cambridge, UK.

77. P. E. Olson, K. F. Reardon, E. A. H. Pilon-Smits, 2003Ecology of rhizosphere bioremediation. In: Phytoremediation: Transformation and Control of Contaminants, McCutcheon, S.C. & Schnoor, J.L. (Ed.), 317 354Wiley: New York.

78. W. Ow, 1996Heavy metal tolerance genes-prospective tools for bioremediation. Resources, Conservation and Recycling, 18 135 149

79. L. A. Papernik, A. S. Bethea, T. E. Singleton, J. V. Magalhaes, D. F. Garvin, L. V. Kochian, 2001Physiological basis of reduced Al tolerance in ditelosomic lines of Chinese Spring wheat. Planta, 212 829 834

80. M. W. Persans, K. Nieman, D. E. Salt, 2001Functional activity and role of cationefflux family members in Ni hyperaccumulation in Thlaspi goesingense. The Proceedings of the National Academy of Sciences USA, 98 9995 1000

81. Pilon-Smits, Phytoremediation, 2005 Annual Review of Plant Biology, 56 15 39

82. Pilon-Smits, M. Pilon, 2002Phytoremediation of metals using transgenic plants. Critical Reviews in Plant Science, 21 439 456

83. Pilon-Smits, 2005Phytoremediation. Annual Reviews of Plant Biology, 56 15 39

84. M. A. Pineros, L. V. Kochian, 2001A patch-clamp study on the physiology of aluminium toxicity and aluminium tolerance in maize. Identification and characterization of Al3+ induced anion channels. Plant Physiology, 125 292 305

85. R. Pires, C. M. Souza, A. A. Silva, S. O. Procópio, L. R. Ferreira, 2003Fitorremediação de solos contaminados com herbicidas. Planta Daninha, 21 2 335 341

86. M. N. V. Prasad, 2004Phytoremediation of metals in the environment for sustainable development. Proceedings of the Indian National Science Academy, 70 71 98

87. M. N. V. Prasad, H. Freitas, 2003Metal hyperaccumulation in plants: Biodiversity prospecting for phytoremediation technology. Electronic Journal of Biotechnology, 6 275 321

88. D. Pulford, C. Watson, 2003Phytoremediation of heavy metal-contaminated land by trees- a review. Environment International, 29 529 540

89. B. H. Robinson, R. R. Brooks, A. W. Howes, J. H. Kirkman, P. E. H. Gregg, 1997The potential of the high biomass nickel hyperaccumulator

Berkheya coddii for phytoremediation and phytomining. Journal of Geochemical Exploration, 60 115 126

90. B. H. Robinson, M. Leblanc, D. Petit, R. R. Brooks, J. H. Kirkman, P. E. H. Gregg, 1998The potential of Thlaspi caerulescens for phytoremediation of contaminated soils. Plant and Soil, 203 47 56

91. R. D. Reeves, A. J. M. Baker, 2000Metal-accumulating plants. In: Phytoremediation of toxic metals using plants to clean up the environment, Raskin, I. & Ensley, B.D. (Ed.) 193 230John Wiley and Sons, New York.

92. C. L. Rugh, S. P. Bizily, R. B. Meagher, 2000Phytoremediation of environmental mercury pollution. In: Phytoremediation of toxic metals using plants to clean up the environment, Raskin, I. & Ensley, B.D. (Ed.) 151 171New York Wiley and Sons, New York.

93. Toppi. L. Sanita di, R. Gabbrielli, 1999Response to cadmium in higher plants. Environmental and Experimental Botany, 41 105 130

94. L. Schnoor, L. A. Licht, S. C. Mc Cutcheon, N. L. Wolf, L. H. Carreira, 1995Phytoremediation of organic and nutrient contaminants. Environmental Science and Technology, 29 318 323

95. S. D. Siciliano, J. J. Germida, K. Banks, C. W. Green, 2003Changes in Microbial Communit Composition and during a polyaromatic hydrocarbon phytoremediation field trial. Applied and Environmental Microbiology, 69 1 483 489

96. O. Siqueira, 1996 Avanços em fundamentos e aplicações de micorrizas. (edition) Lavras: UFLA, 290p.

97. S. M. Ross, 1994 Toxic Metals in Soil-Plant Systems, Wiley & Sons: Chichester, UK.

98. D. E. Salt, R. D. Smith, I. Raskin, 1998Phytoremediation. Annual Review of Plant Physiology and Plant Molecular Biology, 49 643 668

99. H. Sarma, 2011Metal hyperaccumulation in Plants: A Review Focusing on Phytoremediation Technology. Journal of Environmental Science and Technology, 4 2 118 138

100. H. Schat, M. Llugany, R. Voojis, J. Harley-Whitaker, P. M. Bleeker, 2002The role of Phytochelatin in constitutive and adaptive heavy metal

tolerances in hyperaccumulator and non-hyperaccumulator metallophytes. Journal of Experimental Botany, 53 2381 2392

101. L. Schnoor, L. A. Licht, S. C. Mc Cutcheon, N. L. Wolfe, L. H. Carreira, 1995Phytoremediation of contaminated soils and sediments. Environmental Science and Technology, 29

102. Singer, 2006The chemical ecology of pollutants biodegradation. In: Phytoremediation and Rhizoremediation: Theoretical Background, Mackova, M.; Dowling, D. & Macek, K. (Ed.), 5 21Springer.

103. O. V. Singh, S. Labana, G. Pandey, R. Budhiraja, R. K. Jain, 2003Phytoremediation: an overview of metallic ion decontamination from soil. Applied Microbiology and Biotechnology, 61 405 412

104. W. Y. Song, E. Martinoia, J. Lee, D. Kim, D. Y. Kim, E. Vogt, D. Shim, K. S. Choi, I. Hwang, Y. Lee, 2004A novel family of cys-rich membrane proteins mediates cadmium resistance in Arabidopsis. Plant Physiology, 135 1027 1039

105. Taiz, E. Zeiger, 2002 Plant Physiology; Sinauer: Sanderland, M.A., 1 690

106. S. Thomine, R. Wang, J. M. Ward, N. M. Crawford, J. I. Schroeder, 2000Cadmium and iron transport by members of a plant metal transporter family in Arabidopsis with homology to Nramp genes. Proceedings of the National Academy of Sciences USA, 97 4991 4996

107. D. T. Tsao, 2003Overview of phytotechnologies. Advances in Biochemical Engineering and Biotechnology, 78 1 50

108. Y. P. Tong, R. Kneer, Y. G. Zhu, 2004Vacuolar compartimentalization: a second generation approach to engineering plants for phytoremediation. Trends in Plant Science, 9 1 7 9

109. B. Van Aken, 2008Transgenic plants for phytoremediation: helping nature to clean up environmental pollution. Trends in Biotechnology, 26 5 225 227

110. Vassilev, J. P. Schwitzguebel, T. Thewys, D. Van der Lelie, J. Vangronsveld, 2004The use of plants for remediation of metal-contaminated soils. The Scientific World Journal, 4 9 34

111. O. K. Vatamaniuk, S. Mari, Y. P. Lu, P. A. Rea, 1999At PCS1, a phytochelatin synthase from Arabidopsis: isolation and in vitro reconstitution. The Proceedings of the National Academy of Sciences USA, 96 12 7110 7115

112. E. Williams, J. K. Pittman, J. L. Hall, 2000Emerging mechanisms for heavy metal transport in plants. Biochimica et Biophysica Acta, 1465 104 126

113. L. Wolfe, C. F. Hoehamer, 2003Enzymes used by plants and microorganisms to detoxify organic compounds. In: Phytoremediation: Transformation and control of contaminants, Mc-Cutcheon, S.C. & Schnoor, J.L. (Ed.), 158 187Wiley: New York.

114. X. Yang, X. F. Jin, Y. Feng, E. Islam, 2005aMolecular Mechanisms and Genetic bases of Heavy Metal Tolerance/Hyperaccumulation in Plants. Journal of Integrative Plant Biology, 47 9 1025 1035

115. X. Yang, Y. Feng, Z. He, P. Stoffella, 2005bMolecular mechanisms of heavy metal hiperaccumulation and phytoremediation. Journal of Trace Elements in Medicine and Biology, 18 339 353

116. Q. X. Zhou, F. X. Kong, L. Zhu, 2004 Ecotoxicology. Science Press, 454Beijing, China.

117. Y. L. Zhu, E. A. H. Pilon-Smits, L. Jouanin, N. Terry, 1999aOverexpression of glutathione synthetase in Indian mustard enhances cadmium accumulation and tolerance. Plant Physiology, 119 73 79

118. Y. L. Zhu, E. A. H. Pilon-Smits, A. S. Tarun, S. U. Weber, L. Jouanin, N. Terry, 1999bCadmium tolerance and accumulation in Indian mustard is enhanced by overexpressing g-glutamylcysteine synthetase. Plant Physiology, 121 1169 1177

Chapter 10

Identification and characterization of a gene encoding for a nucleotidase from *Phaseolus vulgaris*

Juan Miguel Cabello-Díaz, Gregorio Gálvez-Valdivieso, Cristina Caballo, Rocío Lambert, Francisco Antonio Quiles, Manuel Pineda, Pedro Piedras

Departamento de Botánica, Ecología y Fisiología Vegetal, Grupo de Fisiología Molecular y Biotecnología de Plantas, Campus Rabanales, Edif. Severo Ochoa, 1a Planta, Universidad de Córdoba, 14071 Córdoba, Spain

ABSTRACT

Nucleotidases are phosphatases that catalyze the removal of phosphate from nucleotides, compounds with an important role in plant metabolism. A phosphatase enzyme, with high affinity for nucleotides monophosphate previously identified and purified in embryonic axes from French bean, has been analyzed by MALDI TOF/TOF and two internal peptides have been obtained. The information of these peptide sequences has been used to search in the genome database and only a candidate gene that encodes for the phosphatase was identified (*PvNTD1*). The putative protein contains the conserved domains (motif I–IV) for haloacid dehalogenase-like hydrolases superfamily. The residues involved in the catalytic activity are also conserved. A recombinant protein overexpressed in *Escherichia coli* has shown molybdate resistant phosphatase activity with nucleosides monophosphate as substrate, confirming that the identified gene encodes for the phosphatase with high affinity for nucleotides purified in French bean embryonic axes. The activity of the purified protein was inhibited by adenosine. The expression of *PvNTD1* gene was induced at the specific moment of radicle protrusion in embryonic axes. The gene was also highly expressed in young leaves whereas the level of expression in mature tissues was minimal.

KEYWORDS

Phaseolus vulgaris; Ureide synthesis; Phosphatase; Nucleotidase; Nucleotide monophosphate; Nucleoside; Germination

1. INTRODUCTION

Nucleotides are some of the most crucial molecules for plant growth, development and metabolism. Nucleotides are building blocks for RNA and DNA synthesis and represent the main form for cellular energy currency. Furthermore, nucleotides act as cofactors for enzymes, as secondary messengers as well as they constitute biosynthetic precursors for generation of several cellular polymers. Nucleotide synthesis is performed in mitochondria from heterotrophic cells and in chloroplasts from autotrophic cells (Haferkamp et al., 2011). However, the use of nucleotides is widespread throughout the entire cell and some mechanisms must coordinate its absolute and relative levels in those compartments, and the relative levels of mono-, di-, and triphosphate forms (Zrenner et al., 2006). In this context, efficient transport systems across the different cellular compartments are required, and in fact, transport systems for nucleobases are found in plants in a large number (reviewed in Möhlmann et al., 2010 and Haferkamp et al., 2011).

Nucleotide metabolism has been more studied in animals than in plants due to its medical applications, in the development of anticancer drugs and research carried out in hereditary diseases (Zrenner et al., 2006). In plants, however, the knowledge about the biosynthetic routes, the degradation processes and recycling of nucleotides remains incomplete (Stasolla et al., 2003, Zrenner et al., 2006 and Werner and Witte, 2011). Nucleotide metabolism has a relevant function in some legumes as ureide precursors, which are molecules used for the transport of nitrogen from nodules to aereal parts. In ureide legumes, that include French bean and soybean, the ureides represent almost 90% of total nitrogen transported through xyleme under nitrogen fixing conditions (Todd et al., 2006 and Diaz-Leal et al., 2012).

Nucleotide metabolism can be divided into four processes: de novo synthesis, salvage pathways, nucleotide degradation and phosphotransfer reactions (Zrenner et al., 2006). Whereas the de novo synthesis of nucleotides is very energy consuming, the utilization of preformed bases and nucleosides through the salvage strategy needs less energy (Zrenner et al., 2006). The salvage pathways may be important in certain tissues, especially in non-growing cells that may be able to maintain their pools of nucleotides by salvaging previously used nucleosides with little de novo synthesis. In contrast, de novo synthesis of nucleotides is the main pathway used by growing and dividing cells. Despite its important contribution to plant metabolism, many aspects of nucleotide metabolism are still poorly understood, perhaps because the complexity of the multiple biochemical reactions involved in de novo synthesis, degradation an conversion of nucleotides, nucleosides and nucleobases.

The salvage and de novo synthesis converge at the formation of nucleoside monophosphates, which are the potential substrates of 5′-nucleotidases, enzymes that catalyse the first step in the degradative pathway. Nucleoside kinases catalyse the reverse reactions to 5′-nucleotidases, and these opposite reactions exist for almost all the nucleoside–nucleotide pairs (Bogan and Brenner, 2010). The relative activity of phosphorylation versus dephosphorylation could be the responsible of the balance between synthesis and degradation (Bogan and Brenner, 2010). It has been postulated that 5′-nucleotidase activities are regulated according to the nucleotide–nucleoside ratio of the cells (Bogan and Brenner, 2010). Despite the recent progress in genes and enzymes involved in the synthesis of ureides, the phosphatase activity that catalyses the dephosphorilation of nucleotides in the degradation pathway of nucleotides has not been yet identified. Similarly, the information about genes encoding 5′-nucleotidase in plants is very scarce, despite the availability of the complete genomes of several plant species. The use of databases to identify conserved domains can be the only clue to identify the cellular or molecular function of a protein, because they indicate local or partial similarity to other proteins, some of which may have been experimentally characterized.

The purpose of the present study has been to identify by MALDI TOF/TOF the purified enzyme that catalyses dephosphorylation of purine nucleotides and to study the nucleotidase under different physiological conditions at gene expression level and determination of nucleotidase activity in crude extracts.

2. MATERIALS AND METHODS

2.1. Plant material and growth condition

Common bean (*Phaseolus vulgaris* L. cv. Great Northern) seeds were surface-sterilised in ethanol (30 s) and 0.2% (*w/v*) sodium hypochlorite (5 min) and then washed thoroughly with distilled water. The soaked seeds were allowed to germinate in Petri dishes (120 mm diameter) with wet paper in sterile conditions. The moisture was maintained by regularly adding solutions to the dishes. Three days after germination, seeds were sown on vermiculite:perlite (2:1, *w/w*) pots (16 cm diameter, 18 cm height). Seedlings were inoculated with a fresh suspension of Rhizobia, which had been incubated at low temperatures (28 °C) for less than 30 h. Plants were cultured in a growth chamber under a long-day photoperiod (16 h light, 8 h dark), 300 µmol m^{-2} s^{-1}, and 70% relative humidity at 26–21 °C (day–night temperatures). Inoculated plants were watered three times a week with nitrogen-free nutrient solution (modified form Harper and Gibson, 1984) (0.10 mM KH_2PO_4, pH 6.5; 0.30 mM K_2SO_4; 0.20 mM $MgSO_4$; 0.11 mM $CaCl_2$; 0.25 mM $FeSO_4$; 46 µM H_3BO_3; 12.30 µM $MnSO_4$; 0.77 µM $ZnSO_4$; 0.32 µM $CuSO_4$ and 0.63 µM Na_2O_4Mo). Plant material was collected 3 h into the light period at indicated days after the treatments, frozen with liquid nitrogen and stored at −80 °C until further analysis.

To carry out the analysis during germination and seedling development, the seedlings were maintained in Petri dishes and the embryonic axes were isolated at 1–5 days after the start of imbibition.

To perform the analysis in adult plants, the following samples were collected in plants after 28 days of growth: roots after carefully removal of nodules, mature and immature leaves, shoots after removal of leaves. The pods were obtained from plants in the reproductive stage.

Studies during the development of the leaves were done using plants at the vegetative stage and harvesting only the fourth trifoliate leaves. Leaves were classified into four stages based in the length of the middle leaflet of the fourth trifoliate leaves: A first stage in which the middle leaflet ranged between 5 and 20 mm, a second stage between 20 and 40 mm, a third stage comprising leaves in which the middle leaflet ranged between 40 and 60 mm length; and a fourth stage with leaves completely expanded and developed. All the samples were harvested before the plants reached the reproductive stage.

2.2. Preparation of crude extracts

Crude extracts were obtained from frozen plant material previously stored at −80 °C. Plant material was ground to a fine powder with liquid nitrogen. Plant extracts were obtained by adding 5 ml of extraction buffer 50 mM TES (pH 7.0) and sodium deoxycholate (0.15%, *w/v*) per gram of tissue. The resulting homogenate was centrifuged at 15 000 × *g* for 10 min at 4 °C and the supernatant was considered as the crude extract.

2.3. Identification of a protein with nucleotidase activity by fingerprint

The 5′-nucleotidase from common bean embryonic axes was purified to electrophoretic homogeneity as previously indicated (Cabello-Diaz et al., 2012). The purified protein was subjected to mass spectrometry MALDI–TOF/TOF, at the Proteomic Unit of "Servicio Central de Apoyo a la Investigación" (SCAI, University of Córdoba). The protein purified was separated by SDS-polyacrylamide gel electrophoresis, the gel stained with Coomassie brilliant blue and the corresponding band excised and analysed by using an equipment 4700 Proteomics Analyzer, Applied Biosystems.

The identification of the protein was performed using GPS (Global Protein Server) with data TOF–TOF, combining both types of spectra (MS–MS/MS) and using as search engine MASCOT (MatrixScience, UK). A BLAST search against the Phytozome database with the amino acids sequence from protein identified by fingerprint was carried out.

2.4. RNA isolation and cDNA synthesis

Total RNA was isolated from several tissues using the TRI REAGENT®RNA/DNA/Protein Isolation Reagent (Molecular Research Center). Prior to RT-PCR, total RNA was treated with RNAase-free DNAseI (Promega) at 37 °C for 30 min to eliminate any traces of genomic DNA. First strand cDNA synthesis was performed with 2.5 µg of RNA using a iScript cDNA synthesis kit (Bio-Rad).

2.5. Cloning into the pET30b(+) expression vector

The plasmid vector pGEM-T (Promega) and *Eschirichia coli* DH5αF́ cells were used for standard cloning. To overexpress the recombinant protein, *E. coli* BL21 cells and the plasmid vector pET30b(+) (Novagen) were used. The plasmid pET30b(+), harbors the codons to add to one or both ends of the recombinant protein a histidine tag, which allows purification and identification. The ORF region of PvNTD1 (Phvulv091001267m) was amplified with the primers NTE1 (5'-GTCGACATGGATTCCGGC GTGTGTC TGTTG-3') containing a restriction site for SalI and NTE2 (5'-TGCGGCCGCCTATGGAAT ATAGTACATT-3') that contains a restriction site for Not I. PCR amplification was performed by using a Advantage 2 PCR kit (Clontech). The PCR program used was 95 °C for 2 min, followed by 35 cycles of 95 °C for 1 min, 70 °C for 1 min and 72 °C for 90 s, and 10 min final extension at 72 °C during 10 min. The PCR product was purified using QIAquick kit (Qiagen) and ligated into the cloning vector pGEM-T. The ligation product was used to transform *E. coli* DH5alfaF́ cells. The plasmid pGEM-T containing the cloned sequence and the plasmid pET30b(+) were digested with restriction enzymes NotI and Sal I. After electrophoresis, both DNA fragments were purified using QIAquick kit (Qiagen), ligated and the final construct was inserted into *E. coli* BL21 cells (Novagen). The final construct was sequenced to confirm the absence of mutations.

2.6. Expression and purification of the recombinant protein

The transformed cells were grown in 3 ml LB media supplemented with 50 mg/l kanamycin and grown overnight at 37 °C. After, 1 ml of the overnight grown culture was added to 50 ml LB media supplemented with 50 mg/l kanamycin and incubated at 37 °C until OD_{600} of 0.5–0.6 was achieved. Isopropyl-β-D-thiogalactopyranoside (IPTG) at a final concentration of 0.5 mM was added and the cultures were incubated at 37 °C for 2 h to induce heterologous protein expression. After induction, cells were harvested by centrifugation at 2000 × *g* and 4 °C for 10 min and the pellet stored at −20 °C until purification.

The pellet was resuspended in 4 ml of lysis buffer (50 mM TRIS–HCl pH 7.8; 1 mM PMSF; 0.5 M NaCl and 6 M urea) and cells were lysed by sonication on a Vibra Cell (Sonics and Materials Inc.) using 6 pulses of 5 s each, at all times keeping the samples on ice. After sonication, the homogenate was centrifuged at 12 000 *g* for 15 min at 4 °C. The supernatant was used as crude extract.

The crude extract was passed through a Nickel Chelating Sepharose column (1 ml, GE Healthcare) equilibrated with lysis buffer. The column was washed with 5 column volumes of washing buffer (50 mM TRIS–HCl pH 7.8; 1 mM PMSF; 0.5 M NaCl; 6 M urea and 50 mM imidazole). The protein was eluted with 2 column volumes of elution buffer (50 mM TRIS–HCl pH 7.8; 1 mM PMSF; 0.5 M NaCl; 6 M urea and 0.5 M imidazole).

The eluted protein was sequentially dialysed in 12 h periods of the following dialysis buffers: buffer I (50 mM Tes-NaOH pH 7; 1 mM MgCl; 0.25 M NaCl; 4 M urea; 6 mM cysteine and 10% glycerol), buffer II (50 mM Tes-NaOH pH 7; 1 mM MgCl; 0.25 M NaCl; 2 M urea; 3 mM cysteine and 10% glycerol) and buffer III (50 mM Tes-NaOH pH 7; 1 mM MgCl; 0.25 M NaCl and 2.5% glycerol). The dialysed protein was keep at −20 °C until use.

2.7. Gel electrophoresis, Western blot analysis and protein stain

Proteins were separated in 10% SDS-PAGE gels by using a constant voltage (100 V) in a Mini PROTEAN III system (Bio-Rad). After separation, proteins were electro-transferred to PVDF (polyvinylidene fluoride) membrane (Sigma–Aldrich). To detect His-tagged protein, blots were incubated with a monoclonal anti-polyhistidine antibody (H1029, Sigma–Aldrich) at 1:10000 dilution. Anti-mouse IG, alkaline phosphatase conjugated (A3562, Sigma–Aldrich) was used as secondary antibody at a 1:12500 dilution and the immunoreaction was developed with 5-bromo-4-chloro-3-indoyl phosphate *p*-toluidine salt and nitro-blue tetrazolium chloride as substrates. Coomassie-blue stain was performed as described in Zehr et al. (1989) and silver stain was performed as indicated by Heukeshoven and Dernick (1985).

2.8. Enzymatic activity

Nucleotidase activity was determined by monitoring the phosphate concentration in reaction mixtures according to Katewa and Katyare (2003). Unless otherwise stated, the standard reaction mixture for recombinant protein purified from *E. coli* contained 50 mM MES–HCl buffer (pH 5.5), 1 mM $MgCl_2$, 5 mM substrate and an adequate amount of enzyme. The standard reaction mixture for nucleotidase activity in plant crude extracts contained 50 mM TES–NaOH buffer (pH 7.0), 1 mM $MgCl_2$, 5 mM molybdate, 0.5 mM AMP and an adequate amount of extract. The reactions were initiated by the addition of the enzyme and were performed at 37 °C. Aliquots of 0.2 ml were extracted at several times and placed into tubes containing 0.6 ml of 1 N sulphuric acid. After mixing, 0.1 ml of 2.5% (*w/v*) of ammonium molybdate (prepared in 3 N sulphuric acid) and 0.1 ml of reducing solution (2% (*w/v*) of ascorbic acid and 2% (*w/v*) of hydrazine sulphate, prepared in 0.1 N sulphuric acid) were added. The samples were shaken and incubated at room temperature for 50 min. After, the absorbance at 820 nm was determined.

One unit of enzyme is defined as the amount of enzyme that catalyses the formation of 1 µmol of phosphate per minute.

2.9. Real time PCR

Quantitative RT-PCR (qRT-PCR) was carried out with an iCycler iQ system using the iQ SYBRGreen supermix (Bio-Rad) and the PvNTD1 gene-specific primers NTD-D1 (5′- GAAGTGGTCGGGAATTTGTCG-3′) and NTD-R1 (5′-ATCTCAAACCCGAATCCAATGTC-3). The PCR program consisted of an initial denaturation of 5 min at 95 °C, followed by 40 cycles of 15 s at 95 °C, 30 s at 60 °C and 30 s at 72 °C. After the final cycle, a melting curve analysis was performed over a temperature range of 60–100 °C by increments of 0.5 °C in order to verify the reaction specificity. Analysis of relative gene expression was achieved by using *Actin* -2 (gene ID. EU369189) as reference gene, and Actin-D1 (5′-GGAGAAGATTTGGCATCACACGTT-3′) as forward and Actin-R1 (5′-GTTGGCCTTGGGATTGAGTGGT-3′) as reverse primers. Results were normalized as previously described by Galvez-Valdivieso et al. (2013).

2.10. Analytical determination

Protein concentration was determined by the Bradford method (Bradford, 1976) using the Bio-Rad system and bovine serum albumin as the standard. The concentration of ureides (allantoin and allantoate) was determined by the colorimetric analysis of glyoxylate derivatives as described previously by Quiles et al. (2009).

2.11. Statistical analyses

All results are means of three independent experiments with two technical replicates for experiment. Values are mean ± SE. Statistical analyses were performed with SPSS Statistics, version 17.0, using unifactorial ANOVAs. The homoscedasticity was determined using the Levene test, and least significant difference (LSD) statistics were applied for variables with homogeneity of variance. In the figures, the letters represent significant differences ($P \leq 0.05$) between the studied conditions.

3. RESULTS

3.1. Identification of a cDNA encoding a putative nucleotidase

The molybdate resistant phosphatase purified from French bean seedlings was subjected to SDS-PAGE analysis. The Coomassie stained band was excised from the gel and used for MALDI TOF/TOF analysis. The peptide mass fingerprinting obtained after trypsin digestion did not give positive results in the database search. However, some of the peptides that showed high intensity were further analysed by mass spectrometry. Using the program PEAKS as searching engine, two peptide sequences were identified: AVETNNAGPWER and VPANCVDFVAEYISGE (Supplementary

Figure1A). These peptide sequences were used to search in the *P. vulgaris* database (www.phytozome.org) and the *Phvul.004G174200* gene was identified. We named the gene as*PvNTD1*. The *PvNTD1* full-length cDNA sequence is 1063 bp in nucleotide length, and its ORF was 756 bp long, encoding a 251 amino acids peptide with a predicted molecular weight of 28.5 kDa (Supplementary Figure1B). The phosphatase purified from French bean has a molecular weight of 55 kDa when analysed by SDS-PAGE under reducing conditions. Nevertheless, the treatment of purified protein with 8 M urea and heat for 15 min under reducing conditions resulted in a single band of 28 kDa (Figure1).

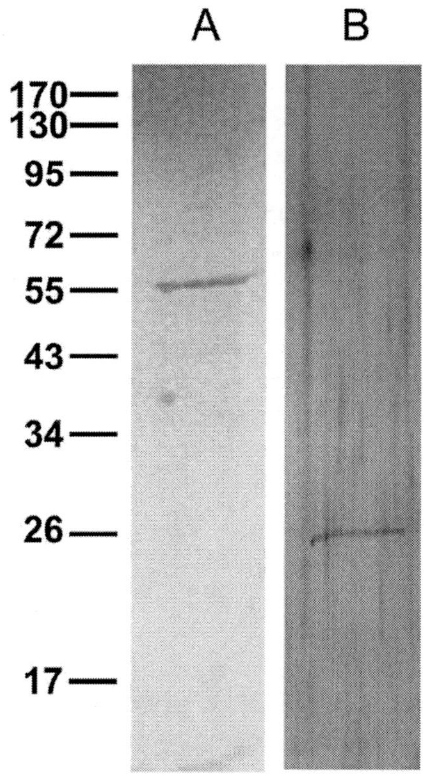

Figure 1. Urea effect of purified protein A) The purified protein was mixed with loading buffer containing SDS, DTT and boiled for 5 min. B) The purified protein was mixed the same loading buffer as above but supplemented with 8 M urea and was boiled for 15 min. Both samples were loaded onto SDS-PAGE and after electrophoresis the gel were silver stained. The positions of the molecular weight markers are showed at the left of the figure.

According to TargetP program, the predicted protein follows the secretory pathway. Further, PvNTD1 contained the domains conserved for haloacid dehalogenase-like hydrolases superfamily, consisted of motifs I–IV (Figure2). The alignment of the deduced amino acid sequence with other homologous sequences identified in the GENBANK database using BLAST searches is shown in Figure2. This alignment revealed that the deduced sequence for *PvNTD1* shares high identity (78%) with the protein from *Glycine max*.

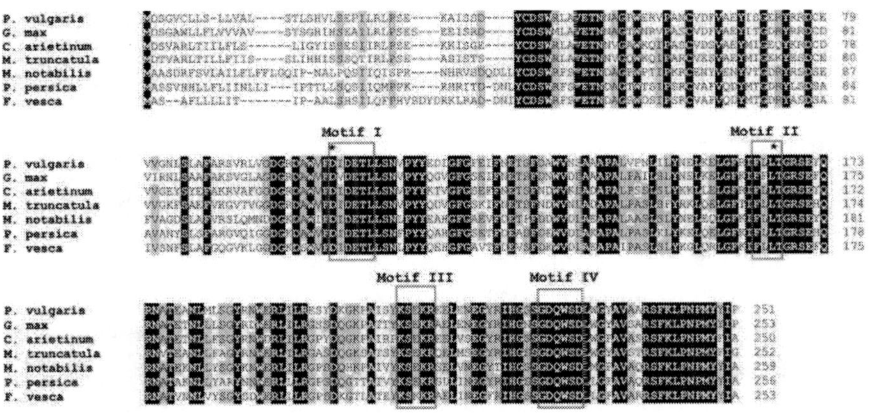

Figure 2. Alignment of amino sequences of nucleotidase identified in *Phaseolus vulgaris* (XP_007152955.1), with the sequences from *Glycine max* (XP_003529064.1), *Cicer arietinum* (XP_004513069.1), *Medicago truncatula* (AFK34288.1), *Morus notabilis* (EXC28686.1), *Prunus persica* (XP_007202474.1), *Fragaria vesca* (XP_004287494). Black boxes indicate the amino acid residues that are identical in all sequences and grey boxes indicate the amino acid residues identical at least in five sequences. Grey boxes mark off the characteristic motifs, M (I), M (II) M (III) and M (IV) of the HAD superfamily (Burroughs et al., 2006). The stars identify the residues associated with the active site. Protein sequences were retrieved from NCBI or Phytozome platforms and aligned using the ClustalW method.

3.2. Heterologous expression

The ORF of *PvNTD1* was cloned into vector pET30b(+) and then transformed into *E. coli*BL21 (DE3) cells. Expression of the recombinant protein was induced by IPTG and the purified protein was analysed by SDS-PAGE and Western blot. These analyses revealed that the overproduced enzyme was mainly in the insoluble fraction (inclusion bodies). Although the amount of protein in the soluble fraction was near negligible, there was an increase in nucleotidase activity in the crude extracts obtained from induced bacterial culture.

The recombinant protein was purified to near homogeneity by a single step consisting of a nickel chelating resin chromatography after adding 6 M urea as denaturing agent through the purification process (Supplementary Figure2). The purified recombinant protein lost the phosphatase activity. However, after a refolding process consisting of several steps of gradual dialysis to remove urea this activity was recovered.

3.3. Characterization of the recombinant protein

The purified protein showed phosphatase activity with all the nucleotides assayed (IMP, GMP, AMP, CMP and UMP) and with pNPP, a non-specific substrate used to assay alkaline and acid phosphatases (Table 1). The resulted activity was similar with all compounds assayed except for AMP when pH 5.5 was used. With AMP as substrate the enzyme showed a different optimum pH, with the maximum activity at pH 7.0. The activity in the presence of the mentioned substrates was not inhibited by molybdate addition to the assay mixtures.

The phosphatase activity of the purified recombinant protein was determined in the presence of several nucleosides (Table 2). The inclusion of adenosine led to the inhibition of the activity independently of the substrate used, and the inhibition was higher than the obtained with other nucleosides (Table 2). Inhibition ratio was similar independently of the nucleotides assayed (data not shown). The recombinant protein was unable to use ADP or ATP as substrate, and these compounds did not show effect on nucleotidase activity assayed with IMP as substrate (data not shown). The inhibition displayed by adenosine followed a dose-dependent curve, with 50% inhibition at 0.2 mM adenosine (Figure3). Only some residual activity was observed at high concentration of adenosine. The same behaviour was observed when nucleotidase purified from embryonic axes was assayed corresponding to an uncompetitive inhibition type (data not shown).

Table 1. Substrate specificity of recombinant phosphatase. Enzymatic activity with the listed substrates at a final concentration of 5 mM was assayed as described in Section 2. Activity with IMP was taken as 100%. Data are mean values (±SE) of four independent determinations. Different letters indicate statistically significant differences according to LSD analysis ($P \leq 0.05$).

Substrate	Activity
IMP	100 ± 12 (ab)
GMP	98 ± 7 (ab)
AMP	60 ± 7 (c)
UMP	92 ± 9 (a)
CMP	88 ± 9 (a)
p-NPP	112 ± 13 (b)

3.4. Expression of the 5′-nucleotidase gene and nucleotidase activity in French bean

Expression of *PvNTD1* gene and nucleotidase activity were analysed in embryonic axes during germination and early seedling development (Figure4). In the assay conditions, the radicle protrusion occurred in the second day after the start of imbibition. The expression was negligible during germination (day 1), meanwhile *PvNTD1* mRNA expression was detected at the moment of radicle protrusion (day 2) reaching the maximum level at 3 days of development (Figure4A). Nucleotidase activity, assayed with AMP as substrate, was detected at 24 h after start of imbibition and the activity increased with axes development (Figure4B), coincidently with the ureide concentration (Figure4C).

Expression level of *PvNTD1* and nucleotidase activity was also analysed in different organs of 28 days old nodulated common bean plants (in the vegetative stage) and in pods of plants in the reproductive stage (Figure 5). *PvNTD1* transcript was detected in all organs analysed, although its expression was particularly high in the developing leaves. A correlation of gene expression with nucleotidase activity and ureide concentration has been found. Since *PvNTD1* expression was particularly high in youngest leaves, tissues with high demand of nutrients, nucleotidase was analysed through leaf ontogeny process (Figure 6). Data corroborated high expression level in youngest leaves with a decrease of gene expression through leaf development (Figure 6). The expression pattern of *PvNTD1*was coincident with the nucleotidase activity and ureide levels (Figure 6).

Table 2. Effect of nucleosides in the recombinant phosphatase. The activity was determined with the indicated substrates in the absence (control) or the presence of the indicated nucleoside. Activity in the absence of nucleoside was taken as 100%. Data are mean values (±SE) of four independent determinations. Different letters indicate statistically significant differences according to LSD analysis ($P \leq 0.05$).

	IMP (5 mM)	CMP (5 mM)	AMP (1 mM)	p-NPP (5 mM)
Control	100 ± 9 (a)	100 ± 8 (a)	100 ± 7 (a)	100 ± 8 (a)
+ Inosine (0.2 mM)	89 ± 10 (ab)	79 ± 9 (b)	92 ± 8 (ab)	104 ± 9 (ab)
+ Adenosine (0.2 mM)	32 ± 4 (c)	34 ± 4 (c)	32 ± 4 (c)	34 ± 6 (c)
+ Guanosine (0.2 mM)	79 ± 12 (b)	81 ± 9 (b)	83 ± 8 (b)	86 ± 9 (b)

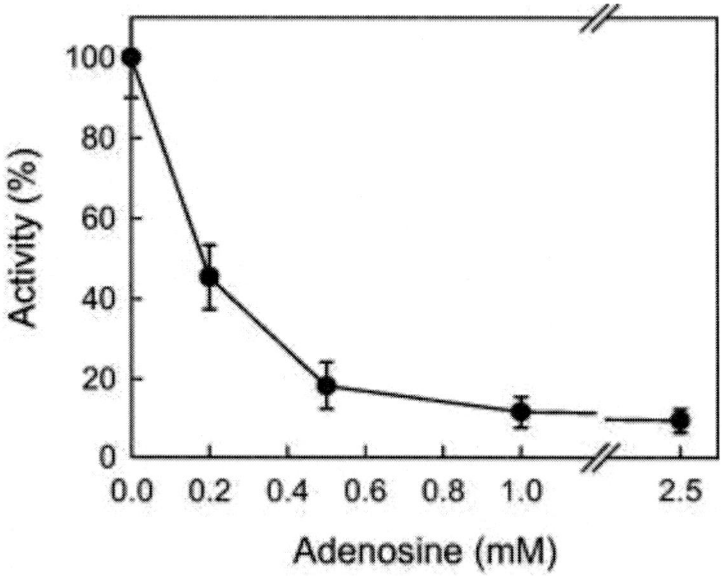

Figure 3. Inhibition by adenosine. Phosphatase activity was determined with 5 mM IMP as substrate in the presence of the amount of adenosine indicated.

4. DISCUSSION

The French bean is a legume that relies on ureide metabolism to obtain its nitrogen when growing under nitrogen fixing conditions. The involvement of ureide in some important physiological processes in French bean has been recently studied (Raso et al., 2007,Quiles et al., 2009, Alamillo et al., 2010 and Diaz-Leal et al., 2012). The first step to ureide synthesis from purines is the removal of the phosphate group. Nevertheless, the enzyme responsible for this step has not been properly identified to date. The difficulty to identify the function of each acid phosphatase is the large variety of phosphorylated substrates that can use these phosphatases (Duff et al., 1994) as well as the large number of enzymes present in plants. Several acid phosphatases that can metabolize nucleotides have been described in ureide legumes (Penheiter et al., 1997, Ferreira et al., 1998, Garcia et al., 2004, Yoneyama et al., 2004 and Liang et al., 2010), although only the phosphatase activity from soybean nodules shows high affinity for nucleotides (Penheiter et al., 1997).

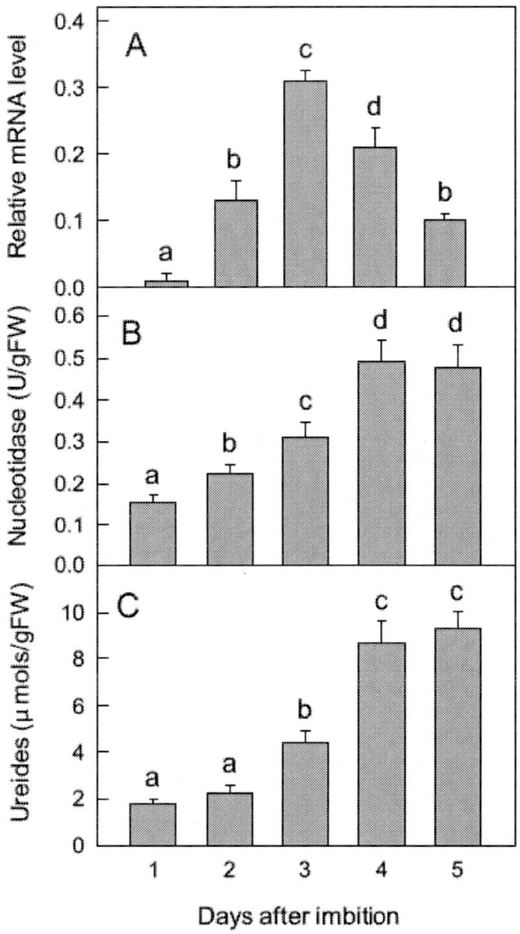

Figure 4. Relative expression levels of *PvNTD1*, nucleotidase activity and ureide concentration in embryonic axes during germination and early seedling development. A) *PvNTD1* gene expression analysis was performed using real-time qRT-PCR on total RNA samples extracted from embryonic axes of French bean at the time after start of imbibition indicated. Each value was normalised with the relative level of expression of the actin-2 gene. Primers used for PCR analyses are indicated in Material and Methods section. B) Nucleotidase activity was determined with AMP as substrate and in the presence of molybdate from crude extracts obtained from the same plant material. C) Ureide concentration was determined in the same plant crude extract. Values are mean ± SE of three independent experiments, with two technical replicates for experiment. Different letters indicate statistically significant differences according to LSD analysis ($P \leq 0.05$).

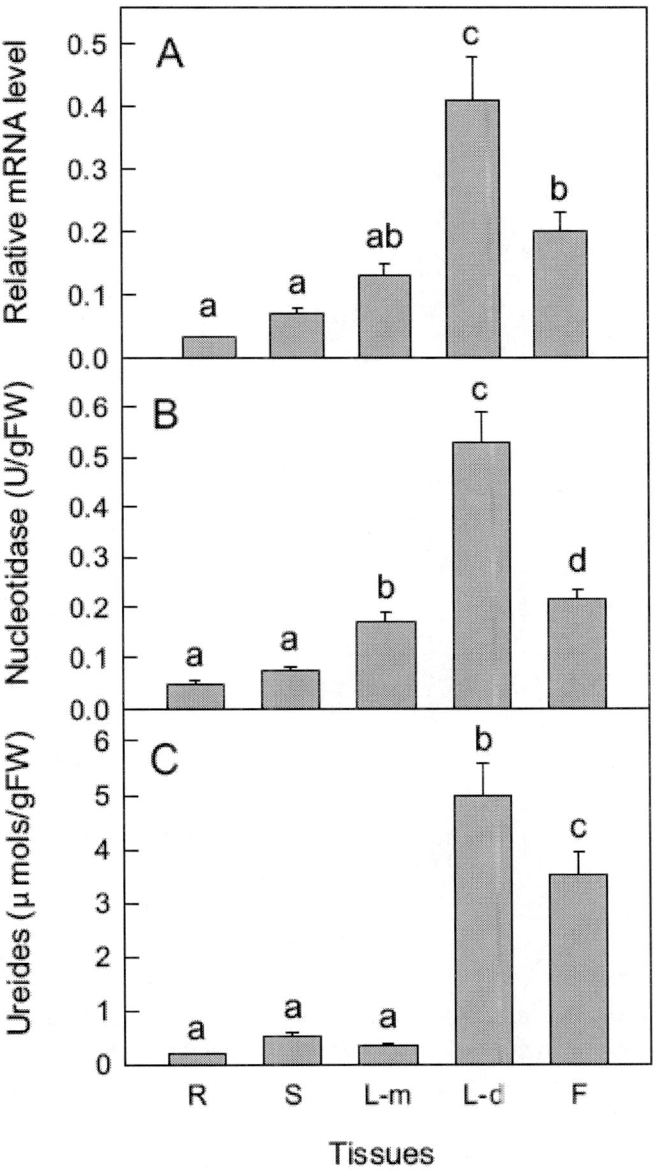

Figure 5. A) Relative expression levels of PvNTD1, B) nucleotidase activity and C) ureide concentration in different tissues of *Phaseolus vulgaris* adult plant were determined as indicated in Figure 4. Different letters indicate statistically significant differences according to LSD analysis ($P \leq 0.05$). Plant material was obtained from roots (R), stems (S), mature leaves (L–m), developing leaves (L–d) and mature pods (P) as indicted in Section 2.

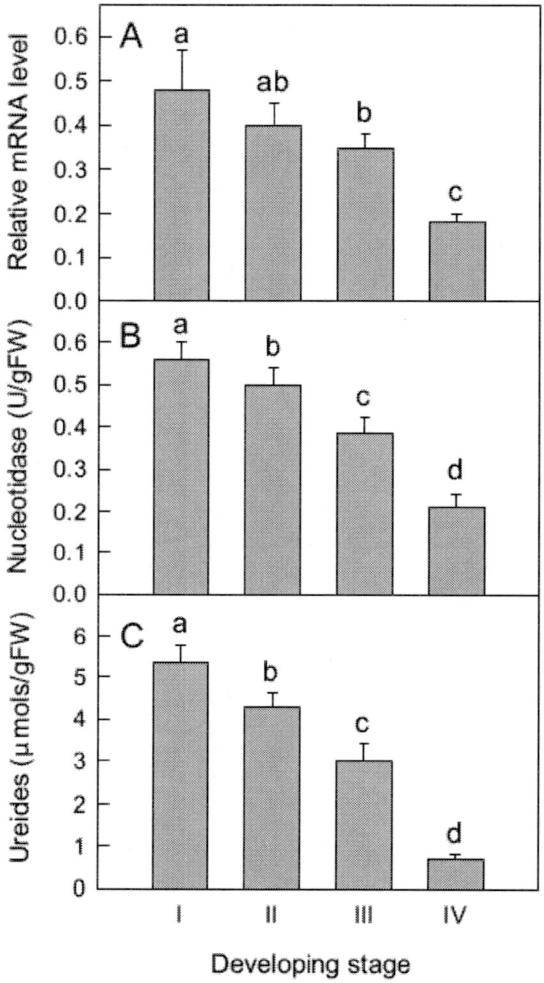

Figure 6. A) Relative expression levels of *PvNTD1*, B) nucleotidase activity and C) ureide concentration in *Phaseolus vulgaris* leaves during leaf ontogeny were determined as indicated in Figure4. Different letters indicate statistically significant differences according to LSD analysis ($P \leq 0.05$). Plant material was obtained from the stages of leaf development as indicated in Section 2.

The information regarding the genes encoding nucleotidases in plants is very scarce, besides availability of whole genome sequences in some species. The use of databases may allow the identification of conserved domains that can be the only evidence to identify the cellular or molecular function of a protein (Marchler-Bauer et al., 2011). In plants, the genes predicted to encode nucleotidases, on the basis of homology to the human or mouse genes, have not been characterized. Therefore, we designed an approach to identify the nucleotidase gene through the analysis of the protein previously purified. We have purified and characterised a phosphatase activity with high affinity for nucleotides (Cabello-Diaz et al., 2012). The purified protein from French bean seedlings has been identified by MALDI–TOF/TOF as an acid phosphatase. However, the purified protein had a molecular weight of 55 kDa (Cabello-Diaz et al., 2012), whereas the identified cDNA codes for a 28 kDa protein. To obtain monomers in the purified protein, 8 M urea was added to the sample buffer before loading the purified protein (Figure2).

The analysis of the amino acid sequence by using conserved domain database and search service (v2.16) revealed the presence of four conserved domains among the superfamily of "Haloacid dehalogenase-like hydrolases". In this group are found enzymes such as hydrolases, phosphatases, nucleotidases and several phosphotransferases (Bogan and Brenner, 2010). All of them contain a nucleophilic aspartate in the phosphate transfer reaction. The deduced amino acid sequence of*PvNTD1* showed the four characteristic domains. The domains M(I) (DxDx[T/V][L/V/I]), M(II) (§ § § [T/S], being § hidrophobic aminoacids), M(III) (K[§ § §], being § amino acids involved in an alpha-helix) and M(IV) (DD or GDx(3-4)D), characteristic of the HAD superfamily (Burroughs et al., 2006). However, neither the S motif (Px(7-8)[R/K]xGF[W/L]) described for human nucleotidases (Rinaldo-Matthis et al., 2002) nor the domains described for UMPH family (5′-nucleotidase for pyrimidines) have been identified in French bean nucleotidase. The *PvNTD1* gene product showed homology with vegetative storage proteins (VSPs) in plants, which function as temporary nitrogen storage in vegetative tissues. The VSP lack phosphatase activity due to the absence of the nucleophilic Asp residue (Leelapon et al., 2004), which is present in PvNTD1 sequence (Figure3). The deduced amino acid for the *PvNTD1* also showed identity to a new phosphatase acid induced by wounding and herbivore attack in hybrid poplar (Veljanovski et al., 2010). The highest identity was found to the gene that codes for the protein purified from soybean nodules (Penheiter et al., 1997).

The whole coding region of *PvNTD1* was over-expressed as recombinant protein in *E. coli*, although most of the protein was associated with inclusion bodies. The denaturation with urea and subsequent refolding yielded an active protein. The recombinant protein showed phosphatase activity with similar properties to those described for the protein purified from embryonic axes, i.e. activity with all the nucleotides, the enzyme showed a optimum pH of 7.0 with AMP whereas with other nucleotides the maximal activity was obtained at pH 5.5, the activity was not unaffected by molybdate or by phosphate, whereas it was inhibited by adenosine. These data indicate that the *PvNTD1* gene encodes a functional nucleotidase.

Nucleotidase activity in plant crude extracts has been determined with AMP as substrate and in the presence of molybdate to inhibit unspecific phosphatases. Among the nucleotides assayed, AMP was hydrolysed more efficiently by the purified protein from embryonic axes since it showed the highest specific constant (Cabello-Diaz et al., 2012). Furthermore, it is of particular interest that the native enzyme showed an optimum pH of 7.0 for the activity with AMP, whereas the pH optima for other nucleotides ranged between 5.5 and 6.0 (Cabello-Diaz et al., 2012). Among nucleotides, adenine pool has some relevant physiological considerations. The adenine nucleotide pool is always the largest (Stasolla et al., 2003). The adenylated compounds play important roles in many metabolic pathways of the plant and therefore, specific carriers are required in nearly every organelle to connect the sites of their production with those of their consumption (Haferkamp et al., 2011). The novo synthesis of purines occurs on the plastids and it is postulated that purines goes to the cytoplasm as AMP, being the precursor of the rest of purine nucleotides (Zrenner et al., 2006).

A physiological process where the nucleotides must be strictly regulated occurs during germination and postgerminative growth (Möhlmann et al., 2010). In fact, the availability of nucleotides during early phases of imbibition seems to be critical for successful embryo germination (Stasolla et al., 2003). Nucleotides are required in bio-energetic processes for the mobilization of storage products and as building blocks for nucleic acid synthesis in developing embryos (Stasolla et al., 2003). Therefore, during germination a tight regulation between de novo, salvage and degradation pathways of both purines and pyrimidines must exits. During the first stages in germination, nucleotides are believed to be obtained from nucleobases and nucleosides via salvage pathways (Möhlmann et al., 2010). With the progression of germination, the nucleotides synthesised de novo acquire relevance, since they are highly required for nucleic acids synthesis in the dividing cells (Bewley, 1997). The identified *PvNTD1* gene is induced during early seedling development (Figure6A) and codes for a phosphatase with high affinity for nucleotides monophosphate (Cabello-Diaz et al., 2012). A nuclease activity with high activity with single stranded DNA or

RNA is induced as well after radicle protrusion (Lambert et al., 2014) in parallel with an accumulation of ureides and an induction of its metabolism (Quiles et al., 2009). The nucleic acid stored during seed formation could be a source for these nucleotides, and therefore could be considered as a potential valuable phosphorous and nitrogen storage compound. This is particularly significant for species with large genomes, and for many plants with high DNA content in seeds as result of endoreduplication, a process during which the nuclei undergo repeated rounds of DNA replication without mitosis. Interestingly, *P. vulgaris* seeds contain endopolyploid nuclei in cotyledons and embryonic axes (Rewers and Sliwinska, 2014).

The level of *PvNTD1* expression and nucleotidase activity correlated in tissues from mature plants with ureide levels (Figure 5). In nitrogen fixing conditions, French bean depends on ureide supply to obtain all the nitrogen required in developing tissues. French bean shows maximum nitrogen fixation levels at 28 days after sowing, just before the onset of flowering (Diaz-Leal et al., 2012). At that point, nucleotidase activity and*PvNTD1* gene expression were very low in all tissues except in developing leaves.*PvNTD1* expression and nucleotidase activity were relatively high in developing fruits as well (Figure 5), tissues with high metabolic rate and high demand of nutrients, supplied in that case both by xylem and phloem. In non-growing cells, the demand of nucleotides is very low and the salvage routes may be able to maintain their pool of nucleotides with little de novo synthesis. However, in developing tissues, the demand for nucleotides increase significantly and de novo synthesis of these compounds could be the main pathway used in growing cells. The high level of *PvNTD1* gene in early developing leaves, could indicate its role in de novo synthesis pathway instead of the degradative pathway. We consider this a plausible hypothesis. The cellular compartment where de novo synthesis of AMP and the synthesis of the other nucleotides might be different. For instance, de novo synthesis of purines in chloroplast ended in MP, which is converted to AMP to be exported to cytosol where other nucleotides are synthesised. In other cases, since nucleosides and nucleobases are more easily transported that nucleotides (Möhlmann et al., 2010 and Haferkamp et al., 2011), the diffusion and transport processes might require dephosphorylation. Furthermore, it has been proposed that even if adenosine is taken by the cells, it may be also hydrolysed to adenine within the cells to be further salvaged (Deng and Ashihara, 2010). The diversity of adenosine salvage in plant species is striking (Deng and Ashihara, 2010) and it has been postulated that plants might use specific purine pathways that differ to those described in other species (Dahncke and Witte, 2013). Detailed studies on the transport processes of nucleobases, nucleosides and nucleotides across membranes as well as the subcellular localization of the enzymes that convert them are needed.

The understanding of nucleotide metabolism in plants remains very limited besides the crucial role of these compounds in plant physiology. We report here the identification of a gene that encodes a nucleotidase representing the main activity able to use nucleotides monophosphate in French bean embryonic axes during early seedling development. This nucleotidase is a acid phosphatase that displays novel properties among its family, such as the lack of inhibition by molybdate and the inhibition by adenosine. The *PvNTD1*gene that codes for the nucleotidase is induced in tissues with high demand for nucleotides. Although we show evidence suggesting its role in the de novo pathway, further investigation needs to be carried out to confirm such statement.

ACKNOWLEDGMENTS

The authors thank Marta Robles for her technical assistance. This work was supported by Ministerio de Ciencia e Innovación (AGL2012-34230), Consejería de Economía, Innovación y Ciencia de la Junta de Andalucía (CVI1761), and Plan Andaluz de Investigación (BIO-115).

REFERENCES

1. Alamillo, J.M., Diaz-Leal, J.L., Sanchez-Moran, M.V., Pineda, M., 2010. Molecularanalysis of ureide accumulation under drought stress in Phaseolus vulgaris L.Plant Cell Environ. 33, 1828–1837.

2. Bewley, J.D., 1997. Seed germination and dormancy. Plant Cell 9, 1055–1066.

3. Bogan, K.L., Brenner, C., 2010. 5_-Nucleotidases and their new roles in NAD(+)andphosphate metabolism. New J. Chem. 34, 845–853.

4. Bradford, M.M., 1976. A rapid and sensitive method for the quantitation ofmicrogram quantities of protein utilizing the principle of protein-dye binding.Anal Biochem. 72, 248–254.

5. Burroughs, A.M., Allen, K.N., Dunaway-Mariano, D., Aravind, L., 2006. Evolutionarygenomics of the HAD superfamily: understanding the structural adaptationsand catalytic diversity in a superfamily of phosphoesterases and alliedenzymes. J. Mol. Biol. 361, 1003–1034.

6. Cabello-Diaz, J.M., Quiles, F.A., Lambert, R., Pineda, M., Piedras, P., 2012.Identification of a novel phosphatase with high affinity for nucleotidesmonophosphate from common bean (Phaseolus vulgaris). Plant Physiol.Biochem. 53, 54–60.

7. Dahncke, K., Witte, C.-P., 2013. Plant purine nucleoside catabolism employs aguanosine deaminase required for the generation of xanthosine in Arabidopsis.Cell 25, 4101–4109, Plant.

8. Deng, W.W., Ashihara, H., 2010. Profiles of purine metabolism in leaves and rootsof Camellia sinensis seedlings. Plant Cell Physiol. 51, 2105–2118.

9. Diaz-Leal, J.L., Galvez-Valdivieso, G., Fernandez, J., Pineda, M., Alamillo, J.M., 2012.Developmental effects on ureide levels are mediated by tissue-specificregulation of allantoinase in Phaseolus vulgaris L. J. Exp. Bot. 63, 4095–4106.

10. Duff, S.M.G., Sarath, G., Plaxton, W.C., 1994. The role of acid phosphatases in plantphosphorus metabolism. Physiol. Plant. 90, 791–800.

11. Ferreira, C.V., Granjeiro, J.M., Taga, E.M., Aoyama, H., 1998. Purification andcharacterization of multiple forms of soybean seed acid phosphatases. PlantPhysiol. Biochem. 36, 487–494

12. Galvez-Valdivieso, G., Alamillo, J.M., Fernandez, J., Pineda, M., 2013. Molecularcharacterization of PVAS3: an asparagine synthetase gene from common beanprevailing in developing organs. J. Plant Physiol. 170, 1484–1490.

13. Garcia, N.A.T., Olivera, M., Iribarne, C., Lluch, C., 2004. Partial purification andcharacterization of a non-specific acid phosphatase in leaves and root nodulesof Phaseolus vulgaris. Plant Physiol. Biochem. 42, 585–591.

14. Haferkamp, I., Fernie, A.R., Neuhaus, H.E., 2011. Adenine nucleotide transport inplants: much more than a mitochondrial issue. Trends Plant Sci. 16, 507–515.

15. Harper, J.E., Gibson, A.H., 1984. Differential nodulation tolerance to nitrate amonglegume species. Crop Sci. 24, 797–801.

16. Heukeshoven, J., Dernick, R., 1985. Simplified method for silver staining of proteinsin polyacrylamide gels and the mechanism of silver staining. Electrophoresis 6,103–112.

17. Katewa, S.D., Katyare, S.S., 2003. A simplified method for inorganic phosphatedetermination and its application for phosphate analysis in enzyme assays.Anal. Biochem. 323, 180–187.

18. Lambert, R., Quiles, F.A., Cabello-Diaz, J.M., Piedras, P., 2014. Purification andidentification of an nuclease activity in embryo axes from French bean. PlantSci. 224, 137–143.

19. Leelapon, O., Sarath, G., Staswick, P.E., 2004. A single amino acid substitution insoybean VSP alpha increases its acid phosphatase activity nearly 20-fold.Planta 219, 1071–1079.

20. Liang, C., Tian, J., Lam, H.-M., Lim, B.L., Yan, X., Liao, H., 2010. Biochemical andmolecular characterization of PvPAP3, a novel purple acid phosphataseisolated from common bean enhancing extracellular ATP utilization. PlantPhysiol. 152, 854–865.

21. Möhlmann, T., Bernard, C., Hach, S., Neuhaus, H.E., 2010. Nucleoside transport andassociate metabolism. Plant Biol. 12, 26–34.

22. Penheiter, A.R., Duff, S.M.G., Sarath, G., 1997. Soybean root nodule acidphosphatase. Plant Physiol. 114, 597–604.

23. Quiles, F.A., Raso, M.J., Pineda, M., Piedras, P., 2009. Ureide metabolism duringseedling development in French bean (Phaseolus vulgaris). Physiol. Plant. 135,19–28.

24. Raso, M.J., Pineda, M., Piedras, P., 2007. Tissue abundance and characterization oftwo purified proteins with allantoinase activity from French bean (Phaseolusvulgaris). Physiol. Plant. 131, 355–366.

25. Rewers, M., Sliwinska, E., 2014. Endoreduplication in the germinating embryo andyoung seedling is related to the type of seedling

establishment but is notcoupled with superoxide radical accumulation. J. Exp. Bot. 65, 4385–4396.

26. Rinaldo-Matthis, A., Rampazzo, C., Reichard, P., Bianchi, V., Nordlund, P., 2002.Crystal structure of a human mitochondrial deoxyribonucleotidase. Nat. Struct.Biol. 9, 779–787.

27. Stasolla, C., Katahira, R., Thorpe, T.A., Ashihara, H., 2003. Purine and pyrimidinenucleotide metabolism in higher plants. J. Plant Physiol. 160, 1271–1295.

28. Todd, C.D., Tipton, P.A., Blevins, D.G., Piedras, P., Pineda, M., Polacco, J.C., 2006.Update on ureide degradation in legumes. J. Exp. Bot. 57, 5–12.

29. Veljanovski, V., Major, I.T., Patton, J.J., Bol, E., Louvet, S., Hawkins, B.J., Constabel,C.P., 2010. Induction of acid phosphatase transcripts, protein and enzymaticactivity by simulated herbivory of hybrid poplar. Phytochemistry 71, 619–626.

30. Werner, A.K., Witte, C.-P., 2011. The biochemistry of nitrogen mobilization: purinering catabolism. Trends Plant Sci. 16, 381–387.

31. Yoneyama, T., Shiozawa, M., Nakamura, M., Suzuki, T., Sagane, Y., Katoh, Y.,Watanabe, T., Ohyama, T., 2004. Characterization of a novel acid phosphatasefrom embryonic axes of kidney bean exhibiting vanadate dependentchloroperoxidase activity. J. Biol. Chem. 279, 37477–37484.

32. Zehr, B.D., Savin, T.J., Hall, R.E., 1989. A one-step, low background coomassiestaining procedure for polyacrylamide gels. Anal. Biochem. 182, 157–159.

33. Zrenner, R., Stitt, M., Sonnewald, U., Boldt, R., 2006. Pyrimidine and purinebiosynthesis and degradation in plants. Annu. Rev. Plant Biol. 57, 805–836.

Chapter 11

Mechanism of Salinity Tolerance in Plants: Physiological, Biochemical, and Molecular Characterization

Bhaskar Gupta[1] and Bingru Huang[2]

[1]Department of Biological Sciences (Section Biotechnology), Presidency University, 86/1 College Street, Kolkata 700073, India

[2]Department of Plant Biology and Pathology, Rutgers University, New Brunswick, NJ 08901, USA

ABSTRACT

Salinity is a major abiotic stress limiting growth and productivity of plants in many areas of the world due to increasing use of poor quality of water for irrigation and soil salinization. Plant adaptation or tolerance to salinity stress involves complex physiological traits, metabolic pathways, and molecular or gene networks. A comprehensive understanding on how plants respond to salinity stress at different levels and an integrated approach of combining molecular tools with physiological and biochemical techniques are imperative for the development of salt-tolerant varieties of plants in salt-affected areas. Recent research has identified various adaptive responses to salinity stress at molecular, cellular, metabolic, and physiological levels, although mechanisms underlying salinity tolerance are far from being completely understood. This paper provides a comprehensive review of major research advances on biochemical, physiological, and molecular mechanisms regulating plant adaptation and tolerance to salinity stress.

1. INTRODUCTION

A major challenge towards world agriculture involves production of 70% more food crop for an additional 2.3 billion people by 2050 worldwide [1]. Salinity is a major stress limiting the increase in the demand for food crops. More than 20% of cultivated land worldwide (~ about 45 hectares) is affected by salt stress and the amount is increasing day by day. Plants on the basis of adaptive evolution can be classified roughly into two major types: the halophytes (that can withstand salinity) and the glycophytes (that cannot withstand salinity and eventually die). Majority of major crop species belong to this second category. Thus salinity is one of the most brutal environmental stresses that hamper crop productivity worldwide [2, 3].

Salinity stress involves changes in various physiological and metabolic processes, depending on severity and duration of the stress, and ultimately inhibits crop production [4–7]. Initially soil salinity is known to represses plant growth in the form of osmotic stress which is then followed by ion toxicity [4, 5]. During the initial phases of salinity stress, water absorption capacity of root systems decreases and water loss from leaves is accelerated due to osmotic stress of high salt accumulation in soil and plants, and therefore salinity stress is also considered as hyperosmotic stress [6]. Osmotic stress in the initial stage of salinity stress causes various physiological changes, such as interruption of membranes, nutrient imbalance, impairs the ability to detoxify reactive oxygen species (ROS), differences in the antioxidant enzymes and decreased photosynthetic activity, and decrease in stomatal aperture [3, 5]. Salinity stress is also considered as a hyperionic stress. One of the most detrimental effects of salinity stress is the accumulation of Na^+ and Cl^- ions in tissues of plants exposed to soils with high NaCl concentrations. Entry of both Na^+ and Cl^- into the cells causes severe ion imbalance and excess uptake might cause significant physiological disorder(s). High Na^+ concentration inhibits uptake of K^+ ions which is an essential element for growth and development that results into lower productivity and may even lead to death [4]. In response to salinity stress, the production of ROS, such as singlet oxygen, superoxide, hydroxyl radical, and hydrogen peroxide, is enhanced [8–12]. Salinity-induced ROS formation can lead to oxidative damages in various cellular components such as proteins, lipids, and DNA, interrupting vital cellular functions of plants.

Genetic variations in salt tolerance exist, and the degree of salt tolerance varies with plant species and varieties within a species. Among major crops, barley (Hordeum vulgare) shows a greater degree of salt tolerance than rice (Oryza sativa) and wheat (Triticum aestivum). The degree of variation is even more pronounced in the case of dicotyledons ranging from Arabidopsis thaliana, which is very sensitive towards salinity, to halophytes such as Mesembryanthemum crystallinum, Atriplex sp., Thellungiella salsuginea (previously known as T. halophila) [3,13, 14]. In the last two decades sumptuous amount of research has been done in order to understand the mechanism of salt tolerance in model plant Arabidopsis [15]. Genetic variations and differential responses to salinity stress in plants differing in stress tolerance enable plant biologists to identify physiological mechanisms, sets of genes, and gene products that are involved in increasing stress tolerance and to incorporate them in suitable species to yield salt tolerant varieties.

The main aim of this review is to discuss research advances on the complex physiological and molecular mechanisms that are involved in plant salinity tolerance.

2. PHYSIOLOGICAL AND BIOCHEMICAL MECHANISMS OF SALT TOLERANCE

Plants develop various physiological and biochemical mechanisms in order to survive in soils with high salt concentration. Principle mechanisms include, but are not limited to, ion homeostasis and compartmentalization, ion transport and uptake, biosynthesis of osmoprotectants and compatible solutes, activation of antioxidant enzyme and synthesis of antioxidant compounds, synthesis of polyamines, generation of nitric oxide (NO), and hormone modulation. Research advances elucidating these mechanisms are discussed below.

2.1. Ion Homeostasis and Salt Tolerance

Maintaining ion homeostasis by ion uptake and compartmentalization is not only crucial for normal plant growth but is also an essential process for growth during salt stress [16–18]. Irrespective of their nature, both glycophytes and halophytes cannot tolerate high salt concentration in their cytoplasm. Hence, the excess salt is either transported to the vacuole or sequestered in older tissues which eventually are sacrificed, thereby protecting the plant from salinity stress [19, 20].

Major form of salt present in the soil is NaCl, so the main focus of research is the study about the transport mechanism of Na^+ ion and its compartmentalization. The Na^+ ion that enters the cytoplasm is then transported to the vacuole via Na^+/H^+ antiporter. Two types of H^+ pumps are present in the vacuolar membrane: vacuolar type H^+-ATPase (V-ATPase) and the vacuolar pyrophosphatase (V-PPase) [21–23]. Of these, V-ATPase is the most dominant H^+ pump present within the plant cell. During nonstress conditions it plays an important role in maintaining solute homeostasis, energizing secondary transport and facilitating vesicle fusion. Under stressed condition the survivability of the plant depends upon the activity of V-ATPase [21]. In a study performed by De Lourdes Oliveira Otoch et al. [22] in hypocotyls of Vigna unguiculata seedlings, it was observed that the activity of V-ATPase pump increased when exposed to salinity stress but under similar conditions, activity of V-PPase was inhibited, whereas in the case of halophyte Suaeda salsa, V-ATPase activity was upregulated and V-PPase played a minor role [23].

Increasing evidence demonstrates the roles of a Salt Overly Sensitive (SOS) stress signalling pathway in ion homeostasis and salt tolerance [24, 25]. The SOS signalling pathway (Figure 1) consists of three major proteins, SOS1, SOS2, and SOS3. SOS1, which encodes a plasma membrane Na^+/H^+ antiporter, is essential in regulating Na^+ efflux at cellular level. It also facilitates long distance transport of Na^+ from root to shoot. Overexpression of this protein confers salt tolerance in plants [26, 27]. SOS2 gene, which encodes a serine/threonine kinase, is activated by salt stress elicited Ca^+ signals. This protein consists of a well-developed N-terminal catalytic domain and a C-terminal regulatory domain [28]. The third type of protein involved in the SOS stress signalling pathway is the SOS3 protein which is a myristoylated Ca^+ binding protein and contains a myristoylation site at its N-terminus. This site plays an essential role in conferring salt tolerance [29]. C-terminal regulatory domain of SOS2 protein contains a FISL motif (also known as NAF domain), which is about 21 amino acid long sequence, and serves as a site of interaction for Ca^{2+} binding SOS3 protein (Figure 1). This interaction between SOS2 and SOS3 protein results in the activation of the kinase [30]. The activated kinase then phosphorylates SOS1 protein thereby increasing its transport activity which was initially identified in yeast [31]. SOS1 protein is characterised by a long cytosolic C-terminal tail, about 700 amino acids long, comprising a putative nucleotide binding motif and an autoinhibitory domain. This autoinhibitory domain is the target site for SOS2 phosphorylation (Figure 1). Besides conferring salt tolerance it also regulates pH homeostasis, membrane vesicle trafficking, and vacuole functions [32, 33]. Thus with the increase in the concentration of Na^+ there is a sharp increase in the intracellular Ca^{2+} level which in turn facilitates its binding with SOS3 protein. Ca^{2+} modulates intracellular Na^+ homeostasis along with SOS proteins. The SOS3 protein then interacts and activates SOS2 protein by releasing its self-inhibition. The SOS3-SOS2 complex is then loaded onto plasma membrane where it phosphorylates SOS1 (Figure 1). The phosphorylated SOS1 results in the increased Na^+ efflux, reducing Na^+ toxicity [34].

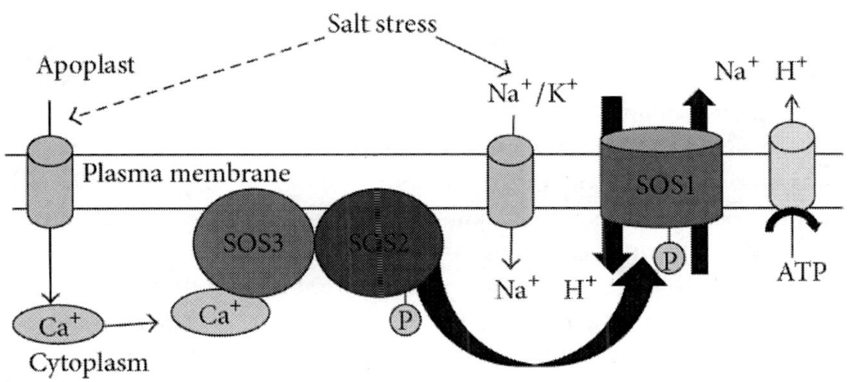

Figure 1. Model of SOS pathway for salinity stress responses.

Many plants have developed an efficient method to keep the ion concentration in the cytoplasm in a low level. Membranes along with their associated components play an integral role in maintaining ion concentration within the cytosol during the period of stress by regulating ion uptake and transport [35]. The transport phenomenon is carried out by different carrier proteins, channel proteins, antiporters and symporters. Maintaining cellular Na^+/K^+ homeostasis is pivotal for plant survival in saline environments. Ma et al. [36] have reported that Arabidopsis NADPH oxidases AtrbohD and AtrbohF function in ROS-dependent regulation of Na^+/K^+ homeostasis in Arabidopsis under salt stress. Plants maintain a high level of K^+ within the cytosol of about 100 mM ideal for cytoplasmic enzyme activities. Within the vacuole K^+ concentration ranges between 10 mM and 200 mM. The vacuole serves as the largest pool of K^+ within the plant cell. K^+ plays a major role in maintaining the turgor within the cell. It is transported into the plant cell against the concentration gradient via K^+ transporter and membrane channels. High affinity K^+ uptake mechanisms are mediated by K^+ transporters when the extracellular K^+ concentration is low, whereas low affinity uptake is carried out by K^+ channels when the extracellular K^+ concentration is high. Thus uptake mechanism is primarily determined by the concentration of K^+ available in the soil. On the other hand a very low concentration of Na^+ ion (about 1 mM or less) is maintained in the cytosol. During salinity stress, due to increased concentration of Na^+ in the soil, Na^+ ion competes with K^+ for the transporter as they both share the same transport mechanism, thereby decreasing the uptake of K^+ [3, 35].

A large number of genes and proteins, such as HKT and NHX, encoding K^+ transporters and channels have been identified and cloned in various plant species. During salt stress expression of some low abundance transcripts is enhanced which are found to be involved in K^+ uptake. This was observed in the halophyteMesembryanthemum crystallinum [37]. Transporters located on the plasma membrane, belonging to the HKT (histidine kinase transporter) family, also play an essential role in salt tolerance by regulating transportation of Na^+ and K^+. Class 1 HKT transporters, that have been identified in Arabidopsis, protect the plant from the adverse effects of salinity by preventing excess accumulation Na^+ in leaves. Similar results were observed in the experiment which was carried out with rice where class 1 HKT transporter removes excess Na^+ from xylem, thus protecting the photosynthetic leaf tissues from the toxic effect of Na^+ [38]. Intracellular NHX proteins are Na^+, K^+/H^+ antiporters involved in K^+ homeostasis, endosomal pH regulation, and salt tolerance. Barragán et al. [39] showed that tonoplast-localized NHX proteins (NHX1 and NHX2: the two major tonoplast-localized NHX isoforms) are essential for active K^+ uptake at the tonoplast, for turgor regulation, and for stomatal function. In fact more such NHX isoforms have been identified and their roles in ion (Na^+, K^+, H^+) homeostasis established from different plant species (e.g., LeNHX3 and LeNHX4 from tomato) [40].

2.2. Compatible Solute Accumulation and Osmotic Protection

Compatible solutes, also known as compatible osmolytes, are a group of chemically diverse organic compounds that are uncharged, polar, and soluble in nature and do not interfere with the cellular metabolism even at high concentration. They mainly include proline [41–45], glycine betaine [46, 47], sugar [48, 49], and polyols [50–53]. Organic osmolytes are synthesised and accumulated in varying amounts amongst different plant species. For example, quaternary ammonium compound beta alanine betaine's accumulation is restricted among few members of Plumbaginaceae [54], whereas accumulation of amino acid proline occurs in taxonomically diverse sets of plants [53]. The concentration of compatible solutes within the cell is maintained either by irreversible synthesis of the compounds or by a combination of synthesis and degradation. The biochemical pathways and genes involved in these processes have been thoroughly studied. As their accumulation is proportional to the external osmolarity, the major functions of these osmolytes are to protect the structure and to maintain osmotic balance within the cell via continuous water influx [24].

Amino acids such as cysteine, arginine, and methionine, which constitute about 55% of total free amino acids, decrease when exposed to salinity stress, whereas proline concentration rises in response to salinity stress [55]. Proline accumulation is a well-known measure adopted for alleviation of salinity stress [53, 56, 57]. Intracellular proline which is accumulated during salinity stress not only provides tolerance towards stress but also serves as an organic nitrogen reserve during stress recovery. Proline is synthesised either from glutamate or ornithine. In osmotically stressed cell glutamate functions as the primary precursor. The biosynthetic pathway comprises two major enzymes, pyrroline carboxylic acid synthetase and pyrroline carboxylic acid reductase. Both these regulatory steps are used to overproduce proline in plants [35]. It functions as an O_2 quencher thereby revealing its antioxidant capability. This was observed in a study carried out by Matysik et al. [56]. Ben Ahmed et al. [57] observed that proline supplements enhanced salt tolerance in olive (Olea europaea) by amelioration of some antioxidative enzyme activities, photosynthetic activity, and plant growth and the preservation of a suitable plant water status under salinity conditions. It has been reported that proline improves salt tolerance in Nicotiana tabacum by increasing the activity of enzymes involved in antioxidant defence system [58]. Deivanai et al. [59] also demonstrated that rice seedlings from seeds pretreated with 1 mM proline exhibited improvement in growth during salt stress.

Glycine betaine is an amphoteric quaternary ammonium compound ubiquitously found in microorganisms, higher plants and animals, and is electrically neutral over a wide range of pH. It is highly soluble in water but also contains nonpolar moiety constituting 3-methyl groups. Because of its unique structural features it interacts both with hydrophobic and hydrophilic domains of the macromolecules, such as enzymes and protein complexes. Glycine betaine is a nontoxic cellular osmolyte that raises the osmolarity of the cell during stress period; thus it plays an important function in stress mitigation. Glycine betaine also protects the cell by osmotic adjustment [60], stabilizes proteins [61], and protects the photosynthetic apparatus from stress damages [62] and reduction of ROS [52, 53]. Accumulation of glycine betaine is found in a wide variety of plants belonging to different taxonomical background. Glycine betaine is synthesised within the cell from either choline or glycine. Synthesis of glycine betaine from choline is a 2-step reaction involving two or more enzymes. In the first step choline is oxidised to betaine aldehyde which is then again oxidised in the next step to form glycine betaine. In higher plants the first conversion is carried out by the enzyme choline monooxygenase (CMO), whereas the next step is catalysed by betaine aldehyde dehydrogenase (BADH) [63]. Another pathway which is observed in some plants, mainly halophytic, demonstrated the synthesis of glycine betaine from glycine. Here glycine betaine is synthesized by three successive N-methylation and the reactions are catalysed by two S-adenosyl methionine dependent methyl transferases, glycine sarcosine N-methyl transferase (GSMT), and sarcosine dimethylglycine N-methyl transferase (SDMT). These two enzymes have overlapping functions as GSMT catalyses the first and the second step while SDMT catalyses the second and third step [63]. Rahman et al. [64] reported the positive effect of glycine betaine on the ultrastructure of Oryza sativa seedlings when exposed to salt stress. Under stressed condition (150 mM NaCl) the ultrastructure of the seedling shows several damages such as swelling of thylakoids, disintegration of grana and intergranal lamellae, and disruption of mitochondria. However, these damages were largely prevented when seedlings were pretreated with glycine betaine. When glycine betaine is applied as a foliar spray in a plant subjected to stress, it led to pigment stabilization and increase in photosynthetic rate and growth [62, 63].

Polyols are compounds with multiple hydroxyl functional groups available for organic reactions. Sugar alcohols are a class of polyols functioning as compatible solutes, as low molecular weight chaperones, and as ROS scavenging compounds [52]. They can be classified into two major types, cyclic (e.g., pinitol) and acyclic (e.g., mannitol). Mannitol synthesis is induced in plants during stressed period via action of NADPH dependent mannose-6-phosphate reductase. These compatible solutes function as a protector or stabilizer of enzymes or membrane structures that are sensitive to dehydration or ionically induced damage. It was found that the transformation with bacterial mltd gene that encodes for mannitol-1-phosphate dehydrogenase in bothArabidopsis and tobacco (Nicotiana tabacum) plants confer salt tolerance, thereby maintaining normal growth and development when subjected to high level of salt stress [65, 66]. Pinitol is accumulated within the plant cell when the plant is subjected to salinity stress. The biosynthetic pathway consists of two major steps, methylation of myo-inositol which results in formation of an intermediate compound, ononitol, which undergoes epimerization to form pinitol. Inositol methyl transferase enzyme encoded by imt gene plays major role in the synthesis of pinitol. Transformation of imt gene in plants shows a result similar to that observed in the case ofmltd gene. Thus it can be said that pinitol also plays a significant role in stress alleviation. Accumulation of polyols, either straight-chain metabolites such as mannitol and sorbitol or cyclic polyols such as myo-inositol and its methylated derivatives, is correlated with tolerance to drought and/or salinity, based on polyol distribution in many species, including microbes, plants, and animals [49].

Accumulations of carbohydrates such as sugars (e.g., glucose, fructose, fructans, and trehalose) and starch occur under salt stress [67]. The major role played by these carbohydrates in stress mitigation involves osmoprotection, carbon storage, and scavenging of reactive oxygen species. It was observed that salt stress increases the level of reducing sugars (sucrose and fructans) within the cell in a number of plants belonging to different species [48]. Besides being a carbohydrate reserve, trehalose accumulation protects organisms against several physical and chemical stresses including salinity stress. They play an osmoprotective role in physiological responses [63]. Sucrose content was found to increase in tomato (Solanum lycopersicum) under salinity due to increased activity of sucrose phosphate synthase [68]. Sugar content, during salinity stress, has been reported to both increase and decrease in various rice genotype [69]. In rice roots it has been observed that starch content decreased in response to salinity while it remained fairly unchanged in the shoot. Decrease in starch content and increase in reducing and nonreducing sugar content were noted in leaves of Bruguiera parviflora [67].

2.3. Antioxidant Regulation of Salinity Tolerance

Abiotic and biotic stress in living organisms, including plants, can cause overflow, deregulation, or even disruption of electron transport chains (ETC) in chloroplasts and mitochondria. Under these conditions molecular oxygen (O_2) acts as an electron acceptor, giving rise to the accumulation of ROS. Singlet oxygen ($^1O_2$), the hydroxyl radical (OH^-), the superoxide radical (), and hydrogen peroxide (H_2O_2) are all strongly oxidizing compounds and therefore potentially harmful for cell integrity [70]. Antioxidant metabolism, including antioxidant enzymes and nonenzymatic compounds, play critical parts in detoxifying ROS induced by salinity stress. Salinity tolerance is positively correlated with the activity of antioxidant enzymes, such as superoxide dismutase (SOD), catalase (CAT), glutathione peroxidise (GPX), ascorbate peroxidase (APX), and glutathione reductase (GR) and with the accumulation of nonenzymatic antioxidant compounds [71, 72]. Gill et al. [73] and Tuteja et al. [74] have recently reported a couple of helicase proteins (e.g., DESD-box helicase and OsSUV3 dual helicase) functioning in plant salinity tolerance by improving/maintaining photosynthesis and antioxidant machinery. Kim et al. [75] showed that silicon (Si) application to rice root zone influenced the hormonal and antioxidant responses under salinity stress. The results showed that Si treatments significantly increased rice plant growth compared to controls under salinity stress. Si treatments reduced the sodium accumulation resulting in low electrolytic leakage and lipid peroxidation compared to control plants under salinity stress. Enzymatic antioxidant (catalase, peroxidase, and polyphenol oxidase) responses were more pronounced in control plants than in Si-treated plants under salinity stress.

Anthocyanin is a flavonoid whose accumulation in plant exposed to salt stress has been largely documented. Van Oosten et al. [76] isolated the anthocyanin-impaired-response-1 (air1) mutant that is unable to accumulate anthocyanins under salt stress. The air1 mutant showed a defect in anthocyanin production in response to salt stress but not to other stresses such as high light, low phosphorous, high temperature, or drought stress. This specificity indicated that air1 mutation did not affect anthocyanin biosynthesis but rather its regulation in response to salt stress. The discovery and characterization of AIR1 opens avenues to dissect the connections between abiotic stress and accumulation of antioxidants in the form of flavonoids and anthocyanins.

Ascorbate is one of the major antioxidants present within the cell. Pea plants grown under saline (150 mM NaCl) stress showed an enhancement of both APX activity and S-nitrosylated APX, as well as an increase of H_2O_2, NO, and S-nitrosothiol (SNO) content that can justify the induction of the APX activity. Proteomic data have shown that APX is one of the potential targets of PTMs mediated by NO-derived molecules [77]. Using recombinant pea cytosolic APX, the impact of peroxynitrite () and S-nitrosoglutathione (GSNO), which are known to mediate protein nitration and S-nitrosylation processes, respectively, was analysed. While peroxynitrite inhibits APX activity, GSNO enhances its enzymatic activity. The results provide new insight into the molecular mechanism of the regulation of APX, which can be both inactivated by irreversible nitration and activated by reversible S-nitrosylation [77]. Exogenous application of ascorbate mitigates the adverse effects of salinity stress in various plant species and promotes plant recovery from the stress [78, 79]. Another antioxidant in stress mitigation is glutathione, which can react with superoxide radical, hydroxyl radical, and hydrogen peroxide, thereby functioning as a free radical scavenger. It can also participate in the regeneration of ascorbate via ascorbate-glutathione cycle [80]. When applied exogenously glutathione helped to maintain plasma membrane permeability and cell viability during salinity stress in Allium cepa [81]. Application of glutathione and ascorbate was found to be effective in increasing the height of the plant, branch number, fresh and dry weight of herbs and flowers, and the content of carbohydrates, phenols, xanthophylls pigment, and mineral ion content when subjected to saline condition [82]. Many studies have found differences in levels of expression or activity of antioxidant enzymes; these differences are sometimes associated with the more tolerant genotype and sometimes with the more sensitive genotype. Munns and Tester [3] suggested that differences in antioxidant activity between genotypes may be due to genotypic differences in degrees of stomatal closure or in other responses that alter the rate of CO_2 fixation and differences that bring into play the processes that avoid photoinhibition and for which the plant has abundant capacity [3]. Roy et al. [83] in their recent review have argued that there are three main traits in plants, which help them in their adaptation to salinity stress: ion exclusion, tissue tolerance, and salinity tolerance. It seems that antioxidants have some role in tissue and salinity tolerance mechanism.

2.4. Roles of Polyamines in Salinity Tolerance

Polyamines (PA) are small, low molecular weight, ubiquitous, polycationic aliphatic molecules widely distributed throughout the plant kingdom. Polyamines play a variety of roles in normal growth and development such as regulation of cell proliferation, somatic embryogenesis, differentiation and morphogenesis, dormancy breaking of tubers and seed germination, development of flowers and fruit, and senescence [84–87]. It also plays a crucial role in abiotic stress tolerance including salinity and increases in the level of polyamines are correlated with stress tolerance in plants [88–91].

The most common polyamines that are found within the plant system are diamine putrescine (PUT), triamine spermidine (SPD), and tetra-amine spermine (SPM) [92–96]. The PA biosynthetic pathway has been thoroughly investigated in many organisms including plants and has been reviewed in details [97–104]. PUT is the smallest polyamine and is synthesised from either ornithine or arginine by the action of enzyme ornithine decarboxylase (ODC) and arginine decarboxylase (ADC), respectively [85, 105]. N-carbamoyl-putrescine is converted to PUT by the enzyme N-carbamoyl-putrescine aminohydrolase [106, 107]. The PUT thus formed functions as a primary substrate for higher polyamines such as SPD and SPM biosynthesis. The triamine SPD and tetramine SPM are synthesized by successive addition of aminopropyl group to PUT and SPD, respectively, by the enzymes spermidine synthase (SPDS) and spermine synthase (SPMS) [108, 109]. ODC pathway is the most common pathway for synthesis of polyamine found in plants. Most of the genes involved in the ODC pathway have been identified and cloned. However there are some plants where ODC pathway is absent; for instance in Arabidopsis polyamines are synthesized via ADC pathway [110–112]. All the genes involved in polyamine biosynthesis pathways have been identified from different plant species including Arabidopsis[113–115]. Polyamine biosynthesis pathway in Arabidopsis involves six major enzymes: ADC encoding genes (ADC1 and ADC2); SPDS (SPDS1 and SPDS2) and SAMDC (SAMDC1, SAMDC2, SAMDC3, SAMDC4) [115–118]. On the contrary, SPM synthase, thermospermine synthase, agmatine iminohydrolase and N-carbamoylputrescine amidohydrolase are represented by single genes only [119, 120].

Increase in endogenous polyamine level has been reported when the plant is exposed to salinity stress. Intracellular polyamine level is regulated by polyamine catabolism. Polyamines are oxidatively catabolised by amine oxidases which include copper binding diamine oxidases and FAD binding polyamine oxidases. These enzymes play a significant role in stress tolerance [121, 122]. The changes in cellular polyamine level due to stress provide possible implications in stress but do not provide evidence of their role in counteracting stress. Hence, to understand whether polyamines actually protect cells from stress-induced damages, exogenous application of polyamines, which is expected to increase endogenous polyamine, has been investigated before or during stress [123, 124]. Application of exogenous polyamine has been found to increases the level of endogenous polyamine during stress; the positive effects of polyamines have been associated with the maintenance of membrane integrity, regulation of gene expression for the synthesis of osmotically-active solutes, reduction in ROS production, and controlling accumulation of Na^+ and Cl^- ion in different organs [123–130]. It was observed that plant deficient in ADC1 and ADC2 is hypersensitive to stress [131]. InArabidopsis, expression of ADC and SPMS increases when exposed to salinity stress. whereas mutants of polyamine biosynthetic genes show sensitivity to salinity [132]. Overproduction of PUT, SPD, and SPM in rice, tobacco, and Arabidopsis enhances salt tolerance [133]. Salt stress regulates polyamine biosynthesis and catabolism by acting as a cellular signal in hormonal pathways thereby regulating abscisic acid (ABA) in response to stress [134]. Additionally, SPM and SPD are regarded as potent inducers of NO which is another important signalling molecule [135] and its involvement in salinity tolerance is discussed below. It has been reported that exogenous application of polyamines could alleviate salt-induced reduction in photosynthetic efficiency, but this effect depends on polyamine concentration and types and level of stress [136]. When the seedling of Sorghum bicolor treated with 0.25 mM SPM is subjected to salt stress it shows improvement in growth and partial increase in the activity of peroxidase and glutathione reductase enzyme with a concomitant decrease in the level of membrane lipid peroxidation [137]. Li et al. [138] performed 2-DE gel electrophoresis and MALDI-TOF/TOF MS with cytosolic proteins to understand the effect of exogenous SPD on proteomic changes under normal and NaCl stress of 3 days old cucumber seedling leaves. Many changes were observed in the levels of proteins involved in energy and metabolic pathways, protein metabolic, stress defense, and other functional proteins. They observed that increased salt tolerance by exogenous SPD would contribute to higher expressions of proteins involved in the SAMs metabolism, protein biosynthesis, and defense mechanisms on antioxidant and detoxification. Li et al. [138] also argued that the regulation of Calvin cycle, protein folding assembly, and the inhibition of protein proteolysis by SPD might play important roles in salt tolerance.

2.5. Roles of Nitric Oxide in Salinity Tolerance

Nitric oxide (NO) is a small volatile gaseous molecule, which is involved in the regulation of various plant growth and developmental processes, such as root growth, respiration, stomata closure, flowering, cell death, seed germination and stress responses, as well as a stress signalling molecule [139–143]. NO directly or indirectly triggers expression of many redox-regulated genes. NO reacts with lipid radicals thus preventing lipid oxidation, exerting a protective effect by scavenging superoxide radical and formation of peroxynitrite that can be neutralised by other cellular processes. It also helps in the activation of antioxidant enzymes (SOD, CAT, GPX, APX, and GR) [144].

Exogenous NO application has been found to play roles in stress mitigation [145–147], but the effects depend on NO concentration. Exogenous application of sodium nitroprusside (SNP), a NO donor, on Lupinus luteusseedlings subjected to salt stress enhanced seed germination and root growth [148]. Seed germination was promoted at concentrations between 0.1 and 800 μM SNP in a dose-dependent manner. The stimulation was most pronounced after 18 and 24 h and ceased after 48 h of imbibition. The promoting effect of NO on seed germination persisted even in the presence of heavy metals (Pb and Cd) and NaCl. Kopyra and Gwóźdź [148] further showed that the pretreatment of L. luteus seedlings for 24 h with 10 μM SNP resulted in efficient reduction of the detrimental effect of the abiotic stressors on root growth and morphology. Pretreatment of maize seedlings with 100 μM SNP increases dry matter of roots and shoots under salinity stress; however, when the concentration of SNP was increased to 1000 μM shoot and root dry weight decreased [149]. Thus, this experiment highlighted both the protective effects of low NO concentration and the toxic effect of high NO concentration on plants.

The positive effects of NO on salinity tolerance or stress mitigation have been attributed to antioxidant activities and modulation of ROS detoxification system [150]. Improved plant growth under salinity stress by exogenous application of NO was associated with increases in antioxidant enzymes such as SOD, CAT, GPX, APX, and GR [151], and suppression of malondialdehyde (MDA) production or lipid peroxidation [152]. Effects of NO on salinity tolerance are also related to its regulation of plasma membrane H^+-ATPase and Na^+/K^+ ratio [143]. NO stimulates H^+-ATPase (H^+-PPase), thereby producing a H^+ gradient and offering the force for Na^+/H^+exchange. Such an increase of Na^+/H^+ exchange may contribute to

K^+ and Na^+ homeostasis [149]. Although NO acts as a signal molecule under salt stress and induces salt resistance by increasing PM H^+-ATPase activity, research results from Zhang et al. [153] with calluses from Populus euphratica also indicated NO cannot activate purified PM H^+-ATPase activity, at least in vitro. They initially hypothesized ABA or H_2O_2 might be downstream signal molecules to regulate the activity of PM H^+-ATPase. Further results indicated H_2O_2 content increased greatly under salt stress. Since H_2O_2 might be the candidate downstream signal molecule, Zhang et al. [153] tested PM H^+-ATPase activity and K to Na ratio in calluses by adding H_2O_2. The results suggested that H_2O_2 inducing an increased PM H^+-ATPase activity resulted in an increased K to Na ratio leading to NaCl stress adaptation.

2.6. Hormone Regulation of Salinity Tolerance

ABA is an important phytohormone whose application to plant ameliorates the effect of stress condition(s). It has long been recognized as a hormone which is upregulated due to soil water deficit around the root. Salinity stress causes osmotic stress and water deficit, increasing the production of ABA in shoots and roots [154–158]. The accumulation of ABA can mitigate the inhibitory effect of salinity on photosynthesis, growth, and translocation of assimilates [158, 159]. The positive relationship between ABA accumulation and salinity tolerance has been at least partially attributed to the accumulation of K^+, Ca^{2+} and compatible solutes, such as proline and sugars, in vacuoles of roots, which counteract with the uptake of Na^+ and Cl^- [160, 161]. ABA is a vital cellular signal that modulates the expression of a number of salt and water deficit-responsive genes. Fukuda and Tanaka [162] demonstrated the effects of ABA on the expression of two genes, HVP1 and HVP10, for vacuolar H^+-inorganic pyrophosphatase, and of HvVHA-A, for the catalytic subunit (subunit A) of vacuolar H^+-ATPase in Hordeum vulgare under salinity stress. ABA treatment in wheat induced the expression of MAPK4-like, TIP 1, and GLP 1 genes under salinity stress [163].

Some other compounds having hormonal properties, such as salicylic acid (SA) and brassinosteroids (BR), also participate in plant abiotic stress responses [164, 165]. Under salinity stress endogenous level of SA increased along with the increase in the activity of salicylic acid biosynthetic enzyme in rice seedling [166]. Jayakannan et al. [167] have recently shown that SA improves salinity tolerance in Arabidopsis by restoring membrane potential and preventing salt-induced K^+ loss via a guard cell outward rectifying K(+) (GORK) channel.Arabidopsis seedling pretreated with SA showed upregulation of H^+-ATPase activity, thereby improving K^+retention during salt stress; SA pretreatment did not prevent accumulation of Na^+ in roots but somehow helped to reduce the concentration of accumulated Na^+ in the shoot [167]. The application of SA also promoted salinity tolerance in barley, as manifested by increases in the content of chlorophyll and carotenoid and maintaining membrane integrity, which was associated with more K^+ and soluble sugar accumulation in the root under saline condition [168]. Nazar et al. [169] have argued that SA alleviates decreases in photosynthesis under salt stress by enhancing nitrogen and sulfur assimilation and antioxidant metabolism differentially in mung bean cultivars. The negative effects of salinity may also be mitigated by BR [170–173]. Application of BR enhanced the activity of antioxidant enzymes (SOD, POX, APX, and GPX) and the accumulation of nonenzymatic antioxidant compounds (tocopherol, ascorbate, and reduced glutathione) [170]. Both BRs and SA are ubiquitous in the plant kingdom, affecting plant growth and development in many different ways, and are known to improve plant stress tolerance. Ashraf et al. [173] have reviewed and discussed the current knowledge and possible applications of BRs and SA that could be used to mitigate the harmful effects of salt stress in plants. They have also discussed the roles of exogenous applications of BRs and SA in the regulation of various biochemical and physiological processes leading to improved salt tolerance in plants.

3. TRANSCRIPTIONAL REGULATION AND GENE EXPRESSION OF SALINITY TOLERANCE

Regulation of gene expression in salinity stress includes a wide array of mechanisms that are used by plants to upregulate or downregulate (increase or decrease) the production of specific gene products (protein or RNA). Various mechanisms of gene regulation have been identified during the central dogma, from transcriptional initiation, to RNA processing, and to the posttranslational modification of a protein.

Transcriptomic analysis provides detailed knowledge about the gene expression at mRNA level, which is widely used to screen candidate genes involved in stress responses. Genomic approaches play a significant role in encoding, cloning, and characterization of important genes. A huge number of salt-responsive transcription factors and genes which are either upregulated or downregulated in response to salinity stress have been identified and characterized using transcriptomic and genomic approaches.

Transcription factors are considered as most important regulators that control gene expressions. Among them, bZIP, WRKY, AP2, NAC, C2H2 zinc finger gene, and DREB families comprise a large number of stress-responsive members. These transcription factor genes are capable of controlling the expression of a broad range of target genes by binding to the specific cis-acting element in the promoters of these genes. Johnson et al. [174] observed that the expression of bZIP genes were upregulated in salt-sensitive wheat cultivar, when exposed to long-term salinity, but decreased in salt-tolerant variety. Overexpression of a NAC transcription factor in both rice and wheat confers salt tolerance, thereby predicting their role in stress mitigation [175]. In rice transcriptional regulators that have been demonstrated to play a significant role in abiotic stress responses involve DREB1/CBF, DREB2, and AREB/ABF [176–180]. Transcriptions factors such as OsNAC5 and ZFP179 show an upregulation under salinity stress, which may regulate the synthesis and accumulation of proline, sugar, and LEA proteins that in turn play an integral role in stress tolerance [181]. In Arabidopsis, salt stress results in the upregulation of AtWRKY8 which directly binds with the promoter of RD29A, suggesting it to be as one of the target genes of AtWRKY8 [182].

A large number of genes and transcription factors are upregulated in response to salinity in different plant species, which serve diverse functions [183–192]. Examples of salt-responsive genes are listed in the Table 1, and these genes are mainly classified into the following functional categories: ion transport or homeostasis (e.g.,SOS genes, AtNHX1, and H^+-ATPase), senescence-associated genes (e.g., SAG), molecular chaperones (e.g.,HSP genes), and dehydration-related transcription factors (e.g., DREB). Among stress-responsive genes, theSOS gene family, which we have already discussed in Section 2.1, is believed to play a very intriguing role in ion homeostasis, thereby conferring salt tolerance [24–37, 190, 191]. Some ROS-scavenging and osmotic-regulating genes are also upregulated by salinity in some plant species. For example, a continuous exposure of rice plants to salinity for about 24 hours resulted in upregulation of glutathione-S-transferase and ascorbate peroxidase, both of which were known to play an active role in ROS scavenging, and with the increase in duration of exposure to salinity stress, upregulation of metallothionein and water channel proteins was also observed [192]. Halophyte plant species Spartina alterniflora when subjected to salt stress exhibits upregulation of 10 genes associated with osmotic regulation [193].

Table 1. Examples of upregulated genes in response to salinity stress.

Species	NaCl concentration (mM)	Gene name	Gene functions	References
Brassica juncea and *Brassica campestris*	25 and 50	*SOS1* *SOS2* *SOS3* *AtNHX1*	(i) Plasma membrane Na^+/K^+ antiporter. (ii) Protein kinase. (iii) Calcium-binding protein. (iv) Vacuolar Na^+/K^+ antiporter.	[183]
Oryza sativa	50	*PRP* *SAG* *HSPC025*	(i) Proline-rich proteins and cell-wall protection. (ii) Senescence associated genes, regulatory processes, and cellular signal transduction. (iii) Heat-shock proteins, protein stabilizing.	[184]
Oryza sativa	100	*OsHSP23.7* *OsHSP71.1,* *OsHSP80.2*	Heat-shock proteins, molecular chaperones, folding, assembling, and transporting proteins.	[185]
Arabidopsis thaliana	150	*AtSKIP*	Transcription factor, transcriptional pre-initiation, splicing, and polyadenylation.	[186]
Oryza sativa	200	*OsHsp17.0,* *OsHsp23.7*	Heat-shock proteins, molecular chaperones, and folding, assembling, and transporting proteins.	[187]
Carrot	300	*DcHsp17.7*	Cell viability and membrane stability under heat stress.	[188]
Arabidopsis thaliana	300	*JcDREB*	Transcription factor	[189]

Recently, Schmidt et al. [194] identified SALT-RESPONSIVE ERF1 (SERF1), a rice (Oryza sativa) transcription factor gene that showed a root-specific induction upon salt and H_2O_2 treatment. Loss of SERF1 impaired the salt-inducible expression of genes encoding members of a mitogen-activated protein kinase (MAPK) cascade and salt tolerance-mediating TFs. Furthermore, they showed that SERF1-dependent genes are H_2O_2 responsive and demonstrated that SERF1 binds to the promoters of MAPK KINASE KINASE6 (MAP3K6), MAPK5, DEHYDRATION-RESPONSIVE ELEMENT BINDING2A (DREB2A), and ZINC FINGER PROTEIN179 (ZFP179) in vitro and in vivo. SERF1 also directly induces its own gene expression. In addition, it was observed that SERF1 is a phosphorylation target of MAPK5, resulting in enhanced transcriptional activity of SERF1 toward its direct target genes. Finally, they demonstrated that the plants deficient for SERF1 are more sensitive to salt stress compared with the wild type, while constitutive overexpression of SERF1 improves salinity tolerance.

There are some transcription factors which are regulated by different kinases and have been found to be significant players of plant adaptation to salinity stress. Serra et al. [195] showed that OsRMC encodes a receptor-like kinase described as a negative regulator of salt stress responses in rice. Two transcription factors, OsEREBP1 and OsEREBP2, belonging to the AP2/ERF family were shown to bind to the same GCC-like DNA motif in OsRMC promoter and to negatively regulate its gene expression. Serra et al. [195] further revealed thatOsEREBP1 transcript level is not significantly affected by salt, ABA, or severe cold (5°C) and is only slightly regulated by drought and moderate cold. On the other hand, the OsEREBP2 transcript level increased after cold, ABA, drought, and high salinity treatments, indicating that OsEREBP2 may play a central role mediating the response to different abiotic stresses. Gene expression analysis in rice varieties with contrasting salt tolerance further suggests that OsEREBP2 is involved in salt stress response in rice. A bZIP class of ABRE binding transcription factor, known as OSBZ8, has also been identified from rice and has been shown to be highly expressed in salt tolerant cultivars than in salt sensitive one [196]. Moreover, OSBZ8 has been shown to be activated/phosphorylated by a SNF-1 group of serine/threonine kinase in the presence of Spd during salinity stress [197].

High-throughput sequencing for transcript profiling in plants has revealed that alternative splicing affects a much higher proportion of the transcriptome than was previously assumed. Alternative splicing is involved in most plant processes and is particularly prevalent in plants exposed to environmental stress. The identification of mutations in predicted splicing factors and spliceosomal proteins that affect cell fate, the circadian clock, plant defense, and tolerance/sensitivity to abiotic stress all points to a fundamental role of splicing/alternative splicing in plant growth, development, and responses to external cues [198]. A suite of Ser/Arg-rich proteins that are key regulators of alternative splicing undergoes alternative themselves in response to various abiotic stresses, such as salt stress [198–200]. PRMT5, a type II protein Arg methyltransferase that symmetrically dimethylates Arg side chains, also impacts splicing/alternative in Arabidopsis. The prmt5 mutant, also known as shk1 kinase binding protein1 (skb1), is sensitive to salt [201]. It was proposed that PRMT5/SKB1 affects plant development and the salt response by altering the methylation status of H4R3sme2 (for symmetric dimethylation of histone H4 arginine 3) and LSm4 and thus linking transcription to pre-mRNA splicing [201]. A nuclear coactivator, At-SKIP (Ski-interacting protein), expression was found to increase in response to salt, mannitol, and ABA treatment, and At-SKIP overexpression or antisense lines show altered tolerance to a plethora of abiotic stress factors [186], and it is likely that a role in alternative splicing contributes to these phenotypes [198].

The small ubiquitin-like modifier (SUMO) is a crucial regulator of signaling proteins in eukaryotes. Attachment of SUMO onto substrates is reversible, and SUMO proteases, which specifically cleave the SUMO-substrate linkages, play a vital regulatory role during SUMOylation. Conti et al. [202] have identified two SUMO proteases, Overly Tolerant To Salt1 (OTS1) and OTS2, which are localized in the nucleus and act redundantly to regulate salt stress responses in Arabidopsis thaliana. Cui et al. [203] have identified an Arabidopsisendoplasmic reticulum (ER-) associated protein degradation (ERAD) component called Ubiquitin conjugase UBC32 that functions in BR-mediated salt stress tolerance. More and more such reports of sumoylation and other ubiquitin like posttranslational modifications during plant salinity stress are coming up.

Downregulated genes are emerging now as essential components of the response to salinity. For example downregulation of -carotene hydroxylase increases -carotene and total carotenoids enhancing salt stress tolerance in transgenic cultured cells of sweet potato [204]. It seems that mutual regulation mechanism exists between different genes and proteins and signals underlying different processes of plant adaptation to abiotic stress.

In addition to protein coding genes, recently discovered microRNAs (miRNAs) and endogenous small interfering RNAs (siRNAs) have emerged as important players in plant stress responses. Initial clues suggesting that small RNAs are involved in plant stress responses stem from studies showing stress regulation of miRNAs and endogenous siRNAs, as well as from target predictions for some miRNAs [205, 206].

4. PROTEOMIC AND METABOLIC RESPONSES TO SALINITY STRESS

Genomics technologies have helped to address the multigenicity of the plant abiotic stress responses. Analysis of genome sequences, and specific transcript collections and their dynamic changes, has provided a more global picture of stress-dependent responses at the cell, tissue, and whole plant level and moved the field from a single-gene approach toward an understanding of interactions between multiple components in cells, facilitating the dissection of abiotic stress circuits and coexpression hubs

[5–8, 207–209]. However, directly focusing on genes may not accurately portray conditions in the cell at a particular state and time during stress due to regulation at the RNA and protein level, including posttranslational regulation. Proteomics, and in particular quantitative proteomics, is emerging as a powerful technique to be applied to the field of crop abiotic stress tolerance research; it has the potential to allow rapid identification and quantification of novel stress- and tolerance-related proteins. Understanding the dynamics of expression and posttranslational modifications of these proteins, and gaining direct insight into their function and interactions, can provide essential information that can be applied to engineer stress-tolerant crops with novel traits through biomarker selection and transgenic strategies [209, 210]. Available data suggest that several common stress responsive proteins are expressed in response to various abiotic stresses in different plant species. About 2171 proteins from 34 plant species have been identified and characterized as salt-responsive proteins, which are either upregulated or downregulated by salinity stress [211]. Based on gene ontology, BLAST alignment and literature information, salt-responsive proteins can be grouped into 14 functional categories. Specifically, there appears to be a general regulation of proteins involved in carbohydrate, nitrogen, and energy metabolisms, with particular emphasis on glycolytic and Krebs cycle enzymes. Moreover, as discussed earlier, salinity and other abiotic stresses lead to metabolic imbalances that lead to ROS generation. Therefore, it is not surprising that plant root or shoot proteomics show the expression of ROS scavenging proteins like SOD, CAT, GPX, APX, and GR [209, 212]. Other proteins that are identified in multiple studies are those involved in protein synthesis, processing, turnover, and degradation, as well as cytoskeleton stability. For photosynthetic processes, there appears to be a general decrease in levels of chlorophyll biosynthesis related proteins but an increase in proteins involved in the light-dependent reactions. Some of the proteins identified are indicative of a general stress-responsive pathway in plants. Less common are proteins identified in the categories of signaling, trafficking, transport, and cell structure [209]. Plant lamina or root membrane proteomics, including that of plasma membrane, mitochondrial, and thylakoid membrane, have revealed the up-/downregulation of a plethora of proteins. These include receptor proteins that perceive the stress, membrane bound signaling, and regulatory proteins that function to relay the stress, vesicle trafficking proteins, and transport proteins that function to maintain ion and water homeostasis, and drive sequestration and/or removal of toxic compounds from the cell, membrane bound kinases, and intrinsic proteins [213–219].

Another significant research approach in plant system biology is the metabolomics which involves the study of metabolome. Higher plants have a remarkable ability to synthesize a vast array of metabolites that differ in chemical complexity and biological functions playing an indispensible role in stress alleviation [229, 230]. Examples of plant metabolites that are involved in salinity tolerance include polyols such as mannitol and sorbitol, dimethylsulfonium compounds, glycine betaine, sugars such as sucrose, trehalose and fructans, or amino acids such as proline that serve as an osmolyte or osmoprotectant [231]. Plants when subjected to salinity stress show an increase in the concentration of these osmolytes thus playing significant role in stress mitigation. We have already discussed their role in Section 2.2. Advances in analytical chemistry, such as MS based methods and NMR, and sophisticated "data processing and mining" techniques, have allowed the plant biologists to venture into hitherto unexplored domains and generate extensive metabolic profiles due to various environmental stimuli including salinity. Results indicate that the metabolic processes are highly specific for given tissues, species, and plant-environment interactions. The clusters of identified compounds not only serve as base in the quest of novel defense compounds but also as markers for the characterization of the plants' defensive state. The latter is especially useful in agronomic applications where meaningful markers are essential for crop protection [232].

5. BIOENGINEERING FOR IMPROVING SALINITY TOLERANCE

Genetic transformation technology enables scientists to achieve gene transfer in precise and predictable manner. Hence genetic engineering approaches would be useful to manipulate the osmoprotectants biosynthetic pathways for accumulating such molecules that act by scavenging ROS, reducing lipid peroxidation, maintaining protein structure and functions [233–235]. Many works on the transformation of plants for improving salinity tolerance focus on genes controlling ion transport, as regulation of Na^+ uptake and compartmentalization is a critically important mechanism for plant survival under salinity stress, and many candidate genes controlling this mechanism have been identified. Engineering plants for overexpression of genes encoding for antiporters is identified as an effective method for generating salt-tolerant plants (Table 2and relative references in Table 2). Gene expression studies using constitutive promoters provide limited biological information compared with the use of inducible promoters or cell type-specific promoters. The choice of promoters can significantly affect the results from a transgenic manipulation. Thus salt tolerant crops could be engineered by successful fine-tuning of the stress response by engineering novel regulatory targets; proper understanding of posttranslational modifications which regulate plant growth performance under stress; overexpression of miRNAs or their targets; maintaining hormone homeostasis to avoid pleiotropic effects under stress; and applying plant synthetic biology approaches to improve genetic engineering strategies [236].

Table 2. Improving plant salt tolerance through engineering genes for various membrane antiporters.

Transgenic host	Gene engineered	Source	Improved functions under salinity stress	References
Arabidopsis	Vacuolar Na^+/H^+ antiporter Ms NHX1	Alfalfa (*Medicago sativa*)	Increased osmotic balance. MDA content rises.	[220]
Rice	Vacuolar Na^+/H^+ Antiporter Pg NHX1	*Pennisetum glaucum*	Elaborate root system.	[221]
Wheat	Vacuolar Na^+/H^+ Antiporter At NHX1	*Arabidopsis thaliana* L.	Increase in grain yield and biomass production. Accumulation of K^+ in leaf. Reduced aggregation of Na^+.	[222]
Tobacco	Vacuolar Na^+/H^+ antiporter GhNHX1	*Gossypium hirsutum*	Na^+ compartmentalization.	[223]
Tomato	Vacuolar Na^+/H^+ antiporter AtNHX1	*Arabidopsis thaliana* L.	Over production of vacuolar Na^+/H^+ antiporter.	[224]
Tobacco	Vacuolar Na^+/H^+ antiporter AlNHXI	*Aeluropus littoralis*	Compartmentalization of Na in roots. Maintenance of K^+/Na^+ ratio in the leaf.	[225]
Brassica	Vacuolar Na^+/H^+ antiporter AtNHX1	*Arabidopsis thaliana* L.	Increased proline content. Improved growth rate. Mitigate the toxic effect of Na^+.	[226]
Arabidopsis	Plasma membrane Na^+/H^+ antiporter SOS1	*Arabidopsis thaliana* L. (wild type)	Improved germination rate, root growth, and chlorophyll content. Reduced accumulation of Na^+.	[227]
Maize	Vacuolar Na^+/H^+ antiporter AtNHX1	*Arabidopsis thaliana* L.	Increased rate of germination.	[228]

6. CONCLUSIONS AND FUTURE RESEARCH PERSPECTIVES

Salinity tolerance involves a complex of responses at molecular, cellular, metabolic, physiological, and whole-plant levels. Extensive research through cellular, metabolic, and physiological analysis has elucidated that among various salinity responses, mechanisms or strategies controlling ion uptake, transport and balance, osmotic regulation, hormone metabolism, antioxidant metabolism, and stress signalling play critical roles in plant adaptation to salinity stress. Taking advantage of the latest advancements in the field of genomic, transcriptomic, proteomic, and metabolomic techniques, plant biologists are focusing on the development of a complete profile of genes, proteins, and metabolites responsible for different mechanisms of salinity tolerance in different plant species. However, there is lack of the integration of information from genomic, transcriptomic, proteomic, and metabolomics studies, and the combined approach is essential for the determination of the key pathways or processes controlling salinity tolerance. In addition, in spite of the significant progress in the understanding of plant stress responses, there is still a large gap in our knowledge of transmembrane ion transport, sensor and receptor in the signalling transduction, molecules in long distance signalling, and metabolites in energy supply. The future focus should be on the study of intercellular and intracellular molecular interaction involved in salinity stress response. Genetic engineering has been proved to be an efficient approach to the development of salinity-tolerant plants, and this approach will become more powerful as more candidate genes associated with salinity tolerance are identified and widely utilized.

ACKNOWLEDGMENTS

Bhaskar Gupta acknowledges the financial help from DBT (Govt. of India) for the DBT-RGYI grant and Government of West Bengal for the WB-DST grant.

REFERENCES

1. FAO, High Level Expert Forum—How to Feed the World in 2050, Economic and Social Development, Food and Agricultural Organization of the United Nations, Rome, Italy, 2009.

2. T. J. Flowers, "Improving crop salt tolerance," Journal of Experimental Botany, vol. 55, no. 396, pp. 307–319, 2004.

3. R. Munns and M. Tester, "Mechanisms of salinity tolerance," Annual Review of Plant Biology, vol. 59, pp. 651–681, 2008.

4. R. A. James, C. Blake, C. S. Byrt, and R. Munns, "Major genes for Na^+ exclusion, Nax1 and Nax2 (wheat HKT1;4 and HKT1;5), decrease Na^+ accumulation in bread wheat leaves under saline and waterlogged conditions," Journal of Experimental Botany, vol. 62, no. 8, pp. 2939–2947, 2011.

5. Rahnama, R. A. James, K. Poustini, and R. Munns, "Stomatal conductance as a screen for osmotic stress tolerance in durum wheat growing in saline soil," Functional Plant Biology, vol. 37, no. 3, pp. 255–263, 2010.

6. R. Munns, "Genes and salt tolerance: bringing them together," New Phytologist, vol. 167, no. 3, pp. 645–663, 2005.

7. J. Rozema and T. Flowers, "Ecology: crops for a salinized world," Science, vol. 322, no. 5907, pp. 1478–1480, 2008.

8. K. Apel and H. Hirt, "Reactive oxygen species: metabolism, oxidative stress, and signal transduction,"Annual Review of Plant Biology, vol. 55, pp. 373–399, 2004.

9. S. Mahajan and N. Tuteja, "Cold, salinity and drought stresses: an overview," Archives of Biochemistry and Biophysics, vol. 444, no. 2, pp. 139–158, 2005.

10. P. Ahmad, "Growth and antioxidant responses in mustard (Brassica juncea L.) plants subjected to combined effect of gibberellic acid and salinity," Archives of Agronomy and Soil Science, vol. 56, no. 5, pp. 575–588, 2010.

11. P. Ahmad and M. N. V. Prasad, Abiotic Stress Responses in Plants: Metabolism, Productivity and Sustainability, Springer, New York, NY, USA, 2012.

12. P. Ahmad and S. Umar, Oxidative Stress: Role of Antioxidants in Plants, Studium Press, New Delhi, India, 2011.

13. Q. Pang, S. Chen, S. Dai, Y. Chen, Y. Wang, and X. Yan, "Comparative proteomics of salt tolerance inArabidopsis thaliana and Thellungiella halophila," Journal of Proteome Research, vol. 9, no. 5, pp. 2584–2599, 2010.

14. E. Ábrahám, I. P. Salamó, C. Koncz, and L. Szabados, "Identification of Arabidopsis and Thellungiellagenes involved in salt tolerance by novel genetic system," Acta Biologica Szegediensis, vol. 55, no. 1, pp. 53–57, 2011.

15. J. L. Zhang and H. Shi, "Physiological and molecular mechanisms of plant salt tolerance," Photosynthesis Research, vol. 115, pp. 1–22, 2013.

16. N. X. Niu Xiaomu, R. A. Bressan, P. M. Hasegawa, and J. M. Pardo, "Ion homeostasis in NaCl stress environments," Plant Physiology, vol. 109, no. 3, pp. 735–742, 1995.

17. R. Serrano, J. M. Mulet, G. Rios et al., "A glimpse of the mechanisms of ion homeostasis during salt stress," Journal of Experimental Botany, vol. 50, pp. 1023–1036, 1999.

18. P. M. Hasegawa, "Sodium (Na^+) homeostasis and salt tolerance of plants," Environmental and Experimental Botany, vol. 92, pp. 19–31, 2013.

19. M. P. Reddy, S. Sanish, and E. R. R. Iyengar, "Photosynthetic studies and compartmentation of ions in different tissues of Salicornia brachiata Roxb. under saline conditions," Photosynthetica, vol. 26, pp. 173–179, 1992.

20. J.-K. Zhu, "Regulation of ion homeostasis under salt stress," Current Opinion in Plant Biology, vol. 6, no. 5, pp. 441–445, 2003.

21. K. J. Dietz, N. Tavakoli, C. Kluge et al., "Significance of the V-type ATPase for the adaptation to stressful growth conditions and its

regulation on the molecular and biochemical level," Journal of Experimental Botany, vol. 52, no. 363, pp. 1969–1980, 2001.

22. M. De Lourdes Oliveira Otoch, A. C. Menezes Sobreira, M. E. Farias De Aragão, E. G. Orellano, M. Da Guia Silva Lima, and D. Fernandes De Melo, "Salt modulation of vacuolar H^+-ATPase and H^+-Pyrophosphatase activities in Vigna unguiculata," Journal of Plant Physiology, vol. 158, no. 5, pp. 545–551, 2001.

23. B. Wang, U. Lüttge, and R. Ratajczak, "Effects of salt treatment and osmotic stress on V-ATPase and V-PPase in leaves of the halophyte Suaeda salsa," Journal of Experimental Botany, vol. 52, no. 365, pp. 2355–2365, 2001.

24. P. M. Hasegawa, R. A. Bressan, J.-K. Zhu, and H. J. Bohnert, "Plant cellular and molecular responses to high salinity," Annual Review of Plant Biology, vol. 51, pp. 463–499, 2000.

25. D. Sanders, "Plant biology: the salty tale of Arabidopsis," Current Biology, vol. 10, no. 13, pp. R486–R488, 2000.

26. H. Shi, M. Ishitani, C. Kim, and J.-K. Zhu, "The Arabidopsis thaliana salt tolerance gene SOS1 encodes a putative Na^+/H^+ antiporter," Proceedings of the National Academy of Sciences of the United States of America, vol. 97, no. 12, pp. 6896–6901, 2000.

27. H. Shi, F. J. Quintero, J. M. Pardo, and J.-K. Zhu, "The putative plasma membrane Na^+/H^+ antiporter SOS1 controls long-distance Na^+ transport in plants," Plant Cell, vol. 14, no. 2, pp. 465–477, 2002.

28. J. Liu, M. Ishitani, U. Halfter, C.-S. Kim, and J.-K. Zhu, "The Arabidopsis thaliana SOS2 gene encodes a protein kinase that is required for salt tolerance," Proceedings of the National Academy of Sciences of the United States of America, vol. 97, no. 7, pp. 3730–3734, 2000.

29. M. Ishitani, J. Liu, U. Halfter, C.-S. Kim, W. Shi, and J.-K. Zhu, "SOS3 function in plant salt tolerance requires N-myristoylation and calcium binding," Plant Cell, vol. 12, no. 9, pp. 1667–1677, 2000.

30. Y. Guo, Q.-S. Qiu, F. J. Quintero et al., "Transgenic Evaluation of Activated Mutant Alleles of SOS2 Reveals a Critical Requirement for Its Kinase Activity and C-Terminal Regulatory Domain for Salt Tolerance in Arabidopsis thaliana," Plant Cell, vol. 16, no. 2, pp. 435–449, 2004.

31. F. J. Quintero, M. Ohta, H. Shi, J.-K. Zhu, and J. M. Pardo, "Reconstitution in yeast of the ArabidopsisSOS signaling pathway for Na^+ homeostasis," Proceedings of the National Academy of Sciences of the United States of America, vol. 99, no. 13, pp. 9061–9066, 2002.

32. F. J. Quintero, J. Martinez-Atienza, I. Villalta et al., "Activation of the plasma membrane Na/H antiporter salt-overly-sensitive 1 (SOS1) by phosphorylation of an auto-inhibitory C-terminal domain," Proceedings of the National Academy of Sciences of the United States of America, vol. 108, no. 6, pp. 2611–2616, 2011.

33. D.-H. Oh, S. Y. Lee, R. A. Bressan, D.-J. Yun, and H. J. Bohnert, "Intracellular consequences of SOS1 deficiency during salt stress," Journal of Experimental Botany, vol. 61, no. 4, pp. 1205–1213, 2010.

34. J. Martínez-Atienza, X. Jiang, B. Garciadeblas et al., "Conservation of the salt overly sensitive pathway in rice," Plant Physiology, vol. 143, no. 2, pp. 1001–1012, 2007.

35. R. K. Sairam and A. Tyagi, "Physiology and molecular biology of salinity stress tolerance in plants,"Current Science, vol. 86, no. 3, pp. 407–421, 2004.

36. L. Ma, H. Zhang, L. Sun et al., "NADPH oxidase AtrbohD and AtrbohF function in ROS-dependent regulation of Na^+/K^+ homeostasis in Arabidopsis under salt stress," Journal of Experimental Botany, vol. 63, no. 1, pp. 305–317, 2012.

37. H. E. Yen, S.-M. Wu, Y.-H. Hung, and S.-K. Yen, "Isolation of 3 salt-induced low-abundance cDNAs from light-grown callus of Mesembryanthemum crystallinum by suppression subtractive hybridization,"Physiologia Plantarum, vol. 110, no. 3, pp. 402–409, 2000.

38. J. I. Schroeder, E. Delhaize, W. B. Frommer et al., "Using membrane transporters to improve crops for sustainable food production," Nature, vol. 497, pp. 60–66, 2013.

39. V. Barragán, E. O. Leidi, Z. Andrés et al., "Ion exchangers NHX1 and NHX2 mediate active potassium uptake into vacuoles to regulate cell

turgor and stomatal function in Arabidopsis," Plant Cell, vol. 24, no. 3, pp. 1127–1142, 2012.

40. F. J. Gálvez, M. Baghour, G. Hao, O. Cagnac, M. P. Rodríguez-Rosales, and K. Venema, "Expression of LeNHX isoforms in response to salt stress in salt sensitive and salt tolerant tomato species," Plant Physiology and Biochemistry, vol. 51, pp. 109–115, 2012.

41. M. A. Hoque, M. N. A. Banu, E. Okuma et al., "Exogenous proline and glycinebetaine increase NaCl-induced ascorbate-glutathione cycle enzyme activities, and proline improves salt tolerance more than glycinebetaine in tobacco Bright Yellow-2 suspension-cultured cells," Journal of Plant Physiology, vol. 164, no. 11, pp. 1457–1468, 2007.

42. P. Ahmad, C. A. Jaleel, M. A. Salem, G. Nabi, and S. Sharma, "Roles of enzymatic and nonenzymatic antioxidants in plants during abiotic stress," Critical Reviews in Biotechnology, vol. 30, no. 3, pp. 161–175, 2010.

43. M. A. Hossain, S. Munemasa, M. Uraji, Y. Nakamura, I. C. Mori, and Y. Murata, "Involvement of endogenous abscisic acid in methyl jasmonate-induced stomatal closure in Arabidopsis," Plant Physiology, vol. 156, no. 1, pp. 430–438, 2011.

44. N. Nounjan, P. T. Nghia, and P. Theerakulpisut, "Exogenous proline and trehalose promote recovery of rice seedlings from salt-stress and differentially modulate antioxidant enzymes and expression of related genes," Journal of Plant Physiology, vol. 169, no. 6, pp. 596–604, 2012.

45. M. A. Tahir, T. Aziz, M. Farooq, and G. Sarwar, "Silicon-induced changes in growth, ionic composition, water relations, chlorophyll contents and membrane permeability in two salt-stressed wheat genotypes," Archives of Agronomy and Soil Science, vol. 58, no. 3, pp. 247–256, 2012.

46. M. A. Khan, I. A. Ungar, and A. M. Showalter, "Effects of sodium chloride treatments on growth and ion accumulation of the halophyte haloxylon recurvum," Communications in Soil Science and Plant Analysis, vol. 31, no. 17-18, pp. 2763–2774, 2000.

47. Y. Wang and N. Nii, "Changes in chlorophyll, ribulose bisphosphate carboxylase-oxygenase, glycine betaine content, photosynthesis and transpiration in Amaranthus tricolor leaves during salt stress,"Journal of Horticultural Science and Biotechnology, vol. 75, no. 6, pp. 623–627, 2000.

48. Kerepesi and G. Galiba, "Osmotic and salt stress-induced alteration in soluble carbohydrate content in wheat seedlings," Crop Science, vol. 40, no. 2, pp. 482–487, 2000.

49. H. J. Bohnert, D. E. Nelson, and R. G. Jensen, "Adaptations to environmental stresses," Plant Cell, vol. 7, no. 7, pp. 1099–1111, 1995.

50. C. W. Ford, "Accumulation of low molecular weight solutes in water-stressed tropical legumes,"Phytochemistry, vol. 23, no. 5, pp. 1007–1015, 1984.

51. M. Dopp, F. Larher, and P. Weigel, "Osmotic adaption in Australian mangroves," Vegetatio, vol. 61, no. 1–3, pp. 247–253, 1985.

52. M. Ashraf and M. R. Foolad, "Roles of glycine betaine and proline in improving plant abiotic stress resistance," Environmental and Experimental Botany, vol. 59, no. 2, pp. 206–216, 2007.

53. S. C. Saxena, H. Kaur, P. Verma, et al., "Osmoprotectants: potential for crop improvement under adverse conditions," in Plant Acclimation to Environmental Stress, pp. 197–232, Springer, New York, NY, USA, 2013.

54. D. Hanson, B. Rathinasabapathi, J. Rivoal, M. Burnet, M. O. Dillon, and D. A. Gage, "Osmoprotective compounds in the Plumbaginaceae: a natural experiment in metabolic engineering of stress tolerance,"Proceedings of the National Academy of Sciences of the United States of America, vol. 91, no. 1, pp. 306–310, 1994.

55. F. El-Shintinawy and M. N. El-Shourbagy, "Alleviation of changes in protein metabolism in NaCl-stressed wheat seedlings by thiamine," Biologia Plantarum, vol. 44, no. 4, pp. 541–545, 2001.

56. Matysik, A. Alia, B. Bhalu, and P. Mohanty, "Molecular mechanisms of quenching of reactive oxygen species by proline under stress in plants," Current Science, vol. 82, no. 5, pp. 525–532, 2002.

57. C. Ben Ahmed, B. Ben Rouina, S. Sensoy, M. Boukhriss, and F. Ben Abdullah, "Exogenous proline effects on photosynthetic performance and antioxidant defense system of young olive tree," Journal of Agricultural and Food Chemistry, vol. 58, no. 7, pp. 4216–4222, 2010.

58. M. A. Hoque, M. N. A. Banu, Y. Nakamura, Y. Shimoishi, and Y. Murata, "Proline and glycinebetaine enhance antioxidant defense and methylglyoxal detoxification systems and reduce NaCl-induced damage in cultured tobacco cells," Journal of Plant Physiology, vol. 165, no. 8, pp. 813–824, 2008.

59. S. Deivanai, R. Xavier, V. Vinod, K. Timalata, and O. F. Lim, "Role of exogenous proline in ameliorating salt stress at early stage in two rice cultivars," Journal of Stress Physiology & Biochemistry, vol. 7, pp. 157–174, 2011.

60. M. A. A. Gadallah, "Effects of proline and glycinebetaine on Vicia faba responses to salt stress," Biologia Plantarum, vol. 42, no. 2, pp. 249–257, 1999.

61. P. Makela, J. Karkkainen, and S. Somersalo, "Effect of glycinebetaine on chloroplast ultrastructure, chlorophyll and protein content, and RuBPCO activities in tomato grown under drought or salinity,"Biologia Plantarum, vol. 43, no. 3, pp. 471–475, 2000.

62. S. Cha-Um and C. Kirdmanee, "Effect of glycinebetaine on proline, water use, and photosynthetic efficiencies, and growth of rice seedlings under salt stress," Turkish Journal of Agriculture and Forestry, vol. 34, no. 6, pp. 517–527, 2010.

63. R. Ahmad, C. J. Lim, and S.-Y. Kwon, "Glycine betaine: a versatile compound with great potential for gene pyramiding to improve crop plant performance against environmental stresses," Plant Biotechnology Reports, vol. 7, pp. 49–57, 2013.

64. S. Rahman, H. Miyake, and Y. Takeoka, "Effects of exogenous glycinebetaine on growth and ultrastructure of salt-stressed rice seedlings (Oryza sativa L.)," Plant Production Science, vol. 5, no. 1, pp. 33–44, 2002.

65. M. L. Binzel, F. D. Hess, R. A. Bressan, and P. M. Hasegawa, "Intracellular compartmentation of ions in salt adapted tobacco cells," Plant Physiology, vol. 86, pp. 607–614, 1988.

66. C. Thomas, M. Sepahi, B. Arendall, and H. J. Bohnert, "Enhancement of seed germination in high salinity by engineering mannitol expression in Arabidopsis thaliana," Plant, Cell and Environment, vol. 18, no. 7, pp. 801–806, 1995.

67. K. Parida, A. B. Das, and P. Mohanty, "Investigations on the antioxidative defence responses to NaCl stress in a mangrove, Bruguiera parviflora: differential regulations of isoforms of some antioxidative enzymes," Plant Growth Regulation, vol. 42, no. 3, pp. 213–226, 2004.

68. Z. Gao, M. Sagi, and S. H. Lips, "Carbohydrate metabolism in leaves and assimilate partitioning in fruits of tomato (Lycopersicon esculentum L.) as affected by salinity," Plant Science, vol. 135, no. 2, pp. 149–159, 1998.

69. N. M. Alamgir and M. Yousuf Ali, "Effect of salinity on leaf pigments, sugar and protein concentrations and chloroplast ATPase activity of rice (Oryza sativa L.)," Bangladesh Journal of Botany, vol. 28, no. 2, pp. 145–149, 1999.

70. F. Groß, J. Durner, and F. Gaupels, "Nitric oxide, antioxidants and prooxidants in plant defence responses," Frontiers in Plant Science, vol. 4, article 419, 2013.

71. Asada, "The water-water cycle in chloroplasts: scavenging of active oxygens and dissipation of excess photons," Annual Review of Plant Biology, vol. 50, pp. 601–639, 1999.

72. K. J. Gupta, M. Stoimenova, and W. M. Kaiser, "In higher plants, only root mitochondria, but not leaf mitochondria reduce nitrite to NO, in vitro and in situ," Journal of Experimental Botany, vol. 56, no. 420, pp. 2601–2609, 2005.

73. S. S. Gill, M. Tajrishi, M. Madan, and N. Tuteja, "A DESD-box helicase functions in salinity stress tolerance by improving photosynthesis and antioxidant machinery in rice (Oryza sativa L. cv. PB1),"Plant Molecular Biology, vol. 82, no. 1-2, pp. 1–22, 2013.

74. Tuteja, R. K. Sahoo, B. Garg, and R. Tuteja, "OsSUV3 dual helicase functions in salinity stress tolerance by maintaining photosynthesis and antioxidant machinery in rice (Oryza sativa L. cv. IR64),"The Plant Journal, vol. 76, no. 1, pp. 115–127, 2013.

75. Y. H. Kim, A. Latif Khan, M. Waqas, et al., "Silicon application to rice root zone influenced the phytohormonal and antioxidant responses under salinity stress," Journal of Plant Growth Regulation, 2013.

76. J. Van Oosten, A. Sharkhuu, G. Batelli, R. A. Bressan, and A. Maggio, "The Arabidopsis thalianamutant air1 implicates SOS3 in the regulation of anthocyanins under salt stress," Plant Molecular Biology, vol. 83, pp. 405–415, 2013.

77. J. C. Begara-Morales, B. Sánchez-Calvo, M. Chaki, et al., "Dual regulation of cytosolic ascorbate peroxidase (APX) by tyrosine nitration and S-nitrosylation," Journal of Experimental Botany, vol. 65, no. 2, pp. 527–538, 2014.

78. S. Agarwal and R. Shaheen, "Stimulation of antioxidant system and lipid peroxidation by abiotic stresses in leaves of Momordica charantia," Brazilian Journal of Plant Physiology, vol. 19, no. 2, pp. 149–161, 2007.

79. Munir and F. Aftab, "Enhancement of salt tolerance in sugarcane by ascorbic acid pretreatment,"African Journal of Biotechnology, vol. 10, no. 80, pp. 18362–18370, 2011.

80. C. H. Foyer, H. Lopez-Delgado, J. F. Dat, and I. M. Scott, "Hydrogen peroxide- and glutathione-associated mechanisms of acclimatory stress tolerance and signalling," Physiologia Plantarum, vol. 100, no. 2, pp. 241–254, 1997.

81. K. H. Aly-Salama and M. M. Al-Mutawa, "Glutathione-triggered mitigation in salt-induced alterations in plasmalemma of onion epidermal cells," International Journal of Agriculture and Biology, vol. 11, no. 5, pp. 639–642, 2009.

82. Rawia Eid, L. S. Taha, and S. M. M. Ibrahiem, "Alleviation of adverse effects of salinity on growth, and chemical constituents of marigold plants by using glutathione and ascorbate," Journal of Applied Sciences Research, vol. 7, pp. 714–721, 2011.

83. S. J. Roy, S. Negrão, and M. Tester, "Salt resistant crop plants," Current Opinion in Biotechnology, vol. 26, pp. 115–124, 2014.

84. J. M. Knott, P. Römer, and M. Sumper, "Putative spermine synthases from Thalassiosira pseudonana andArabidopsis thaliana synthesize thermospermine rather than spermine," FEBS Letters, vol. 581, no. 16, pp. 3081–3086, 2007.

85. K. Gupta, A. Dey, and B. Gupta, "Plant polyamines in abiotic stress responses," Acta Physiologiae Plantarum, vol. 35, no. 7, pp. 2015–2036, 2013.

86. M. Panicot, E. G. Minguet, A. Ferrando et al., "A polyamine metabolon involving aminopropyl transferase complexes in Arabidopsis," Plant Cell, vol. 14, no. 10, pp. 2539–2551, 2002.

87. W. Galston, R. Kaur-Sawhney, T. Altabella, and A. F. Tiburcio, "Plant polyamines in reproductive activity and response to abiotic stress," Botanica Acta, vol. 110, no. 3, pp. 197–207, 1997.

88. K. Gupta, A. Dey, and B. Gupta, "Polyamines and their role in plant osmotic stress tolerance," in Climate Change and Plant Abiotic Stress Tolerance, N. Tuteja and S. S. Gill, Eds., pp. 1053–1072, Wiley-VCH, Weinheim, Germany, 2013.

89. J. Yang, J. Zhang, K. Liu, Z. Wang, and L. Liu, "Involvement of polyamines in the drought resistance of rice," Journal of Experimental Botany, vol. 58, no. 6, pp. 1545–1555, 2007.

90. M. D. Groppa and M. P. Benavides, "Polyamines and abiotic stress: recent advances," Amino Acids, vol. 34, no. 1, pp. 35–45, 2008.

91. Z. Kovács, L. Simon-Sarkadi, A. Szucs, and G. Kocsy, "Differential effects of cold, osmotic stress and abscisic acid on polyamine accumulation in wheat," Amino Acids, vol. 38, no. 2, pp. 623–631, 2010.

92. R. Alcázar, J. C. Cuevas, J. Planas et al., "Integration of polyamines in the cold acclimation response,"Plant Science, vol. 180, no. 1, pp. 31–38, 2011.

93. J. Martin-Tanguy, "Metabolism and function of polyamines in plants: recent development (new approaches)," Plant Growth Regulation, vol. 34, no. 1, pp. 135–148, 2001.

94. V. V. Kuznetsov and N. I. Shevyakova, "Polyamines and stress tolerance of plants," Plant Stress, vol. 1, pp. 50–71, 2007.

95. S. S. Hussain, M. Ali, M. Ahmad, and K. H. M. Siddique, "Polyamines: natural and engineered abiotic and biotic stress tolerance in plants," Biotechnology Advances, vol. 29, no. 3, pp. 300–311, 2011.

96. S. Shu, S. R. Guo, and L. Y. Yuan, "A review: polyamines and photosynthesis," in Advances in Photosynthesis—Fundamental Aspects, M. M. Najafpour, Ed., pp. 439–464, InTech, Rijeka, Croatia, 2012.

97. I. Alet, D. H. Sánchez, J. C. Cuevas et al., "New insights into the role of spermine in Arabidopsis thaliana under long-term salt stress," Plant Science, vol. 182, no. 1, pp. 94–100, 2012.

98. R. Alcázar, J. L. García-Martínez, J. C. Cuevas, A. F. Tiburcio, and T. Altabella, "Overexpression of ADC2 in Arabidopsis induces dwarfism and late-flowering through GA deficiency," Plant Journal, vol. 43, no. 3, pp. 425–436, 2005.

99. C. Bortolotti, A. Cordeiro, R. Alcázar et al., "Localization of arginine decarboxylase in tobacco plants,"Physiologia Plantarum, vol. 120, no. 1, pp. 84–92, 2004.

100. C. Illingworth, M. J. Mayer, K. Elliott, C. Hanfrey, N. J. Walton, and A. J. Michael, "The diverse bacterial origins of the Arabidopsis polyamine biosynthetic pathway," FEBS Letters, vol. 549, no. 1–3, pp. 26–30, 2003.

101. J. M. Knott, P. Römer, and M. Sumper, "Putative spermine synthases from Thalassiosira pseudonana andArabidopsis thaliana synthesize thermospermine rather than spermine," FEBS Letters, vol. 581, no. 16, pp. 3081–3086, 2007.

102. T. Kusano, K. Yamaguchi, T. Berberich, and Y. Takahashi, "The polyamine spermine rescues Arabidopsisfrom salinity and drought stresses (Plant Signaling and Behavior)," Plant Signaling and Behavior, vol. 2, no. 4, pp. 251–252, 2007.

103. J. Martin-Tanguy, "Metabolism and function of polyamines in plants: recent development (new approaches)," Plant Growth Regulation, vol. 34, no. 1, pp. 135–148, 2001.

104. J. L. Rambla, F. Vera-Sirera, M. A. Blázquez, J. Carbonell, and A. Granell, "Quantitation of biogenic tetraamines in Arabidopsis thaliana," Analytical Biochemistry, vol. 397, no. 2, pp. 208–211, 2010.

105. M. Hasanuzzaman, K. Nahar, and M. Fujita, "Regulatory role of polyamines in growth, development and abiotic stress tolerance in plants," in Plant Adaptation to Environmental Change: Significance of Amino Acids and Their Derivatives, pp. 157–193, 2014.

106. R. Alcázar, J. Planas, T. Saxena et al., "Putrescine accumulation confers drought tolerance in transgenicArabidopsis plants over-expressing the homologous Arginine decarboxylase 2 gene," Plant Physiology and Biochemistry, vol. 48, no. 7, pp. 547–552, 2010.

107. Bouchereau, A. Aziz, F. Larher, and J. Martin-Tanguy, "Polyamines and environmental challenges: recent development," Plant Science, vol. 140, no. 2, pp. 103–125, 1999.

108. R. Alcázar, F. Marco, J. C. Cuevas et al., "Involvement of polyamines in plant response to abiotic stress,"Biotechnology Letters, vol. 28, no. 23, pp. 1867–1876, 2006.

109. R. Fluhr and A. K. Mattoo, "Ethylene-biosynthesis and perception," Critical Reviews in Plant Sciences, vol. 15, pp. 479–523, 1996.

110. T. Kusano, K. Yamaguchi, T. Berberich, and Y. Takahashi, "Advances in polyamine research in 2007,"Journal of Plant Research, vol. 120, no. 3, pp. 345–350, 2007.

111. K. D. Allen, "Assaying gene content in Arabidopsis," Proceedings of the National Academy of Sciences of the United States of America, vol. 99, pp. 9568–9572, 2002.

112. C. Hanfrey, S. Sommer, M. J. Mayer, D. Burtin, and A. J. Michael, "Arabidopsis polyamine biosynthesis: absence of ornithine decarboxylase and the mechanism of arginine decarboxylase activity," Plant Journal, vol. 27, no. 6, pp. 551–560, 2001.

113. K. Urano, Y. Yoshiba, T. Nanjo et al., "Characterization of Arabidopsis genes involved in biosynthesis of polyamines in abiotic stress responses and developmental stages," Plant, Cell and Environment, vol. 26, no. 11, pp. 1917–1926, 2003.

114. C. Ge, X. Cui, Y. Wang et al., "BUD2, encoding an S-adenosylmethionine decarboxylase, is required forArabidopsis growth and development," Cell Research, vol. 16, no. 5, pp. 446–456, 2006.

115. T. Janowitz, H. Kneifel, and M. Piotrowski, "Identification and characterization of plant agmatine iminohydrolase, the last missing link in polyamine biosynthesis of plants," FEBS Letters, vol. 544, no. 1–3, pp. 258–261, 2003.

116. M. B. Watson and R. L. Malmberg, "Regulation of Arabidopsis thaliana (L.) Heynh arginine decarboxylase by potassium deficiency stress," Plant Physiology, vol. 111, no. 4, pp. 1077–1083, 1996.

117. Y. Hanzawa, A. Imai, A. J. Michael, Y. Komeda, and T. Takahashi, "Characterization of the spermidine synthase-related gene family in Arabidopsis thaliana," FEBS Letters, vol. 527, no. 1–3, pp. 176–180, 2002.

118. T. Hashimoto, K. Tamaki, K.-I. Suzuki, and Y. Yamada, "Molecular cloning of plant spermidine synthases," Plant and Cell Physiology, vol. 39, no. 1, pp. 73–79, 1998.

119. K. Urano, Y. Yoshiba, T. Nanjo, T. Ito, K. Yamaguchi-Shinozaki, and K. Shinozaki, "Arabidopsis stress-inducible gene for arginine decarboxylase AtADC2 is required for accumulation of putrescine in salt tolerance," Biochemical and Biophysical Research Communications, vol. 313, no. 2, pp. 369–375, 2004.

120. Y. Hanzawa, A. Imai, A. J. Michael, Y. Komeda, and T. Takahashi, "Characterization of the spermidine synthase-related gene family in Arabidopsis thaliana," FEBS Letters, vol. 527, no. 1–3, pp. 176–180, 2002.

121. T. Takahashi and J.-I. Kakehi, "Polyamines: ubiquitous polycations with unique roles in growth and stress responses," Annals of Botany, vol. 105, no. 1, pp. 1–6, 2010.

122. Cona, G. Rea, R. Angelini, R. Federico, and P. Tavladoraki, "Functions of amine oxidases in plant development and defence," Trends in Plant Science, vol. 11, no. 2, pp. 80–88, 2006.

123. Tisi, R. Angelini, and A. Cona, "Wound healing in plants: cooperation of copper amine oxidase and flavin-containing polyamine oxidase," Plant Signaling and Behavior, vol. 3, no. 3, pp. 204–206, 2008.

124. E. Navakoudis, C. Lütz, C. Langebartels, U. Lütz-Meindl, and K. Kotzabasis, "Ozone impact on the photosynthetic apparatus and the protective role of polyamines," Biochimica et Biophysica Acta—General Subjects, vol. 1621, no. 2, pp. 160–169, 2003.

125. Roy, K. Niyogi, D. N. SenGupta, and B. Ghosh, "Spermidine treatment to rice seedlings recovers salinity stress-induced damage of plasma membrane and PM-bound H^+-ATPase in salt-tolerant and salt-sensitive rice cultivars," Plant Science, vol. 168, no. 3, pp. 583–591, 2005.

126. Roychoudhury, S. Basu, and D. N. Sengupta, "Amelioration of salinity stress by exogenously applied spermidine or spermine in three varieties of indica rice differing in their level of salt tolerance," Journal of Plant Physiology, vol. 168, no. 4, pp. 317–328, 2011.

127. X. Wang, G. Shi, Q. Xu, and J. Hu, "Exogenous polyamines enhance copper tolerance of Nymphoides peltatum," Journal of Plant Physiology, vol. 164, no. 8, pp. 1062–1070, 2007.

128. M. Ali, "Role of putrescine in salt tolerance of Atropa belladonna plant," Plant Science, vol. 152, no. 2, pp. 173–179, 2000.

129. Afzal, F. Munir, C. M. Ayub, S. M. A. Basra, A. Hameed, and A. Nawaz, "Changes in antioxidant enzymes, germination capacity and vigour of tomato seeds in response of priming with polyamines,"Seed Science and Technology, vol. 37, no. 3, pp. 765–770, 2009.

130. J.-C. Yiu, L.-D. Juang, D. Y.-T. Fang, C.-W. Liu, and S.-J. Wu, "Exogenous putrescine reduces flooding-induced oxidative damage by increasing the antioxidant properties of Welsh onion," Scientia Horticulturae, vol. 120, no. 3, pp. 306–314, 2009.

131. Hussain, K. Nawaz, A. Majeed et al., "Role of exogenous salicylic acid applications for salt tolerance in violet (Viola Odorata L.)," Sarhad Journal of Agriculture, vol. 27, pp. 171–175, 2011.

132. Yamaguchi, Y. Takahashi, T. Berberich et al., "The polyamine spermine protects against high salt stress in Arabidopsis thaliana," FEBS Letters, vol. 580, no. 30, pp. 6783–6788, 2006.

133. Roy and R. Wu, "Overexpression of S-adenosylmethionine decarboxylase gene in rice increases polyamine level and enhances sodium chloride-stress tolerance," Plant Science, vol. 163, no. 5, pp. 987–992, 2002.

134. I. Shevyakova, L. I. Musatenko, L. A. Stetsenko, et al., "Effects of abscisic acid on the contents of polyamines and proline in common bean plants under salt stress," Russian Journal of Plant Physiology, vol. 60, pp. 200–211, 2013.

135. P. N. Moschou, K. A. Paschalidis, and K. A. Roubelakis-Angelakis, "Plant polyamine catabolism: the state of the art," Plant Signaling and Behavior, vol. 3, no. 12, pp. 1061–1066, 2008.

136. J. Duan, J. Li, S. Guo, and Y. Kang, "Exogenous spermidine affects polyamine metabolism in salinity-stressed Cucumis sativus roots and enhances short-term salinity tolerance," Journal of Plant Physiology, vol. 165, no. 15, pp. 1620–1635, 2008.

137. Y. Y. Chai, C. D. Jiang, L. Shi, T. S. Shi, and W. B. Gu, "Effects of exogenous spermine on sweet sorghum during germination under salinity," Biologia Plantarum, vol. 54, no. 1, pp. 145–148, 2010.

138. Li, L. He, S. Guo, et al., "Proteomics reveal cucumber Spd-responses under normal condition and salt stress," Plant Physiology and Biochemistry, vol. 67, pp. 7–14, 2013.

139. Delledonne, Y. Xia, R. A. Dixon, and C. Lamb, "Nitric oxide functions as a signal in plant disease resistance," Nature, vol. 394, no. 6693, pp. 585–588, 1998.

140. L. Lamattina, C. García-Mata, M. Graziano, and G. Pagnussat, "Nitric oxide: the versatility of an extensive signal molecule," Annual Review of Plant Biology, vol. 54, pp. 109–136, 2003.

141. Besson-Bard, A. Pugin, and D. Wendehenne, "New insights into nitric oxide signaling in plants,"Annual Review of Plant Biology, vol. 59, pp. 21–39, 2008.

142. M.-G. Zhao, L. Chen, L.-L. Zhang, and W.-H. Zhang, "Nitric reductase-dependent nitric oxide production is involved in cold acclimation and freezing tolerance in Arabidopsis," Plant Physiology, vol. 151, no. 2, pp. 755–767, 2009.

143. M. Crawford, "Mechanisms for nitric oxide synthesis in plants," Journal of Experimental Botany, vol. 57, no. 3, pp. 471–478, 2006.

144. Bajgu, "Nitric oxide: role in plants under abiotic stress," in Physiological Mechanisms and Adaptation Strategies in Plants Under Changing Environment, pp. 137–159, Springer, 2014.

145. H. Sung and J. K. Hong, "Sodium nitroprusside mediates seedling development and attenuation of oxidative stresses in Chinese cabbage," Plant Biotechnology Reports, vol. 4, no. 4, pp. 243–251, 2010.

146. K. K. Hossain, R. D. Itoh, G. Yoshimura et al., "Effects of nitric oxide scavengers on thermoinhibition of seed germination in Arabidopsis thaliana," Russian Journal of Plant Physiology, vol. 57, no. 2, pp. 222–232, 2010.

147. J. Xiong, G. Fu, L. Tao, and C. Zhu, "Roles of nitric oxide in alleviating heavy metal toxicity in plants,"Archives of Biochemistry and Biophysics, vol. 497, no. 1-2, pp. 13–20, 2010.

148. M. Kopyra and E. A. Gwóźdź, "Nitric oxide stimulates seed germination and counteracts the inhibitory effect of heavy metals and salinity on root growth of Lupinus luteus," Plant Physiology and Biochemistry, vol. 41, no. 11-12, pp. 1011–1017, 2003.

149. Y. Zhang, L. Wang, Y. Liu, Q. Zhang, Q. Wei, and W. Zhang, "Nitric oxide enhances salt tolerance in maize seedlings through increasing activities of proton-pump and Na^+/H^+ antiport in the tonoplast,"Planta, vol. 224, no. 3, pp. 545–555, 2006.

150. Mishra, A. B. Jha, and R. S. Dubey, "Arsenite treatment induces oxidative stress, upregulates antioxidant system, and causes phytochelatin synthesis in rice seedlings," Protoplasma, vol. 248, no. 3, pp. 565–577, 2011.

151. L. Zhao, F. Zhang, J. Guo, Y. Yang, B. Li, and L. Zhang, "Nitric oxide functions as a signal in salt resistance in the calluses from two ecotypes of reed," Plant Physiology, vol. 134, no. 2, pp. 849–857, 2004.

152. M. Nalousi, S. Ahmadiyan, A. Hatamzadeh, and M. Ghasemnezhad, "Protective role of exogenous nitric oxide against oxidative stress induced by salt stress in bell-pepper (Capsicum annum L.),"American-Eurasian Journal of Agricultural & Environmental Science, vol. 12, no. 8, pp. 1085–1090, 2012.

153. F. Zhang, Y. Wang, Y. Yang, H. Wu, D. Wang, and J. Liu, "Involvement of hydrogen peroxide and nitric oxide in salt resistance in the calluses from Populus euphratica," Plant, Cell and Environment, vol. 30, no. 7, pp. 775–785, 2007.

154. He and G. R. Cramer, "Abscisic acid concentrations are correlated with leaf area reductions in two salt-stressed rapid-cycling Brassica species," Plant and Soil, vol. 179, no. 1, pp. 25–33, 1996.

155. Cabot, J. V. Sibole, J. Barceló, and C. Poschenrieder, "Abscisic acid decreases leaf Na^+ exclusion in salt-treated Phaseolus vulgaris L," Journal of Plant Growth Regulation, vol. 28, no. 2, pp. 187–192, 2009.

156. G. R. Cramer and S. A. Quarrie, "Abscisic acid is correlated with the leaf growth inhibition of four genotypes of maize differing in their response to salinity," Functional Plant Biology, vol. 29, no. 1, pp. 111–115, 2002.

157. D.-J. Kang, Y.-J. Seo, J.-D. Lee et al., "Jasmonic acid differentially affects growth, ion uptake and abscisic acid concentration in salt-tolerant and salt-sensitive rice cultivars," Journal of Agronomy and Crop Science, vol. 191, no. 4, pp. 273–282, 2005.

158. L. P. Popova, Z. G. Stoinova, and L. T. Maslenkova, "Involvement of abscisic acid in photosynthetic process in Hordeum vulgare L. during salinity stress," Journal of Plant Growth Regulation, vol. 14, no. 4, pp. 211–218, 1995.

159. W. D. Jeschke, A. D. Peuke, J. S. Pate, and W. Hartung, "Transport, synthesis and catabolism of abscisic acid (ABA) in intact plants of castor bean (Ricinus communis L.) under phosphate deficiency and moderate

salinity," Journal of Experimental Botany, vol. 48, no. 314, pp. 1737–1747, 1997.

160. Chen, J. Li, S. Wang, A. Hüttermann, and A. Altman, "Salt, nutrient uptake and transport, and ABA ofPopulus euphratica; a hybrid in response to increasing soil NaCl," Trees—Structure and Function, vol. 15, no. 3, pp. 186–194, 2001.

161. R. Gurmani, A. Bano, S. U. Khan, J. Din, and J. L. Zhang, "Alleviation of salt stress by seed treatment with abscisic acid (ABA), 6-benzylaminopurine (BA) and chlormequat chloride (CCC) optimizes ion and organic matter accumulation and increases yield of rice (Oryza sativa L.)," Australian Journal of Crop Science, vol. 5, no. 10, pp. 1278–1285, 2011.

162. Fukuda and Y. Tanaka, "Effects of ABA, auxin, and gibberellin on the expression of genes for vacuolar H^+-inorganic pyrophosphatase, H^+-ATPase subunit A, and Na^+/H^+ antiporter in barley," Plant Physiology and Biochemistry, vol. 44, no. 5-6, pp. 351–358, 2006.

163. C. Keskin, A. T. Sarikaya, B. Yüksel, and A. R. Memon, "Abscisic acid regulated gene expression in bread wheat (Triticum aestivum L.)," Australian Journal of Crop Science, vol. 4, no. 8, pp. 617–625, 2010.

164. Fragnire, M. Serrano, E. Abou-Mansour, J.-P. Métraux, and F. L'Haridon, "Salicylic acid and its location in response to biotic and abiotic stress," FEBS Letters, vol. 585, no. 12, pp. 1847–1852, 2011.

165. S. D. Clause and J. M. Sasse, "Brassinosteroids: essential regulators of plant growth and development,"Annual Review of Plant Biology, vol. 49, pp. 427–451, 1998.

166. H. Sawada, I.-S. Shim, and K. Usui, "Induction of benzoic acid 2-hydroxylase and salicylic acid biosynthesis-Modulation by salt stress in rice seedlings," Plant Science, vol. 171, no. 2, pp. 263–270, 2006.

167. M. Jayakannan, J. Bose, O. Babourina, et al., "Salicylic acid improves salinity tolerance in Arabidopsis by restoring membrane potential and preventing salt-induced K^+ loss via a GORK channel," Journal of Experimental Botany, vol. 64, no. 8, pp. 2255–2268, 2013.

168. M. A. El-Tayeb, "Response of barley grains to the interactive effect of salinity and salicylic acid," Plant Growth Regulation, vol. 45, no. 3, pp. 215–224, 2005.

169. R. Nazar, N. Iqbal, S. Syeed, and N. A. Khan, "Salicylic acid alleviates decreases in photosynthesis under salt stress by enhancing nitrogen and sulfur assimilation and antioxidant metabolism differentially in two mungbean cultivars," Journal of Plant Physiology, vol. 168, no. 8, pp. 807–815, 2011.

170. A. A. El-Mashad and H. I. Mohamed, "Brassinolide alleviates salt stress and increases antioxidant activity of cowpea plants (Vigna sinensis)," Protoplasma, vol. 249, no. 3, pp. 625–635, 2012.

171. S. Anuradha and S. Seeta Ram Rao, "Effect of brassinosteroids on salinity stress induced inhibition of seed germination and seedling growth of rice (Oryza sativa L.)," Plant Growth Regulation, vol. 33, no. 2, pp. 151–153, 2001.

172. Krishna, "Brassinosteroid-Mediated Stress Responses," Journal of Plant Growth Regulation, vol. 22, no. 4, pp. 289–297, 2003.

173. M. Ashraf, N. A. Akram, R. N. Arteca, and M. R. Foolad, "The physiological, biochemical and molecular roles of brassinosteroids and salicylic acid in plant processes and salt tolerance," Critical Reviews in Plant Sciences, vol. 29, no. 3, pp. 162–190, 2010.

174. R. Johnson, R. L. Wagner, S. D. Verhey, and M. K. Walker-Simmons, "The abscisic acid-responsive kinase PKABA1 interacts with a seed-specific abscisic acid response element-binding factor, TaABF, and phosphorylates TaABF peptide sequences," Plant Physiology, vol. 130, no. 2, pp. 837–846, 2002.

175. K. Nakashima, L.-S. P. Tran, D. Van Nguyen et al., "Functional analysis of a NAC-type transcription factor OsNAC6 involved in abiotic and biotic stress-responsive gene expression in rice," Plant Journal, vol. 51, no. 4, pp. 617–630, 2007.

176. J. Mizoi, K. Shinozaki, and K. Yamaguchi-Shinozaki, "AP2/ERF family transcription factors in plant abiotic stress responses," Biochimica et Biophysica Acta—Gene Regulatory Mechanisms, vol. 1819, no. 2, pp. 86–96, 2012.

177. Y. Fujita, M. Fujita, K. Shinozaki, and K. Yamaguchi-Shinozaki, "ABA-mediated transcriptional regulation in response to osmotic stress in plants," Journal of Plant Research, vol. 124, no. 4, pp. 509–525, 2011.

178. RoyChoudhury, B. Gupta, and D. N. Sengupta, "Trans-acting factor designated OSBZ8 interacts with both typical abscisic acid responsive elements as well as abscisic acid responsive element-like sequences in the vegetative tissues of indica rice cultivars," Plant Cell Reports, vol. 27, no. 4, pp. 779–794, 2008.

179. Y. Ito, K. Katsura, K. Maruyama et al., "Functional analysis of rice DREB1/CBF-type transcription factors involved in cold-responsive gene expression in transgenic rice," Plant and Cell Physiology, vol. 47, no. 1, pp. 141–153, 2006.

180. Y. Fujita, T. Yoshida, and K. Yamaguchi-Shinozaki, "Pivotal role of the AREB/ABF-SnRK2 pathway in ABRE-mediated transcription in response to osmotic stress in plants," Physiologia Plantarum, vol. 147, no. 1, pp. 15–27, 2013.

181. S.-Y. Song, Y. Chen, J. Chen, X.-Y. Dai, and W.-H. Zhang, "Physiological mechanisms underlying OsNAC5-dependent tolerance of rice plants to abiotic stress," Planta, vol. 234, no. 2, pp. 331–345, 2011.

182. Y. Hu, L. Chen, H. Wang, L. Zhang, F. Wang, and D. Yu, "Arabidopsis transcription factor WRKY8 functions antagonistically with its interacting partner VQ9 to modulate salinity stress tolerance," The Plant Journal, vol. 74, pp. 730–745, 2013.

183. K. Chakraborty, R. K. Sairam, and R. C. Bhattacharya, "Differential expression of salt overly sensitive pathway genes determines salinity stress tolerance in Brassica genotypes," Plant Physiology and Biochemistry, vol. 51, pp. 90–101, 2012.

184. P. Roshandel and T. Flowers, "The ionic effects of NaCl on physiology and gene expression in rice genotypes differing in salt tolerance," Plant and Soil, vol. 315, no. 1-2, pp. 135–147, 2009.

185. J. Zou, A. Liu, X. Chen et al., "Expression analysis of nine rice heat shock protein genes under abiotic stresses and ABA treatment," Journal of Plant Physiology, vol. 166, no. 8, pp. 851–861, 2009.

186. G.-H. Lim, X. Zhang, M.-S. Chung et al., "A putative novel transcription factor, AtSKIP, is involved in abscisic acid signalling and confers salt and osmotic tolerance in Arabidopsis," New Phytologist, vol. 185, no. 1, pp. 103–113, 2010.

187. J. Zou, C. Liu, A. Liu, D. Zou, and X. Chen, "Overexpression of OsHsp17.0 and OsHsp23.7 enhances drought and salt tolerance in rice," Journal of Plant Physiology, vol. 169, no. 6, pp. 628–635, 2012.

188. N.-H. Song and Y.-J. Ahn, "DcHsp17.7, a small heat shock protein in carrot, is tissue-specifically expressed under salt stress and confers tolerance to salinity," New Biotechnology, vol. 28, no. 6, pp. 698–704, 2011.

189. M. Tang, X. Liu, H. Deng, and S. Shen, "Over-expression of JcDREB, a putative AP2/EREBP domain-containing transcription factor gene in woody biodiesel plant Jatropha curcas, enhances salt and freezing tolerance in transgenic Arabidopsis thaliana," Plant Science, vol. 181, no. 6, pp. 623–631, 2011.

190. J. Liu, M. Ishitani, U. Halfter, C.-S. Kim, and J.-K. Zhu, "The Arabidopsis thaliana SOS2 gene encodes a protein kinase that is required for salt tolerance," Proceedings of the National Academy of Sciences of the United States of America, vol. 97, no. 7, pp. 3730–3734, 2000.

191. H. Shi, M. Ishitani, C. Kim, and J.-K. Zhu, "The Arabidopsis thaliana salt tolerance gene SOS1 encodes a putative Na^+/H^+ antiporter," Proceedings of the National Academy of Sciences of the United States of America, vol. 97, no. 12, pp. 6896–6901, 2000.

192. Kawasaki, C. Borchert, M. Deyholos et al., "Gene expression profiles during the initial phase of salt stress in rice," Plant Cell, vol. 13, no. 4, pp. 889–905, 2001.

193. N. Baisakh, P. K. Subudhi, and N. P. Parami, "cDNA-AFLP analysis reveals differential gene expression in response to salt stress in a halophyte Spartina alterniflora Loisel," Plant Science, vol. 170, no. 6, pp. 1141–1149, 2006.

194. Schmidt, D. Mieulet, H.-M. Hubberten, et al., "SALT-RESPONSIVE ERF1 regulates reactive oxygen species-dependent signaling during the

initial response to salt stress in rice," The Plant Cell, vol. 25, no. 6, pp. 2115–2131, 2013.

195. Serra, D. D. Figueiredo, A. M. Cordeiro, et al., "OsRMC, a negative regulator of salt stress response in rice, is regulated by two AP2/ERF transcription factors," Plant Molecular Biology, vol. 82, pp. 439–455, 2013.

196. K. Mukherjee, A. R. Choudhury, B. Gupta, et al., "An ABRE-binding factor, OSBZ8, is highly expressed in salt tolerant cultivars than in salt sensitive cultivars of indica rice," BMC Plant Biology, vol. 6, article 18, no. 1, 2006.

197. B. Gupta, K. Gupta, and D. N. Sengupta, "Spermidine-mediated in vitro phosphorylation of transcriptional regulator OSBZ8 by SNF1-type serine/threonine protein kinase SAPK4 homolog in indica rice," Acta Physiologiae Plantarum, vol. 34, no. 4, pp. 1321–1336, 2012.

198. Staiger and J. W. Brown, "Alternative splicing at the intersection of biological timing, development, and stress responses," The Plant Cell, vol. 25, no. 10, pp. 3640–3656, 2013.

199. S. G. Palusa, G. S. Ali, and A. S. N. Reddy, "Alternative splicing of pre-mRNAs of Arabidopsisserine/arginine-rich proteins: regulation by hormones and stresses," Plant Journal, vol. 49, no. 6, pp. 1091–1107, 2007.

200. N. Tanabe, K. Yoshimura, A. Kimura, Y. Yabuta, and S. Shigeoka, "Differential expression of alternatively spliced mRNAs of Arabidopsis SR protein homologs, atSR30 and atSR45a, in response to environmental stress," Plant and Cell Physiology, vol. 48, no. 7, pp. 1036–1049, 2007.

201. Z. Zhang, S. Zhang, Y. Zhang et al., "Arabidopsis floral initiator SKB1 confers high salt tolerance by regulating transcription and pre-mRNA splicing through altering histone H4R3 and small nuclear ribonucleoprotein LSM4 methylation," Plant Cell, vol. 23, no. 1, pp. 396–411, 2011.

202. L. Conti, G. Price, E. O'Donnell, B. Schwessinger, P. Dominy, and A. Sadanandom, "Small ubiquitin-like modifier proteases overly tolerant

to Salt1 and -2 regulate salt stress responses in Arabidopsis," Plant Cell, vol. 20, no. 10, pp. 2894–2908, 2008.

203. Cui, L. Liu, Q. Zhao et al., "Arabidopsis ubiquitin conjugase UBC32 is an ERAD component that functions in brassinosteroid-mediated salt stress tolerance," Plant Cell, vol. 24, no. 1, pp. 233–244, 2012.

204. S. H. Kim, Y. O. Ahn, M.-J. Ahn, H.-S. Lee, and S.-S. Kwak, "Down-regulation of β-carotene hydroxylase increases β-carotene and total carotenoids enhancing salt stress tolerance in transgenic cultured cells of sweetpotato," Phytochemistry, vol. 74, pp. 69–78, 2012.

205. R. Sunkar, V. Chinnusamy, J. Zhu, and J.-K. Zhu, "Small RNAs as big players in plant abiotic stress responses and nutrient deprivation," Trends in Plant Science, vol. 12, no. 7, pp. 301–309, 2007.

206. Xie, C. N. Stewart Jr., F. A. Taki, et al., "High-throughput deep sequencing shows that microRNAs play important roles in switchgrass responses to drought and salinity stress," Plant Biotechnology Journal, 2013.

207. J. Bohnert, Q. Gong, P. Li, and S. Ma, "Unraveling abiotic stress tolerance mechanisms—getting genomics going," Current Opinion in Plant Biology, vol. 9, no. 2, pp. 180–188, 2006.

208. J. Weston, L. E. Gunter, A. Rogers, and S. D. Wullschleger, "Connecting genes, coexpression modules, and molecular signatures to environmental stress phenotypes in plants," BMC Systems Biology, vol. 2, article 16, 2008.

209. B. Barkla, T. Castellanos-Cervantes, J. L. de León, et al., "Elucidation of salt stress defense and tolerance mechanisms of crop plants using proteomics—current achievements and perspectives," Proteomics, vol. 13, no. 12-13, pp. 1885–1900, 2013.

210. B. Gupta, A. Sengupta, J. Saha, and K. Gupta, "Plant abiotic stress: "Omics" approach," Journal of Plant Biochemistry & Physiology, vol. 1, article e108, 2013.

211. Zhang, B. Han, T. Wang et al., "Mechanisms of plant salt response: insights from proteomics," Journal of Proteome Research, vol. 11, no. 1, pp. 49–67, 2012.

212. S. S. Gill and N. Tuteja, "Reactive oxygen species and antioxidant machinery in abiotic stress tolerance in crop plants," Plant Physiology and Biochemistry, vol. 48, no. 12, pp. 909–930, 2010.

213. B. J. Barkla, R. Vera-Estrella, and O. Pantoja, "Progress and challenges for abiotic stress proteomics of crop plants," Proteomics, vol. 13, no. 12-13, pp. 1801–1815, 2013.

214. Y. Osakabe, K. Yamaguchi-Shinozaki, K. Shinozaki, and L. S. Tran, "Sensing the environment: key roles of membrane-localized kinases in plant perception and response to abiotic stress," Journal of Experimental Botany, vol. 64, no. 2, pp. 445–458, 2013.

215. G. Friso, L. Giacomelli, A. J. Ytterberg et al., "In-depth analysis of the thylakoid membrane proteome ofArabidopsis thaliana chloroplasts: new proteins, new functions, and a plastid proteome database," Plant Cell, vol. 16, no. 2, pp. 478–499, 2004.

216. Marmagne, M.-A. Roue, M. Ferro et al., "Identification of new intrinsic proteins in Arabidopsis plasma membrane proteome," Molecular and Cellular Proteomics, vol. 3, no. 7, pp. 675–691, 2004.

217. Takahashi, B. Li, T. Nakayama, Y. Kawamura, and M. Uemura, "Plant plasma membrane proteomics for improving cold tolerance," Frontiers in Plant Science, vol. 4, article 90, 2013.

218. D. Ghosh and J. Xu, "Abiotic stress responses in plant roots: a proteomics perspective," Frontiers in Plant Science, vol. 5, article 6, 2014.

219. S. T. Kim, S. G. Kim, G. K. Agrawal, S. Kikuchi, R. Rakwal, et al., "Rice proteomics: a model system for crop improvement and food security," Proteomics, vol. 14, no. 4-5, pp. 593–610, 2014.

220. B.-Y. An, Y. Luo, J.-R. Li, W.-H. Qiao, X.-S. Zhang, and X.-Q. Gao, "Expression of a vacuolar Na^+/H^+antiporter gene of alfalfa enhances salinity tolerance in transgenic Arabidopsis," Acta Agronomica Sinica, vol. 34, no. 4, pp. 557–564, 2008.

221. D. Verma, S. L. Singla-Pareek, D. Rajagopal, M. K. Reddy, and S. K. Sopory, "Functional validation of a novel isoform of Na^+/H^+ antiporter from Pennisetum glaucum for enhancing salinity tolerance in rice,"Journal of Biosciences, vol. 32, no. 3, pp. 621–628, 2007.

222. Z.-Y. Xue, D.-Y. Zhi, G.-P. Xue, H. Zhang, Y.-X. Zhao, and G.-M. Xia, "Enhanced salt tolerance of transgenic wheat (Tritivum aestivum L.) expressing a vacuolar Na^+/H^+ antiporter gene with improved grain yields in saline soils in the field and a reduced level of leaf Na^+," Plant Science, vol. 167, no. 4, pp. 849–859, 2004.

223. C.-A. Wu, G.-D. Yang, Q.-W. Meng, and C.-C. Zheng, "The cotton GhNHX1 gene encoding a novel putative tonoplast Na^+/H^+ antiporter plays an important role in salt stress," Plant and Cell Physiology, vol. 45, no. 5, pp. 600–607, 2004.

224. H.-X. Zhang and E. Blumwald, "Transgenic salt-tolerant tomato plants accumulate salt in foliage but not in fruit," Nature Biotechnology, vol. 19, no. 8, pp. 765–768, 2001.

225. G.-H. Zhang, Q. Su, L.-J. An, and S. Wu, "Characterization and expression of a vacuolar Na^+/H^+antiporter gene from the monocot halophyte Aeluropus littoralis," Plant Physiology and Biochemistry, vol. 46, no. 2, pp. 117–126, 2008.

226. H.-X. Zhang, J. N. Hodson, J. P. Williams, and E. Blumwald, "Engineering salt-tolerant Brassica plants: characterization of yield and seed oil quality in transgenic plants with increased vacuolar sodium accumulation," Proceedings of the National Academy of Sciences of the United States of America, vol. 98, no. 22, pp. 12832–12836, 2001.

227. Shi, B.-H. Lee, S.-J. Wu, and J.-K. Zhu, "Overexpression of a plasma membrane Na^+/H^+ antiporter gene improves salt tolerance in Arabidopsis thaliana," Nature Biotechnology, vol. 21, no. 1, pp. 81–85, 2003. · ·

228. X.-Y. Yin, A.-F. Yang, K.-W. Zhang, and J.-R. Zhang, "Production and analysis of transgenic maize with improved salt tolerance by the introduction of AtNHX1 gene," Acta Botanica Sinica, vol. 46, no. 7, pp. 854–861, 2004.

229. Saito and F. Matsuda, "Metabolomics for functional genomics, systems biology, and biotechnology,"Annual Review of Plant Biology, vol. 61, pp. 463–489, 2010.

230. R. A. Dixon and D. Strack, "Phytochemistry meets genome analysis, and beyond," Phytochemistry, vol. 62, no. 6, pp. 815–816, 2003.

231. Shulaev, D. Cortes, G. Miller, and R. Mittler, "Metabolomics for plant stress response," Physiologia Plantarum, vol. 132, no. 2, pp. 199–208, 2008.

232. D. Balmer, V. Flors, G. Glauser, and B. Mauch-Mani, "Metabolomics of cereals under biotic stress: current knowledge and techniques," Frontiers in Plant Science, vol. 4, article 82, 2013.

233. M. Ashraf, "Biotechnological approach of improving plant salt tolerance using antioxidants as markers,"Biotechnology Advances, vol. 27, no. 1, pp. 84–93, 2009.

234. M. Abogadallah, "Antioxidative defense under salt stress," Plant Signaling and Behavior, vol. 5, no. 4, pp. 369–374, 2010.

235. S. Saxena, H. Kaur, P. Verma, et al., "Osmoprotectants: potential for crop improvement under adverse conditions," in Plant Acclimation to Environmental Stress, N. Tuteja and S. Singh Gill, Eds., pp. 197–232, Springer, New York, NY, USA, 2013.

236. V. Cabello, A. F. Lodeyro, and M. D. Zurbriggen, "Novel perspectives for the engineering of abiotic stress tolerance in plants," Current Opinion in Biotechnology, vol. 26, pp. 62–70, 2014.

Chapter 12

Lectin domains at the frontiers of plant defense

Nausicaä Lannoo and Els J. M. Van Damme*

Laboratory of Biochemistry and Glycobiology, Department of Molecular Biotechnology, Ghent University, Ghent, Belgium

Plants are under constant attack from pathogens and herbivorous insects. To protect and defend themselves, plants evolved a multi-layered surveillance system, known as the innate immune system. Plants sense their encounters upon perception of conserved microbial structures and damage-associated patterns using cell-surface and intracellular immune receptors. Plant lectins and proteins with one or more lectin domains represent a major part of these receptors. The whole group of plant lectins comprises an elaborate collection of proteins capable of recognizing and interacting with specific carbohydrate structures, either originating from the invading organisms or from damaged plant cell wall structures. Due to the vast diversity in protein structures, carbohydrate recognition domains and glycan binding specificities, plant lectins constitute a very diverse protein superfamily. In the last decade, new types of nucleocytoplasmic plant lectins have been identified and characterized, in particular lectins expressed inside the nucleus and the cytoplasm of plant cells often as part of a specific plant response upon exposure to different stress factors or changing environmental conditions. In this review, we provide an overview on plant lectin motifs used in the constant battle against pathogens and predators during plant defenses.

1. INTRODUCTION

In nature, plants are constantly exposed to a plethora of different pathogens including bacteria, viruses, and fungi. Whereas the interaction with some of these organisms can be beneficial, most microbial infection is harmful for the plant (Dangl et al., 2013). In order to resist pathogen colonization, plants developed a highly sophisticated, multilayered system enabling the plant to recognize invading pathogens and mount rapidly efficient defense responses (Muthamilarasan and Prasad, 2013; Wirthmueller et al., 2013).

Microbial entry into the host tissue is a critical step in causing infection. Pathogens can enter plants through natural openings such as stomata, hydathodes, lateral roots, or through accidental wounds, but can also form specialized structures such as haustoria to penetrate directly into the plant surface (Melotto et al., 2006;Gudesblat et al., 2009). Many phytopathogens also produce lytic enzymes to damage the plant cell wall in favor of pathogen invasion. Perception of the invading pathogen is the first step in the plant's defense and is governed by cell surface transmembrane receptors. These pattern recognition receptors or PRRs are able to recognize two types of molecules.

The first group encompasses the damage-associated molecular patterns (DAMPs) which are produced in the plant apoplast as a consequence of pathogen entry. Examples include cell wall fragments such as oligogalacturonides and cellulose fragments, cutin monomers, and peptides such as systemin, defensins, and phytosulfokines (Ryan, 2000; Nühse, 2012; Albert, 2013). PRRs are also able to recognize conserved microbial structures, known as pathogen- or microbe-associated molecular patterns (PAMPs/MAMPs), which are essential for the microbial physiology and the pathogen's fitness (Newman et al., 2013; Wirthmueller et al., 2013). Examples of PAMPs/MAMPs include lipopolysaccharides (LPS) of Gram-negative bacteria, peptidoglycan (PGN) of Gram-positive bacteria, bacterial flagellins, eubacterial elongation factors (EF-Tu), and fungal cell wall derived glucans, chitins, and proteins.

Upon PAMP/MAMP and DAMP perception by the PRRs, the so-called PAMP/MAMP-triggered immunity (PTI/MTI) response is activated which gives rise to downstream intracellular signaling events such as activation of mitogen-activated protein kinases, production of reactive oxygen species and transcriptional reprogramming ultimately leading to a complex output response of the plant that limits microbial growth (Wirthmueller et al., 2013).

However, successful pathogens have elaborated a counter defense response to overcome PTI by means of expression of specific elicitors or effector proteins [also known as avirulence (Avr) proteins; Grant et al., 2006]. Pathogenic bacteria typically inject these effectors directly into the cytoplasm of the plant host cell through type III secretion mechanisms to suppress and/or block PRR-dependent signaling, to facilitate nutrient acquisition and to contribute to the pathogen's dispersal which can lead to effector-triggered susceptibility (ETS; Block et al., 2013; Cui et al., 2013).

As a counter move, plants have co-evolved a second layer of defense, known as effector-triggered immunity (ETI) which, in contrast to PTI/MTI acts mostly inside the plant cell. In ETI, specific resistance (*R*) genes become expressed upon recognition of an effector to produce defense proteins. The majority of the R proteins include nucleotide-binding leucine-rich repeat (NB-LRR)-containing proteins. The outcome of PTI/MTI and ETI can lead to programmed cell death of the host cell via (local) activation of a hypersensitive response (HR), but can also initiate systemic acquired resistance (SAR) to activate defenses in distal, non-infected parts of plants in order to establish a heightened state of immunity throughout the plant (Thomma et al., 2011).

2. PATHOGEN RECOGNITION RECEPTORS (PRRS)

The cell wall confers the first tier of the plant's immunity (Malinovsky et al., 2014). The extracellular PRRs are able to detect pathogen determinants (the so-called PAMPs/MAMPs), DAMPs and effectors at the surface of the plant cell and are used to translocate the extracellular message of 'danger' to the intracellular environment to trigger appropriate defense mechanisms (Figure1A; Dubery et al., 2012). The PRR family encompasses two groups of plasma membrane-localized proteins: the receptor-like kinases (RLKs) and the receptor-like proteins (RLPs). RLKs are single-pass transmembrane proteins with an extracellular domain that is responsible for the perception of the P/M/DAMPs and an intracellular serine/threonine kinase domain that activates the downstream signaling responses. RLPs possess a similar structure but, because they only have a short cytosolic domain without an obvious signaling module, they depend on the association with kinases for signaling. However, there is emerging evidence that upon ligand binding

RLKs also form homodimers or heterodimers with other kinases and RLPs and as such function in multiprotein complexes to initiate plant immunity (Boller and Felix, 2009;Böhm et al., 2014; Han et al., 2014; Macho and Zipfel, 2014).

Thus far, only a limited number of RLKs and RLPs that may function in plant immunity have been functionally characterized. Matching these proteins to their ligands is still a challenging study. The majority of the known PRR ectodomains contains LRRs for direct/indirect recognition of pathogenic effector proteins (Table 1). In addition, a large diversity of membrane-bound and soluble PRRs have been described to carry lectin domains that are implicated in the recognition of carbohydrate structures from microbial organisms or derived from plant cell wall damage (Tables 2 and 3).

3. PATHOGEN RECOGNITION BASED ON PROTEIN–PROTEIN INTERACTIONS

The study of plant–pathogen interactions has focused on those PRRs which use protein–protein interactions to recognize invading pathogens. Phytopathogens are recognized upon perception of characteristic epitopes present on their surface and essential for the pathogen's survival. These epitopes are mostly recognized by plant cell surface receptors carrying LRR motifs in their ectodomain structures and a kinase domain in their intracellular domain, collectively named the LRR-RLKs. Since these protein–protein interactions have been the subject of several recent overview papers, we only briefly summarize some plant LRR-RLKs and LRR-RLPs (Figure 1B and Table 1).

Amongst the plant PRRs of the LRR-RLK type, the *Arabidopsis*LRR-RLK *At*FLS2 (Flagellin Sensing 2) is the best-studied protein, containing 28 extracellular LRRs. This FLS2 recognizes bacterial flagellin via perception of the conserved 22-amino acid epitope flg22. *At*FLS2 directly interacts with flg22 resulting in phosphorylation of *At*FLS2 and immediate dimerization with its co-receptor BAK1/SERK, another LRR-RLK. Transphosphorylation of the kinase domain of BAK1 enables conformational changes and subsequent release of phosphorylated BAK1 leading to activation of downstream MAPK defense signaling (Gómez-Gómez et al., 2001;Chinchilla et al., 2006, 2007; Schulze et al., 2010). In the absence of PAMP recognition, BAK1 itself interacts with the pseudokinase BIR2 (also LRR-RLK-type) to prevent FLS2-BAK1 heterodimerization (Halter et al., 2014). After flg22 perception,*At*FLS2 is subject to endocytosis and degradation by the E3 ubiquitin ligase PUB12/13 to prevent continuous defense signaling. Newly synthesized *At*FLS2 is incorporated in the plasma membrane at later times (Smith et al., 2014). In turn, virulent *Pseudomonas syringae* pathovars produce effector proteins, such as AvrPto, AvrPtoB, and AvrPphB to destabilize *At*FLS2 and thus compromise host immunity (Block and Alfano, 2011).

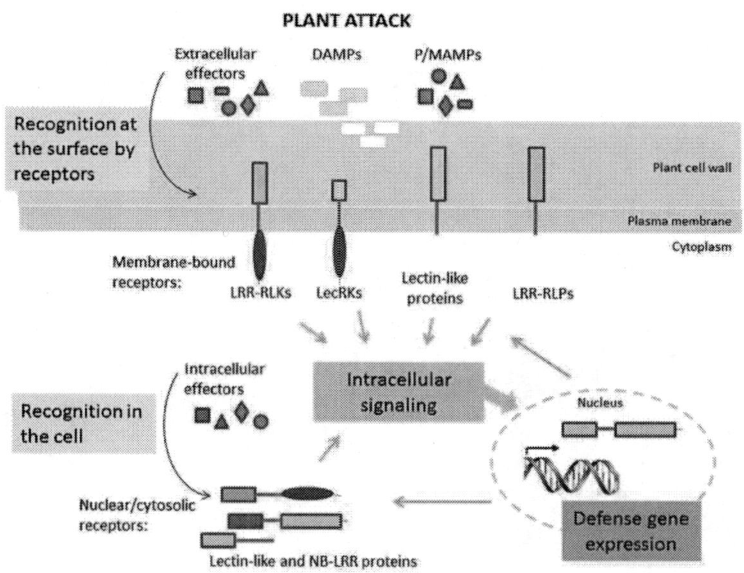

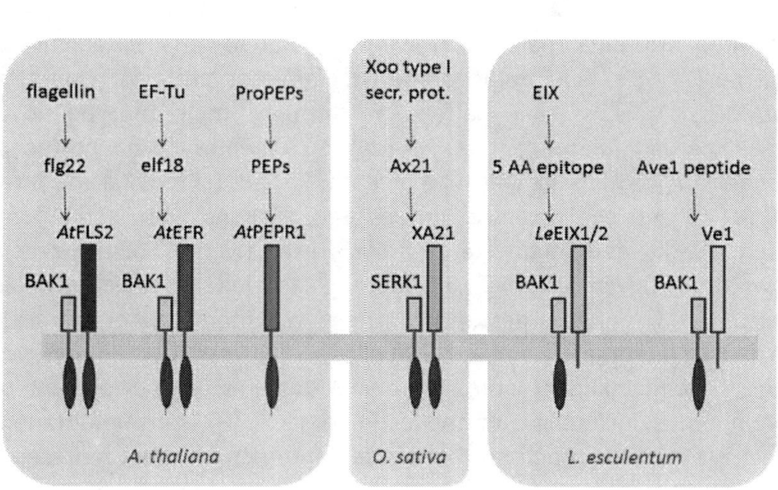

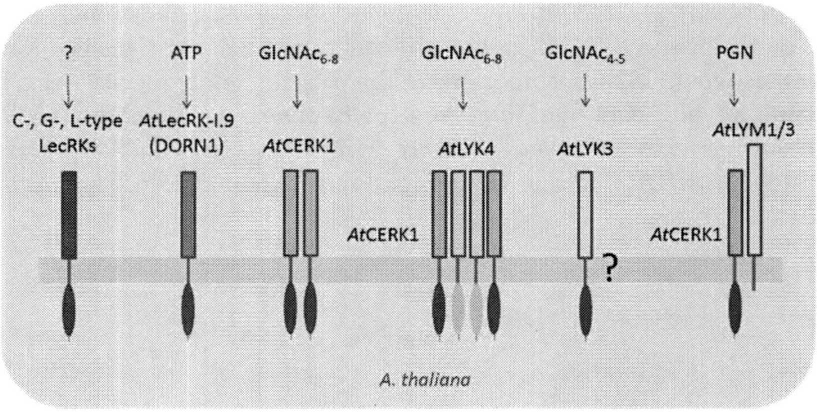

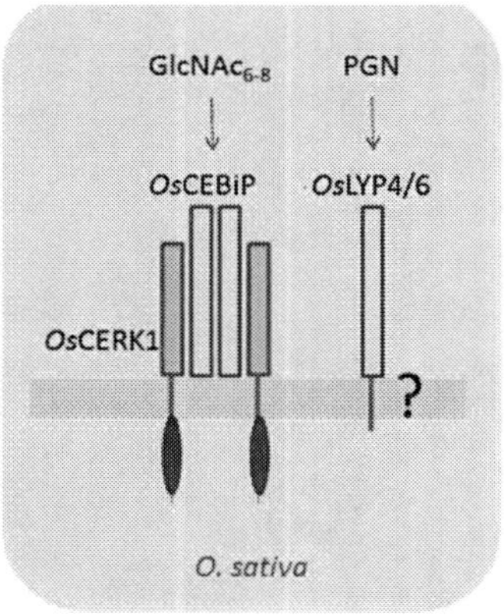

Figure 1. Plant innate immunity. (A) Perception of pathogen/microbe-associated molecular patterns (P/MAMPs), damage-associated molecular patterns (DAMPs) and pathogen-derived effector proteins. Plants sense P/MAMPs, DAMPs, and effectors through membrane-bound and intracellular (soluble) receptors. Four types of membrane-bound receptors can be distinguished: the LRR-type receptor kinases (LRR-RLKs) and proteins (LRR-RLPs), and the receptor kinases and proteins with lectin domains (called LecRKs and lectin-like proteins, respectively). Soluble receptors known thus far include NB-LRR proteins as well as nucleocytoplasmic lectins. Upon 'danger' perception, these receptors trigger intracellular signals which ultimately will result in altered expression of defense-related genes. Legend: ellipses represent kinase domains; bars represent other protein motifs, including LRRs and lectin domains. **(B)** Transmembrane PRRs detect P/MAMPs through protein–protein interactions. Bars represent LRR

domains, red ellipses indicate functional kinase domains. **(C)** Transmembrane PRRs with lectin domains identified in*Arabidopsis thaliana*. In the case of the LysM domain evidence supports protein–carbohydrate interactions to detect P/M/DAMPs. Bars represent lectin domains, including C-, G-, and L-type (brown) and LysM (yellow) domains. Ellipses represent kinase domains (red = functional, pink = non-functional, purple = putative kinase domain). **(D)** Transmembrane PRRs with lectin domains identified in *Oryza sativa*. The LysM domain recognizes P/M/DAMPs through specific binding of chitin fragments. Bars represent the LysM lectin domains; ellipses represent functional kinase domains.

The transmembrane protein *At*EFR represents another Arabidopsis LRR-RLK-type receptor involved in bacterial PAMP signaling (Zipfel et al., 2006). The ectodomain of *At*EFR consists of 24 LRRs and is involved in the perception of the elf18 peptide, a conserved N-terminal fragment of bacterial elongation factor Tu. Many of the signaling compounds downstream of *At*EFR are shared with*At*FLS2, and *At*EFR also requires dimerization with *At*BAK1 for signaling. However, the action of *At*EFR is independent of flagellin perception and unlike *At*FLS2, *At*EFR requires N-glycosylation to become functional. Indeed, a single *N*-glycan is crucial for receptor abundance and ligand recognition between the pathogen and the plant cell surface (Häweker et al., 2010).

Rice plants use the transmembrane XA21 receptor kinase to confer immunity toward a number of *Xanthomonas oryzae* pv *oryzae*(*Xoo*) isolates, which cause leaf blight in rice. The XA21 receptor recognizes Ax21, a sulfated 17-amino acid peptide derived from the*Xoo* type I secreted protein (Lee et al., 2009). Also here, the action of XA21 is tightly regulated. Without PAMP, XA21 is kept in an inactive state through binding with and autophosphorylation by the ATPase XB24. Upon binding of Ax21 to XA21, the XB24/XA21 dimer dissociates and the XA21 kinase domain is released and translocated to the cell nucleus for subsequent immune signaling (Park and Ronald, 2012). Chen et al. (2014) recently reported that XA21 can also be found in a constitutive heteromeric complex with a BAK1 ortholog, named OsSERK2, and undergoes bidirectional transphosphorylation to confer resistance to the *Xanthomonas*bacterium.

Table 1. TABLE 1. LRR-type PRRs involved in plant defense signaling.

PRR	Plant species	Ligand	Reference
LRR-RLK type			
*At*FLS2	*Arabidopsis thaliana*	Flagellin (Flg22)	Chinchilla et al. (2006)
*At*EFR	*A. thaliana*	Ef-TU	Zipfel et al. (2006)
XA21	*Oryza sativa*	Activator of XA21 (Ax21)	Lee et al. (2009)
XA3/XA26		Epitope derived from *Xanthomonas oryzae* pv *oryzae*	Sun et al. (2004)
SR160	*Lycopersicon peruvianum*	(pro)systemin	Scheer and Ryan (2002)
PEPR1	*A. thaliana*	PEPR1	Krol et al. (2010)
NORK	*Medicago truncatula*	?	Endre et al. (2002)
SYMRK	*Lotus japonicus*	?	Stracke et al. (2002)
LRR-RLP type			
LeEIX2	*Lycopersicon esculentum*	Xylanase (EIX)	Ron and Avni (2004)
Ve1	*L. esculentum*	Ave1 peptide	Fradin et al. (2009)

Table 2. Membrane-bound lectin-type PRRs involved in plant defense signaling and symbiosis.

PRR	Plant species	Ligand	Reference
LysM-RLK type			
AtCERK1	*Arabidopsis thaliana*	Chitin	Miya et al. (2007), Brotman et al. (2012)
AtLYK3	*A. thaliana*	Chitin	Paparella et al. (2014)
AtLYK4	*A. thaliana*	Chitin	Wan et al. (2012)
OsCERK1	*Oryza sativa*	Chitin (when in combination with OsCEBiP)	Kaku et al. (2006)
NFR1	*Lotus japonicus*	Lipochitooligosaccharide Nod factors	Radutoiu et al. (2003)
LYK3	*Medicago truncatula*	Lipochitooligosaccharide Nod factors	Knogge and Scheel (2006)
NFR5	*L. japonicus*	Lipochitooligosaccharide Nod factors	Radutoiu et al. (2003), Madsen et al. (2003)
SYM10	*Pisum sativum*	Lipochitooligosaccharide Nod factors?	Madsen et al. (2003)
LYK4	*M. truncatula*	Lipochitooligosaccharide Nod factors	Limpens et al. (2003)
NFP	*M. truncatula*	Lipochitooligosaccharide Nod factors	Mulder et al. (2006)
LysM-RLP type			
LYM1/AtLYP2, LYM3/AtLYP3	*A. thaliana*	Peptidoglycan	Willmann et al. (2011), Tanaka et al. (2013)
OsCEBiP	*O. sativa*	Chitin	Shimizu et al. (2010)
OsLYP4, OsLYP6	*O. sativa*	Chitin + Peptidoglycan	Liu et al. (2012a)

Table 3. Nucleocytoplasmic lectin domains involved in plant defense signaling.

Lectin domain	Carbohydrate Specificity	Subcellular localisation	Examples
Amaranthin domain	GalNAc, T-antigen	nucleus, cytosol	Amaranthin, Hfr-2
EUL domain	Galactosides, high-mannose *N*-glycans	nucleus, cytosol	EEA, ArathEULS3
Jacalin domain	Mannose-specific subgroup / galactose-specific subgroup	nucleus, cytosol / vacuole	Orysata, TaVER2, TaHfr-1, TaJA-1
Nictaba domain	$(GlcNAc)_n$, high-mannose *N*-glycans, complex *N*-glycans	nucleus, cytosol	Nictaba, PPL
Ricin-B domain	Gal/GalNAc, Siaα2,6Gal/GalNAc	Vacuole, nucleus, cytosol	Ricin, abrin, SNA-I, SNA-V

Tomato plants encode several cell-surface LRR-RLPs such as LeEIX1/EIX2 and Ve1 which confer resistance toward *Trichoderma*and race 1 strains of *Verticillium* pathogens, respectively (Ron and Avni, 2004; Fradin et al., 2009). The ethylene-inducing xylanase EIX is a fungal β-1–4-endoxylanase that is used by *Trichoderma viride* to enter tomato and tobacco plants. The epitope that is recognized by the plants to elicit defense responses constitutes five amino acids that are not involved in the enzymatic activity (Rotblat et al., 2002). Both LeEIX1/EIX2 can bind EIX, but only LeEIX2 transmits the signal to activate immune responses (Ron and Avni, 2004). The ligand of the Ve1 receptor is the Ave1 peptide, a peptide conserved in several fungi and phytopathogenic bacteria. BAK1 signaling is involved in induced defense responses for both LeEIX1 and Ve1 (Fradin et al., 2009; Bar et al., 2010).

4. PATHOGEN RECOGNITION BASED ON PROTEIN–CARBOHYDRATE INTERACTIONS

4.1 The Carbohydrates

Major part of the P/M/DAMPs that are perceived in the plant as 'danger' molecules include carbohydrate structures which are either present at the cell surface of the invading pathogen or originate from the plant itself, when released from cell wall degradation caused by pathogen entry. These structures comprise bacterial LPS and PGNs and fungal chitin molecules as well as plant-derived oligogalacturonides and cellulose fragments. Also arabinogalactan proteins residing in the plant cell wall have been reported to be involved in plant immune responses (Newman et al., 2013).

Lipopolysaccharides are large outer membrane glycoconjugates found in Gram-negative bacteria that are composed of a lipid, a core oligosaccharide and an *O*-antigen polysaccharide chain. The lipid, called lipid A, is embedded in the bacterial membrane and is linked to the core oligosaccharide by the KDO sugar (3-deoxy-D-mannose-2-octulosonate). The core sugar ends in the *O*-antigen which is composed of oligorhamnans in many phytopathogens.

Peptidoglycans are essential cell wall components of Gram-positive and Gram-negative bacteria, and comprise alternating β(1–4) linked *N*-acetylmuramic acid (MurNAc) and *N*-acetylglucosamine (GlcNAc) residues, with a short peptide chain attached to MurNAc.

Chitin is a long-chain polymer of β(1–4) linked GlcNAc residues and is the main component of the fungal cell wall and the exoskeleton of insects. In case of fungi, chitin is cross-linked to β-glucan.

Oligogalacturonides are oligomers of α(1,4) linked galacturonosyl residues that are released from plant cell walls upon partial degradation of homogalacturonan (i.e., the major component of pectin) by pathogen attack and also upon mechanical damage.

Cellulose is an important component of the plant cell wall, built up of hundreds of β(1–4) linked glucose residues which form long polymer chains. These chains are packed into microfibrils which give strength and flexibility to the plant cell wall.

Arabinogalactan proteins are a distinct class of complex, extensively glycosylated hydroxyproline-rich proteins (the so-called proteoglycans), widely distributed among plant species. They consist of a rather small core protein backbone which is *O*-glycosylated by type II arabinogalactan glycans and often contain an N-terminal GPI anchor. AGPs are located near the cell surface, including the plasma membrane, the apoplast, the cell wall, and the intercellular matrix, and have been implicated in many aspects of plant growth and development, such as cell expansion, proliferation, and differentiation. These AGPs are not only involved in establishing a connection between the cell wall and the plasma membrane, but would also extend to the cytoplasm, establishing a continuum between intracellular and extracellular compartments.

4.2 Lectin Domains Involved in Plant Immunity

Lectins are proteins that contain at least one non-catalytic domain which enables them to selectively recognize and bind in a reversible way to specific glycans that are either present in a free form or are part of glycoproteins and glycolipids. Plants express a huge number of highly diverse lectins, exhibiting different molecular structures and binding specificities toward endogenous (plant) glycans as well as to glycans from exogenous (non-plant) origin (Van Damme et al., 2008, 2011).

A lot of plant lectins are constitutively expressed in high amounts in seeds and vegetative storage tissues where they have been shown to play a role in plant defense (Peumans and Van Damme, 1995). In addition, plants also express minute amounts of specific lectins as particular responses toward environmental stresses and pathogen attack. In the absence of plant stress, the inducible lectins are not expressed at detectable levels. Most of the constitutively expressed lectins are synthesized with a signal peptide, and are sequestered in the vacuole or secreted to the extracellular space. In contrast, most of the inducible plant lectins reside in the nucleus and the cytoplasm of a plant cell (Lannoo and Van Damme, 2010).

The majority of the known plant lectins are built up of one or more lectin-like domains coupled to un-related domains such as aerolysin, AIG1, chitinase, dirigent, F-box, Kelch, kinase, LRR, NB-ARC, PAG, or TIR domains (Van Damme et al., 2008). Up till now, most attention of the scientific community dealing with plant innate immunity has been given to transmembrane receptor proteins containing one or more ectopic lectin domains coupled to an intracellular kinase domain. Amongst these lectin receptor kinases (LecRKs), those comprising LysM-type lectin domains are the most studied ones (Table 2; Singh and Zimmerli, 2013). However, plants use a broad variety of lectin domains to counteract pathogen attack (Table 3).

4.3 Membrane-bound proteins with a lectin domain

Lectin receptor kinases (LecRKs). Typically, LecRKs are two-domain proteins composed of an N-terminal extracellular lectin domain and a C-terminal cytosolic Ser/Thr kinase domain, separated by a transmembrane region (Figure 1C). Based on their lectin domain LecRKs are classified into 4 types; G-, C-, L-, and LysM-type (Singh and Zimmerli, 2013; Vaid et al., 2013). Although these LecRKs consist of at least one domain that shows striking sequence similarity with a lectin motif, very little information is available with respect to the ability of this domain to recognize and interact with specific carbohydratestructures.

G-type LecRKs contain an extracellular lectin domain which resembles the *Galanthus nivalis* agglutinin (GNA). However, it remains to be shown whether this sugar binding domain is indeed involved in ligand binding. Based on genome-wide analyses, 32 G-type LecRKs have been identified in *Arabidopsis thaliana* and 100 G-type LecRKs in rice (Vaid et al., 2012). G-type LecRKs function in self-incompatibility reactions in flowering plants (the so-called SRKs) and in plant defense to biotic stress as well as to abiotic stress (Sherman-Broyles et al., 2007; Kim et al., 2009; Sun et al., 2013).

C-type (calcium-dependent) LecRKs are mostly found in mammalian proteins that mediate immune responses and play a role in pathogen recognition. In plants, C-type RLKs are rather rare. At present only one C-type LecRK encoding gene has been identified in *A. thaliana* (At1g52310), though its function has not been elucidated yet (Cambi et al., 2005; Bouwmeester and Govers, 2009).

L-type (legume-like) LecRKs represent a more abundant group of LecRKs. Thus far, 45 L-type LecRKs have been identified in *A. thaliana*. Based on phylogenetic relationships the genes encoding*Arabidopsis* L-type LecRK can be classified into nine clusters and nine clades (designated with the Roman numerals I to IX). These genes showed variable expression patterns in different tissues and developmental stages in response to stimuli (Bouwmeester and Govers, 2009). Some LecRKs were indeed reported to be involved in plant resistance to pathogens, e.g., *At*LecRK-I.9 is involved in sensing cell wall integrity and defense response to *Phytophthora infestans* (Bouwmeester et al., 2011). *At*LecRK-VI.2 is critical for resistance against *Pseudomonas syringae* and *Pectobacterium carotovorum* (Singh et al., 2012; Huang et al., 2014) while*At*LecRK-IV.3 induces resistance against *Botrytis cinerea* (Huang et al., 2013). Some *At*LecRKs have also been reported to act in hormone signaling (ABA) and stomatal immunity (e.g., *At*LecRK-VI.2 and *At*LecRK-V.5; Singh et al., 2012). L-type LecRKs have also been identified in other plants. For instance, tobacco plants express L-type LecRKs with a role in plant immunity against pathogens and insects (Kanzaki et al., 2008; Gilardoni et al., 2011). In turn,*Medicago* plants contain L-type LecRKs that are involved in symbiosis (Navarro-Gochicoa et al., 2003).

At present, it is not yet clear whether the L-type LecRKs possess lectin activity since the amino acids important for interaction of the legume lectin domain with its specific carbohydrate ligand are poorly conserved. In contrast the hydrophobic site present in the legume-type lectin domain is preserved, suggesting that LecRKs may act in the recognition of small hydrophobic ligands (Huang et al., 2013; Choi et al., 2014). Recently, evidence was obtained that the plasma membrane localized DORN1, encoded by the *AtLecRK-I.9* gene, plays an important role as a receptor for extracellular ATP (Figure 1C; Choi et al., 2014). DORN1 lacks the conserved Ca^{2+} and Mn^{2+} binding residues that are critical for carbohydrate binding activity of legume lectins. Early studies also suggested the ability of the legume lectin domain to bind adenine, a component of ATP (Roberts and Goldstein, 1983). However, since adenine was unable to compete with ATP for binding to DORN1, the exact ATP binding site in DORN1 remains to be determined. These data are in good agreement with the fact that extracellular ATP is now perceived as a central signaling molecule in plant stress responses (Cao et al., 2014; Choi et al., 2014).

LysM LecRKs are the most studied LecRKs (Figures 1C,D and Table2). They contain ectopic lysin motifs, referred to as LysM domains, which are considered to mediate binding to various types of bacterial PGN and fungal chitin, upon recognition of the GlcNAc moieties (Buist et al., 2008; Gust et al., 2012). The lysine motif, approximately 40 amino acids in length, is a ubiquitous protein domain found in most living organisms except the Archaea. It can be used as a single domain, but is also present in the form of two or occasionally three repeats in a large number of proteins. In most cases LysM motifs are coupled to other protein domains exhibiting some enzymatic activity, such as GlcNAc modification in the case of microbial hydrolases or intracellular signaling for plant kinases.

The *Arabidopsis* chitin elicitor receptor kinase 1 (*At*CERK1, also known as LYK1/RLK1) is the major chitin receptor found in *A. thaliana* (Miya et al., 2007; Petutschnig et al., 2010; Tanaka et al., 2013). It is a membrane-anchored protein with an extracellular domain containing three LysM motifs coupled to an intracellular kinase domain. This kinase domain contains a canonical RD (Arginine-Aspartate) motif in its catalytic loop and possesses autophosphorylation activity, unlike the non-RD kinase domain of typical PRRs. *At*CERK1 was reported to be involved in fungal resistance. The protein directly binds to fungal chitooligosaccharides ($GlcNAc_n$ with $n >$ 2) through its LysM domains, but only longer oligomers ($n > 4$) trigger immune responses. Liu et al. (2012b) demonstrated that binding of chitin oligomers ($n = 8$) to *At*CERK1 induces homodimerization of the receptor, which is essential for the activation of downstream intracellular signaling, most likely by phosphorylation of both CERK1 cytoplasmic kinase domains.

*At*CERK1 can also mediate perception of PGN when part of a complex with *At*LYM1 and *At*LYM3, two other transmembrane LysM containing proteins lacking an intracellular kinase domain (the so-called LYP proteins; Willmann et al., 2011). Next to*At*CERK1 *A. thaliana* contains four additional LysM RLKs, designated *At*LYK2–5. Since *At*LYK4 and *At*LYK5 can also recognize and bind to chitin molecules but have a non-functional pseudokinase domain it has been suggested they should form a complex with *At*CERK1 to compose a functional chitin receptor (Wan et al., 2012). *At*LYK3 possesses a functional kinase domain, but was suggested be involved in ABA signaling rather than in chitin recognition (Paparella et al., 2014).

Unlike *At*CERK1, the rice ortholog *Os*CERK1 cannot bind directly to chitooligosaccharides. *Os*CERK1, a LysM LecRK with a single extracellular LysM domain, requires heterodimerization with its co-receptor *Os*CEBiP (chitin elicitor binding protein) for chitin binding and subsequent activation of innate immunity (Figure 1D).*Os*CEBiP is a transmembrane LysM receptor protein containing three LysM domains and lacking a kinase domain, resembling the*Arabidopsis* proteins *At*LYM1 and *At*LYM3. However, whereas*At*LYM1 and *At*LYM3 are involved in PGN binding through receptor formation with *At*CERK1, *Os*CEBiP seems to play a major role in fungal chitin perception. Upon binding to the fungal $GlcNAc_{6\text{-}8}$ oligomers, *Os*CEBiP homo-dimerizes at the plasma membrane of the plant cell. After this ligand-induced dimerization, the *Os*CEBiP sandwich-like structure forms a heteromeric complex with two *Os*CERK1 proteins, to activate intracellular defenses, including ROS signaling, callose deposition, and defense gene expression (Shimizu et al., 2010; Shinya et al., 2012; Hayafune et al., 2014; Kouzai et al., 2014). *Os*LYP4 and *Os*LYP6 are two additional rice LysM proteins lacking a kinase domain which presumably mediate perception of both PGN and chitin, but their action and transmembrane signal transfer remain unclear (Liu et al., 2012a).

Plants not only use LysM LecRKs to recognize pathogenic organisms, they also use them to perceive beneficial organisms such as mycorrhizal fungi and rhizobacteria implicating a dual role of LysM in both innate immunity and symbiosis (Table 2; Gust et al., 2012). Examples include the *Lotus japonicus* NRF1 and NRF5 (Radutoiu et al., 2003) and the *Medicago truncatula* LYK3 and LYK4 (Knogge and Scheel, 2006) which can recognize rhizobial lipochitin-oligosaccharide signals or Nod factors.

4.4 Soluble proteins with a lectin domain

An overview of soluble proteins with a lectin domain involved in plant defense signaling is given in Table 3.

Amaranthins. The Amaranthin family groups all proteins related to amaranthin, a lectin present in the seeds of *Amaranthus caudatus*. Native amaranthin is a homodimeric protein built of two 33 kDa subunits, each comprising two tandem-arrayed homologous amaranthin domains. Amaranthin specifically recognizes the T-antigen disaccharide Galβ(1,3) GalNAc but also interacts with GalNAc. Interestingly, the amaranthin domain itself possesses no sugar binding site(s), but the specific head-to-tail arrangement of two amaranthin subunits is necessary to establish the T-antigen disaccharide binding site (Van Damme et al., 2008). Up till now, only amaranthins originating from *Amaranthus* species have been purified and biochemically characterized. This nucleocytoplasmic lectin was reported to enhance the plant's resistance against aphids when ectopically expressed in transgenic tobacco, potato, and cotton by affecting growth and development of the invading aphids (Wu et al., 2006; Xin et al., 2011).

Screening of the publicly available genome databases revealed that the amaranthin domain is widespread throughout the plant kingdom. Several chimeric proteins containing N-terminal amaranthin domain(s) coupled to unrelated protein domains have been identified (Van Damme et al., 2011). Columbine plants (*Aquilegia formosa*× *Aquilegia pubescens*) encode a protein in which two amaranthin domains are coupled to a kinase domain. Since this protein does not have a transmembrane domain, it is suggested to reside inside the cell. Cucumber, maize, and wheat plants were found to contain genes that encode proteins with amaranthin domain(s) coupled to an aerolysin domain. Aerolysins are cytolytic toxins that are mostly produced by the bacterium*Aeromonas* and can kill host cells upon pore formation in the plasma membrane (Degiacomi et al., 2013). In wheat, a chimeric protein called Hfr-2 (Hessian fly responsive-2) is up-regulated in the leaf sheaths after feeding of virulent Hessian fly larvae and armyworms, and enhances wheat tolerance against Hessian fly larvae (Puthoff et al., 2005).

Calreticulin/calnexin. Calreticulin (CRT) and calnexin (CNX) are glucose binding lectins residing in the endoplasmic reticulum (ER) of eukaryotic cells. Both CRT and CNX act as molecular chaperones and are essential ER components ensuring proper folding and quality control of newly synthesized secretory and membrane-bound glycoproteins before ER release (Ellgaard and Frickel, 2003;Kapoor et al., 2004; Williams, 2006). While CNX is a type-I integral membrane protein, CRT is a soluble protein. Both CRT and CNX act together with their co-chaperones ERp57 and PDI, two soluble thiol-disulfide oxidoreductases. Whereas the classical chaperones associate with the peptide moiety of their substrates, CNX and CRT bind to their glycoprotein substrates primarily through specific recognition and binding to the oligosaccharide intermediates $Glc_1Man_{7-9}GlcNAc_2$ present on nascent glycoproteins.

In the ER quality control system, a growing polypeptide initially gets N-glycosylated with the core glycan $Glc_3Man_9GlcNAc_2$. By successive action of glucosidase I and II, the outer glucoses are trimmed resulting in a monoglucosylated glycoprotein which then serves as the substrate of CNX/CRT for proper folding. Once the glycoprotein is correctly folded, the terminal glucose of its oligosaccharide is cleaved by glucosidase II and the glycoprotein is released from the CNX/CRT/ERp57 complex for further processing. If the glycoprotein is not correctly folded, it is recognized by the UDP-glucose:glycoprotein glucosyltransferase enzyme, which acts as a folding sensor, and gets re-glucosylated to promote its renewed association with CNX/CRT. As such, de- and re-glucosylation of glycoproteins facilitates their correct folding. When folding ultimately fails, the misfolded glycoproteins are sorted out of the ER and are degraded by the proteasome, a system known as ER associated degradation or ERAD. Under adverse environmental conditions, the demand for protein folding exceeds the capacity of the system resulting in the accumulation of misfolded proteins in the ER, giving rise to so-called ER stress (Howell, 2013; Liu and Li, 2014).

The correct folding of membrane-bound PRRs is a critical step in plant immunity. The ER quality control system not only regulates the abundance and quality of transmembrane receptors, it also affects downstream signaling of the receptor (Tintor and Saijo, 2014). The LRR-RLKs *At*EFR and *At*FLS2 from *A. thaliana* as well as the LysM LecRK NFP from *M. truncatula* require proper *N*-glycosylation for accurate functioning. Nevertheless, *At*EFR and*At*FLS2 production are coordinated by different ER components (Li et al., 2009; Häweker et al., 2010). *Arabidopsis* plants contain two types of CRTs: CRT1/2 and CRT3 isoforms (Thelin et al., 2011). Whereas CRT1 is a key chaperone in plant ER stress, CRT3 is typically involved in the quality control of *At*EFR and the brassinosteroid receptor BRI1, but is not essential for *At*FLS2 biogenesis (Jin et al., 2009). CRT2 appears to have a dual regulatory role in plant defense against the biotrophic pathogen *P. syringae* pv tomato DC3000 (Qiu et al., 2012). Upon pathogen invasion, CRT2 is involved in the up-regulation of SA-dependent immune signaling through its Ca^{2+} buffering capacity. However, CRT2 negatively influences these SA-dependent responses through its chaperone activity, resulting in the overall suppression of plant resistance toward *P. syringae* pv tomato DC3000.

EUL-related lectins. The family of EUL-related lectins groups all nucleocytoplasmic proteins that comprise at least one *Euonymus*lectin (EUL) domain. The prototype of this family is the so-called*Euonymus europaeus* agglutinin (EEA) which is present at very high concentrations in the arillus tissue of the spindle tree (*E. europaeus*). EEA is a non-glycosylated homodimeric protein composed of 17 kDa subunits, and recognizes two structurally different classes of glycans. Glycans with carbohydrate epitopes containing galactose, such as Galα1–3Gal and Galα1–3Galβ1–4GlcNAc, blood group B [Galα1–3(Fucα1-2)Gal-], and O (Fucα1–2Gal-) epitopes are bound with a higher affinity compared to high-mannose *N*-glycans. Based on inhibition studies, it was suggested that the EUL domain might contain two different binding sites (Fouquaert et al., 2008).

Sequences with an EUL domain are present in almost all sequenced plant genomes from Embryophyta, ranging from liverworts to flowering plants. The sequence of the EUL domain is well conserved suggesting that the corresponding EUL proteins fulfill an essential role in plants. Based on the overall protein domain architecture, the EUL family can be divided into two groups containing proteins either composed of a single EUL domain (S-type EUL proteins) or of two tandemly arrayed EUL domains separated by a spacer sequence (D-type EUL proteins; Fouquaert and Van Damme, 2012). The majority of the EUL sequences encode chimeric proteins, in which the EUL domain is linked to other unknown domains. Whereas most dicot species encode one or two EUL S-type proteins, monocot, and lower plant species contain a whole set of S- and D-type EUL sequences.

In contrast to EEA, which is expressed at high concentrations in the arilli of spindle tree seeds, the EUL proteins from *Oryza sativa* and*A. thaliana* are very low abundant proteins. In both plants the amount of EUL transcripts is enhanced after the plant was subjected to different abiotic (such as dehydration, salinity, osmotic stress, and ABA treatment) and biotic (such as bacterial and fungal infection) stresses (Fouquaert and Van Damme, 2012; Al Atalah et al., 2014a).

Next to the differential regulation of gene expression, the EUL proteins show different carbohydrate binding properties. ArathEULS3 from *A. thaliana* preferentially interacts with *N*-glycans containing galactosylated structures such as Lewis X [Galβ1–4(Fucα1–3) GlcNAc], Lewis Y [Fucα1–2Galβ1–4(Fucα1–3) GlcNAc] and lactosamine (Galβ1–4GlcNAc) motifs (Van Hove et al., 2011). Similarly, both EUL domains composing the two-domain EUL protein from rice, OrysaEULD1A, show carbohydrate specificity toward galactose containing glycans (Al Atalah et al., 2014b). In contrast, the rice protein OrysaEULS2 preferably binds mannosylated N-glycan structures (Al Atalah et al., 2012). All these EUL proteins are located in the nucleus and the cytoplasm of the plant cell. In search for interacting proteins, Li et al. (2014) recently reported that ArathEULS3 interacts with the nuclear/cytosolic ABA receptor RCAR1. Furthermore, Berendzen et al. (2012) demonstrated interaction of ArathEULS3 with CPK3, a Ca^{2+} dependent kinase involved in the ABA response in guard cells, supporting a role for ArathEULS3 in ABA signaling and stomatal closure.

Jacalin-related lectins (JRL). The family of jacalin-related lectins is named after jacalin, a 18 kDa T-antigen disaccharide-binding lectin domain first isolated from the seeds of jackfruit (*Artocarpus integrifolia*). Based on differences in molecular structure, subcellular localization, and carbohydrate binding properties, the large group of jacalins can be subdivided into two subgroups, further referred to as the galactose binding and mannose binding JRLs, residing in the vacuolar and nucleocytoplasmic compartment of the plant cell, respectively.

Galactose-specific JRLs have been reported mainly within the family Moraceae, whereas the mannose-specific JRLs are widespread in Viridiplantae. Furthermore, recent studies have shown that the jacalin domain is not restricted to plant proteins, since a similar domain has been reported in eukaryotes outside the plant kingdom as well as in some prokaryotes (Van Damme et al., 2008; Kanagawa et al., 2014). In plants, chimeric proteins comprising one or more jacalin domains linked to an unrelated domain are widely distributed. For instance, in *A. thaliana*, sequences composed of one or two jacalin domains C-terminally linked to five in tandem arranged Kelch domains are present. In addition, multiple *Arabidopsis* genes encode a putative F-box protein with a C-terminal jacalin domain (Nagano et al., 2008). Several Poaceae species (wheat, rice, maize) express proteins consisting of an N-terminal dirigent domain (also called disease-response domain) C-terminally fused to a jacalin domain. In rice, additional types of chimerolectins were identified, either composed of an N-terminal tyrosine kinase domain coupled to two or three jacalin domains or an N-terminal NB-ARC motif coupled to a LRR and a C-terminal jacalin domain (Van Damme et al., 2008).

Many jacalin-related lectin genes have been shown to be associated with disease resistance, abiotic stress signaling, wounding, insect damage or multiple stresses (Song et al., 2014). Especially the jacalin proteins with a dirigent domain are functionally involved in plant defense. To our knowledge, these chimeric proteins have only been reported in Poaceae species. In wheat, nearly half of the jacalin-related lectin genes encode dirigent domain-containing jacalin-related proteins. Several of these proteins have been studied in some detail, amongst them TaVER2, TaHfr-1, and TaJA-1 (Song et al., 2014). Interestingly all these mannose binding lectins are expressed as a response toward plant stress. TaVER2 is specifically expressed during vernalization in wheat (Yong et al., 2003) but is also up-regulated upon jasmonate and ABA treatment (Feng et al., 2009). TaHfr-1 is up-regulated after herbivory of Hessian fly larvae (Williams et al., 2002) and TaJA-1 is specifically accumulating after jasmonate treatment (Ma et al., 2013). The mannose specific TaHfr-1 was shown to effectively inhibit Hessian fly larval feeding resulting in the delay of larval development and premature death of the pest insects (Subramanyam et al., 2008). Transgenic tobacco plants overexpressing Ta-JA1 revealed increased resistance to bacterial, fungal, and viral pathogens (Ma et al., 2010). TaVER2 homologs have been found in maize and sorghum (β-glucosidase aggregating factor) and in rice (OsJAC1; Esen and Blanchard, 2000; Jiang et al., 2006; Kittur et al., 2009). Transgenic rice plants overexpressing OsJAC1 indicated the importance of OsJAC1 for rice growth and development (Jiang et al., 2007).

Similar to the chimerolectins also jacalin-related proteins composed only of jacalin domains are up-regulated in plant tissues subjected to certain stress treatments. For instance, Orysata was first reported as a salt inducible mannose binding lectin in the leaves of *O. sativa* (Zhang et al., 2000). Later it was shown that Orysata is also expressed upon JA and ABA treatment, after infection with an incompatible *Magnaporthe grisea* strain as well as during senescence (Lee et al., 2001; de Souza Filho et al., 2003;Qin et al., 2003). Glycan array analyses revealed that Orysata preferentially interacts with high-mannose and some more complex*N*-glycans (Al Atalah et al., 2011). In recent years, several orthologs of Orysata have been identified in Gramineae species but also in other plants, such as *Helianthus tuberosus* and *Ipomoea batatas*. The mannose specific wheat protein TaJRLL1 consisting of two jacalin-like domains is considered a component of SA and JA dependent plant defense signaling mechanisms and is activated upon fungal infection (*Fusarium graminearum* and *Blumeria graminis*; Xiang et al., 2011). However, not all jacalin-related defense proteins depend on hormone signaling. For example, the*Arabidopsis* jacalin-related JAX1 confers broad but specific resistance to potex viruses by inhibition of viral RNA accumulation, independent of hormone signaling (Yamaji et al., 2012). Other*Arabidopsis* jacalin-related proteins interact with proteins of ER bodies, i.e., ER-derived organelles presumably involved in defense against herbivores and/or pathogens (Nagano et al., 2008). Recently, the jacalin-related protein from sunflower seedlings named Helja was reported as a lectin with antifungal properties toward some pathogenic fungi of the genus *Candida*. Helja induces morphological changes as well as ROS production in yeast cells. Furthermore, lectin treatment also altered the membrane permeability of the cells (Regente et al., 2014).

Nictaba-related lectins. The family of Nictaba-related lectins was named after the *Nicotiana tabacum* agglutinin, abbreviated as Nictaba, a 19 kDa lectin domain originally discovered in tobacco leaves (Chen et al., 2002a) after jasmonate treatment. Though the lectin was first reported as a chito-oligosaccharide binding protein, glycan array analyses revealed that Nictaba also reacts with the inner core structure $(Man)_3$β1–4GlcNAcβ1–4GlcNAcβ-N-Asn of high-mannose and complex *N*-glycans. Biochemical assays confirmed that Nictaba can interact in a sugar-inhibitable way with many N-glycosylated proteins (Lannoo et al., 2006, 2007). A nuclear proteomics approach revealed the interaction of Nictaba with the core histone proteins H2A, H2B, and H4 through their *O*-GlcNAc modification (Schouppe et al., 2011), which was later confirmed at the microscopical level (Delporte et al., 2014a).

An extensive survey of the genome/transcriptome databases indicated that Nictaba-like domains are widespread among the Embryophyta but are absent from other eukaryotes and prokaryotes. Few proteins belonging to the Nictaba-like family consist of a single Nictaba domain. Furthermore, numerous sequences were identified encoding chimeric proteins comprising the Nictaba domain fused to unrelated N-terminal domains (e.g., F-box domain) or a Nictaba domain C-terminally fused to an N-terminal TIR (toll/interleukin-1 receptor) domain (found in*Arabidopsis* and tomato), an AIG1 (avrRpt2-induced gene) domain (identified in *Arabidopsis*) or a kinase domain (found in rice;Delporte et al., 2014b).

At present, only few Nictaba-related proteins have been studied for their biological properties and physiological role. A comparative analysis of the carbohydrate binding properties of Nictaba from *N. tabacum*, the Cucurbitaceae phloem lectin PPL, the *A. thaliana*homolog PP2-A1 and the *A. thaliana* F-box-Nictaba protein encoded by the gene *At2g02360* revealed that despite the sequence similarity and the presence of conserved amino acids in the carbohydrate binding site, different Nictaba domains can interact with different glycan motifs (Delporte et al., 2014b), suggesting different biological roles.

Since insect herbivory by Lepidopteran insects also triggers the JA pathway the tobacco lectin also accumulates after caterpillar attack. Furthermore, feeding assays with (transgenic) tobacco lines demonstrated the entomotoxic activity of Nictaba. It was suggested that the entomotoxic effect of Nictaba is caused by interaction of the lectin with glycoconjugates present in the digestive tract of the insect (Vandenborre et al., 2010, 2011). Within the plant cell, insect herbivory results in enhanced Nictaba expression in the cytoplasm, followed by partial translocation of the lectin to the nucleus, where it can interact with core histone proteins through their *O*-GlcNAc modification. It is hypothesized that Nictaba binding to chromatin results in enhanced transcription of defense related genes (Lannoo and Van Damme, 2010).

The Cucurbitaceae phloem lectins are a group of Nictaba-related lectins that are found in phloem exudates of different Cucurbitaceae species. Unlike Nictaba, the PP2 lectins are exclusively and continuously expressed in the companion cells of the phloem and then translocated into the phloem sap. The pumpkin lectin PPL and the PP2-like protein from *A. thaliana*, PP2-A1, show high binding affinity for chitin oligomers ($GlcNAc_{3\text{-}6}$). Similar to Nictaba, PP2-A1 also binds with the $Man_3GlcNAc_2$core of high-mannose *N*-glycans (Beneteau et al., 2010). Interestingly, the expression of PP2-A1 was enhanced by ethylene treatment and *Pseudomonas* infection. Furthermore, PP2-A1 represses phloem feeding of the green peach aphid *Myzus persicae*(Zhang et al., 2011), displayed antifungal activity against various fungal strains (Lee et al., 2014), and negatively affects Cucurbit aphid borne yellow virus transmission (Bencharki et al., 2010), strongly supporting a role in plant defense.

Several F-box Nictaba proteins are encoded in the *A. thaliana*genome. Glycan array analyses revealed the F-box protein encoded by *At2g02360* exhibits carbohydrate binding activity toward *N*- and *O*-glycans with *N*-acetyllactosamine (LacNAc; Galβ1–3GlcNAc and Galβ1–4GlcNAc) and poly-LacNAc structures as well as with Lewis A (Galβ1–3(Fucα1–4)GlcNAc), Lewis X (Galβ1–4(Fuca1–3)GlcNAc), Lewis Y (Fucα1–2Galβ1–4(Fucα1–3)GlcNAc), and type-1 B motifs (Galα1–3(Fucα1–2)Galβ1–3GlcNAc. Since these glycan structures have been reported in bacteria, viruses, and animals rather than in plants the physiological importance of this glycan interaction remains enigmatic (Lannoo et al., 2008; Stefanowicz et al., 2012). Furthermore, the same *Arabidopsis* F-box Nictaba protein was shown to interact with core members of an E3-type ubiquitin ligase complex which resulted in the hypothesis assuming a role of the Nictaba domain in glycoprotein degradation (Takahashi et al., 2004; Arabidopsis Interactome Mapping Consortium, 2011).

Ricin-B lectins. The ricin-B lectin family is one of the most widespread families of carbohydrate binding proteins in nature. The most famous member of this family is ricin, a toxic protein from castor bean (*Ricinus communis* L.) seeds, which was the very first lectin discovered in plants by Peter Hermann Stillmark in 1888. Ricin is a chimeric protein consisting of an A chain with enzymatic activity linked through a disulfide bridge with a B chain with lectin activity. This B chain consists of a duplicated ricin-B domain, responsible for the carbohydrate binding activity of the protein toward galactosylated structures. The enzymatic activity of ricin involves RNA *N*-glycosidase activity and is responsible for the removal of a highly conserved adenine residue from the sarcin/ricin loop of the 28S ribosomal RNA. As a result, the ribosomes are no longer able to bind elongation factor 2 and protein synthesis is blocked. Because of their catalytic activity these chimeric proteins are also referred to as type 2 ribosome-inactivating proteins (RIPs), and are considered as toxic proteins if they succeed in entering the cell. The uptake of the protein by the host cell is aided by their lectinic B-chain which can specifically interact with glycoconjugate structures on the cell surface. Except for ricin and abrin (from the jequirity bean *Abrus precatorius*), most type 2 RIPs are only moderately or even weakly toxic (Van Damme et al., 2001; Stirpe and Battelli, 2006).

The family of ricin-B related lectins is widespread in the plant kingdom and has been characterized in detail for what concerns its biological activity and toxicity in several plant species, especially *R. communis* (castor bean), *Abrus precatorius* (jequirity bean), *Viscum album* (mistletoe), and *Sambucus nigra* (elderberry). Unlike the other soluble lectins described above, most ricin-B related proteins accumulate in the plant vacuole or are secreted to the extracellular space (Van Damme et al., 2008). Over the years the ricin-B domain was identified in numerous plants, animals, fungi, and bacteria. All these proteins with ricin-B domains are also classified as the R-type lectins (Cummings and Etzler, 2009).

Within the genus *Sambucus* (elderberry) an extended family of ricin-B related proteins, including several chimerolectins and hololectins has been identified (Shang and Van Damme, 2014). Detailed hapten inhibition assays and glycan array studies revealed that all these *S. nigra* proteins exhibit different carbohydrate binding properties, and allowed to classify the *Sambucus* lectins into three groups. A first group covers the lectins SNA-II and SNA-IV as well as the type 2 RIP SNA-V with specificity toward Gal/GalNAc and Gal/GalNAc-containing glycan structures. The second group comprises only the type 2 RIP SNA-I, which specifically interacts with terminal sialic acid residues (Neu5Ac; α2–6) linked to Gal/GalNAc. Finally, the type 2 RIP SNLRP represents a third specificity group with strong interaction with GlcNAc oligomers (Shang and Van Damme, 2014).

Several lines of evidence support the idea that ricin-B related lectins play a role in plant defense against pathogens (Chen et al., 2002b; Vandenbussche et al., 2004a,b) and insects (Wei et al., 2004; Shahidi-Noghabi et al., 2009). Over-expressing SNA-I′ or SNA-V from *S. nigra* in transgenic tobacco (Samsun NN) plants enhances the tobacco plant's resistance against infection with tobacco mosaic virus. Though the antiviral effect is clearly related to the amount of protein expressed it cannot be related to an increased expression of pathogenesis-related proteins (Chen et al., 2002b; Vandenbussche et al., 2004a, b). Furthermore, no clear correlation was observed between *in planta* antiviral activity of the transgenic tobacco lines and the *in vitro N*-glycosidase activity of the proteins toward genomic RNA of the tobacco mosaic virus, suggesting that the *in planta* antiviral activity of these RIPs may rely on a direct interaction with the virus (Vandenbussche et al., 2004a). At present the importance of the lectin activity in the antiviral activity of the proteins remains unclear.

The first evidence for the insecticidal activity of ricin-B related proteins came from feeding assays with ricin and cinnamomin (from *Cinnamomum camphora* tree). Ricin showed strong toxicity to several insects such as cowpea weevil (*Callosobruchus macultatus*), cotton boll weevil (*Anthonomus grandis*), house fly (*Musca domestica*), and larvae of the silkworm *Bombyx mori* (Wei et al., 2004). The differences in activity between ricin and cinnamomin could not be related to the enzymatic activity but rather were attributed to differences in the activity of the lectin chain of the proteins (Wei et al., 2004). Shahidi-Noghabi et al. (2009) reported the enhanced resistance of transgenic tobacco plants overexpressing SNA-I or its isoform SNA-I′ toward different pest insect species including aphids and caterpillars. Since mutation of the carbohydrate binding site can abolish or reduce the toxic effect, the entomotoxic properties of the proteins can be linked to their carbohydrate binding activity (Shahidi-Noghabi et al., 2008). In addition, the cytotoxic effects of *S. nigra* RIPs toward insect cells was accompanied by caspase 3-like protease-induced apoptosis (Shahidi-Noghabi et al., 2008, 2011). More research is needed to identify the interacting proteins for the *Sambucus* lectins on the cell surface.

5. CONCLUSION

Plant genomes encode a plethora of RLKs, RLPs, and lectins to protect themselves against the vast array of pathogenic bacteria, viruses, fungi, oomycetes, and pest insects. A key feature of the plant's innate immunity is the ability to recognize D/P/MAMPs of potential pathogens through PRRs, and subsequently respond in a highly sensitive and specific manner. Many advances have been made in the understanding how different proteins function in plant innate immunity. As more structural and biochemical data become available, common themes are emerging on receptor organization, ligand perception and binding, receptor activation, and intracellular defense signaling. It is clear now that PRRs are ultradynamic multiprotein structures which often use phosphorylation to activate downstream signaling.

A first interaction between the pathogen and the plant occurs at the level of the cell wall and the plasma membrane where extracellular effectors, DAMPs, and P/MAMPs will be recognized by membrane-bound receptors (Figure 1; Tables 1 and 2), among which a large group of RLPs and RLKs some of which carry an extracellular lectin domain. Though different lectin motifs have been recognized, only the LysM domain was unambiguously shown to be dependent on carbohydrate interactions for recognition of fungal and bacterial components and subsequent signal transmission into the plant cell. Interestingly, the LysM motif shows high specificity for chito-oligosaccharides, an abundant component in different pathogens but absent from plants. Other (lectin) receptor kinases/proteins probably depend on protein–protein interactions to recognize specific ligands. Though our knowledge on receptor kinase function and signaling has greatly improved, several issues still remain with respect to the potential ligands for pathogen recognition and the downstream signaling events, especially for those receptors lacking the kinase domain.

In addition to the lectin motifs present at the level of the cell wall/plasma membrane, plants synthesize well-defined soluble lectins upon exposure to multiple abiotic and biotic stresses (Table3). Most of these inducible lectins reside in the nucleus and the cytosol of the plant cell and evidence is emerging for their role in signal transduction inside the plant cell as part of multiple plant defense pathways. Hence protein–carbohydrate interactions should not only be envisaged at the level of the interaction between the pathogen and the plant cell, but also play an important role inside the plant cell as part of the intracellular signaling resulting from the recognition of plant attackers. At least for some cytoplasmic lectins (Nictaba-related proteins, EUL-related proteins, amaranthins) it was shown that they are also translocated inside the plant nucleus. The Nictaba-related proteins in particular have been shown to interact with glycosylated histones, and therefore are suggested to act as chromatin remodelers, enabling altered gene expression as a result of stress signaling. Surveying the plant genome sequences also revealed that most lectin domains are part of larger proteins, consisting of one or more lectin domains linked to un-related protein domains, most often with unknown functions. Future challenges include the characterization of the ligands for these soluble lectins or lectin domains present as part of a larger protein to elucidate the biological relevance of these interactions. Large-scale experiments integrating genomics, biochemistry, cell biology, structural biology, and bioinformatics will enable to elucidate the physiological importance of the lectin motifs in protein–carbohydrate interactions in signal transduction and plant defense.

ACKNOWLEDGMENTS

This work was supported by the Fund for Scientific Research – Flanders and the Research Council of Ghent University. N. Lannoo acknowledges the receipt of a Postdoctoral fellowship from the Fund for Scientific Research – Flanders.

REFERENCES

1. Al Atalah, B., De Vleesschauwer, D., Xu, J., Fouquaert, E., Höfte, M., and Van Damme, E. J. (2014a). Transcriptional behavior of EUL-related rice lectins toward important abiotic and biotic stresses. *J. Plant Physiol.* 171, 986–992. doi: 10.1016/j.jplph.2014.04.004

2. Al Atalah, B., Vanderschaeghe, D., Bloch, Y., Proost, P., Plas, K., Callewaert, N., et al. (2014b). Characterization of a type D1A EUL-related lectin from rice expressed in*Pichia pastoris*. *Biol. Chem.* 395, 413–424.

3. Al Atalah, B., Fouquaert, E., Vanderschaeghe, D., Proost, P., Balzarini, J., Smith, D. F., et al. (2011). Expression analysis of the nucleocytoplasmic lectin “Orysata” from rice in *Pichia pastoris*. *FEBS J.* 278, 2064–2079.

4. Al Atalah, B., Rougé, P., Smith, D. F., Proost, P., Lasanajak, Y., and Van Damme, E. J. M. (2012). Expression analysis of a type S2 EUL-related lectin from rice in *Pichia pastoris*. *Glycoconj. J.* 29, 467–479.

5. Albert, M. (2013). Peptides as triggers of plant defence. *J. Exp. Bot.* 64, 5269–5279.

6. Arabidopsis Interactome Mapping Consortium. (2011). Evidence for network evolution in an *Arabidopsis* interactome map. *Science* 333, 601– 607.

7. Bar, M., Sharfman, M., Ron, M., and Avni, A. (2010). BAK1 is required for the attenuation of ethylene-inducing xylanase (Eix)-induced defense responses by the decoy receptor LeEix1. *Plant J.* 63, 791–800.

8. Bencharki, B., Boissinot, S., Revollon, S., Ziegler-Graff, V., Erdinger, M., Wiss, L., et al. (2010). Phloem protein partners of Cucurbit aphid borne yellows virus: possible involvement of phloem proteins in virus transmission by aphids. *Mol. Plant Microbe Interact.* 23, 799–810.

9. Beneteau, J., Renard, D., Marché, L., Douville, E., Lavenant, L., Rahbé, Y., et al. (2010). Binding properties of the N-acetylglucosamine and high-mannose N-glycan PP2-A1 phloem lectin in *Arabidopsis*. *Plant Physiol.* 153, 1345–1361.

10. Berendzen, K. W., Böhmer, M., Wallmeroth, N., Peter, S., Vesić, M., Zhou, Y.,et al. (2012). Screening for in planta protein–protein interactions combining bimolecular fluorescence complementation with flow cytometry. *Plant Methods* 8:25.

11. Block, A., and Alfano, J. R. (2011). Plant targets for *Pseudomonas syringae* type III effectors: virulence targets or guarded decoys? *Curr. Opin. Microbiol.* 14, 39–46.

12. Block, A., Toruño, T. Y., Elowsky, C. G., Zhang, C., Steinbrenner, J., Beynon, J., et al. (2013). The *Pseudomonas syringae* type III effector HopD1 suppresses effector-triggered immunity, localizes to the endoplasmic reticulum, and targets the*Arabidopsis* transcription factor NTL9. *New Phytol.* 201, 1358–1370.

13. Böhm, H., Albert, I., Fan, L., Reinhard, A., and Nürnberger, T. (2014). Immune receptor complexes at the plant cell surface. *Curr. Opin. Plant Biol.* 20, 47–54.

14. Boller, T., and Felix, G. (2009). A renaissance of elicitors: perception of microbe-associated molecular patterns and danger signals by pattern-recognition receptors.*Annu. Rev. Plant Biol.* 60, 379–406.

15. Bouwmeester, K., de Sain, M., Weide, R., Gouget, A., Klamer, S., Canut, H., et al. (2011). The lectin receptor kinase LecRK-I.9 is a novel Phytophthora resistance component and a potential host target for a RXLR effector. *PLoS Pathog.* 7:e1001327.

16. Bouwmeester, K., and Govers, F. (2009). *Arabidopsis* L-type lectin receptor kinases: phylogeny, classification, and expression profiles. *J. Exp. Bot.* 60, 4383–4396.

17. Brotman, Y., Landau, U., Pnini, S., Lisec, J., Balazadeh, S., Mueller-Roeber, B.,et al. (2012). The LysM receptor-like kinase LysM RLK1 is required to activate defense and abiotic-stress responses induced by overexpression of fungal chitinases in *Arabidopsis*plants. *Mol. Plant* 5, 1113–1124.

18. Buist, G., Steen, A., Kok, J., and Kuipers, O. P. (2008). LysM, a widely distributed protein motif for binding to (peptido)glycans. *Mol. Microbiol.* 68, 838–847.

19. Cambi, A., Koopman, M., and Figdor, C. G. (2005). How C-type lectins detect pathogens. *Cell. Microbiol.* 7, 481–488.

20. Cao, Y., Tanaka, K., Nguyen, C. T., and Stacey, G. (2014). Extracellular ATP is a central signaling molecule in plant stress responses. *Curr. Opin. Plant Biol.* 20, 82–87.

21. Chen, X., Zuo, S., Schwessinger, B., Chern, M., Canlas, P. E., Ruan, D.,et al. (2014). An XA21-associated kinase (OsSERK2) regulates immunity mediated by the XA21 and XA3 immune receptors. *Mol. Plant* 7, 874–892.

22. Chen, Y., Peumans, W. J., Hause, B., Bras, J., Kumar, M., Proost, P.,et al. (2002a). Jasmonic acid methyl ester induces the synthesis of a cytoplasmic/nuclear chitooligosaccharide-binding lectin in tobacco leaves. *FASEB J.* 16, 905–907.

23. Chen, Y., Peumans, W. J., and Van Damme, E. J. M. (2002b). The *Sambucus nigra*type-2 ribosome-inactivating protein SNA-I′ exhibits in planta antiviral activity in transgenic tobacco. *FEBS Lett.* 516, 27–30.

24. Chinchilla, D., Bauer, Z., Regenass, M., Boller, T., and Felix, G. (2006). TheArabidopsis receptor kinase FLS2 binds flg22 and determines the specificity of flagellin perception. *Plant Cell* 18, 465–476.

25. Chinchilla, D., Zipfel, C., Robatzek, S., Kemmerling, B., Nürnberger, T., Jones, J. D., et al. (2007). A flagellin-induced complex of the receptor FLS2 and BAK1 initiates plant defence. *Nature* 448, 497–500.

26. Choi, J., Tanaka, K., Cao, Y., Qi, Y., Qiu, J., Liang, Y.,et al. (2014). Identification of a plant receptor for extracellular ATP. *Science* 343, 290–294.

27. Cui, F., Wu, S., Sun, W., Coaker, G., Kunkel, B., He, P.,et al. (2013). The *Pseudomonas syringae* type III effector AvrRpt2 promotes pathogen virulence via stimulating*Arabidopsis* auxin/indole acetic acid protein turnover. *Plant Physiol.* 162, 1018–1029.

28. Cummings, R. D., and Etzler, M. E. (2009). "R-type lectins," in *Essentials of Glycobiology*, 2nd Edn, Chap. 28, eds E. Varki, R. D. Cummings, J. D. Esko, H. H. Freeze, P. Stanley, C. R. Bertozzi.,et al. (Cold Spring Harbor, NY: Cold Spring Harbor Laboratory Press).

29. Dangl, J. L., Horvath, D. M., and Staskawicz, B. J. (2013). Pivoting the plant immune system from dissection to deployment. *Science* 341, 746–751.

30. Degiacomi, M. T., Iacovache, I., Pernot, L., Chami, M., Kudryashev, M., Stahlberg, H., et al. (2013). Molecular assembly of the aerolysin pore reveals a swirling membrane-insertion mechanism. *Nat. Chem. Biol.* 9, 623–629.

31. Delporte, A., De Vos, W. H., and Van Damme, E. J. M. (2014a). In vivo interaction between the tobacco lectin and the core histone proteins. *J. Plant Physiol.* 171, 986–992.

32. Delporte, A., Van Holle, S., Lannoo, N., and Van Damme, E. J. M. (2014b). The tobacco lectin, prototype of the family of Nictaba-related proteins. *Curr. Prot. Pept. Sci.* (in press).

33. de Souza Filho, G. A., Ferreira, B. S., Dias, J. M. R., Queiroz, K. S., Branco, A. T., Bressan-Smith, R. E.,et al. (2003). Accumulation of SALT protein in rice plants as a response to environmental stresses. *Plant Sci.* 164, 623–628.

34. Dubery, I. A., Sanabria, N. M., and Huang, J.-C. (2012). "Chapter 6: nonself perception in plant innate immunity," in *Self and Nonself*, ed. C. López-Larrea (Austin, TX: Landes Bioscience; New York, NY: Springer Science+Business Media).

35. Ellgaard, L., and Frickel, E. M. (2003). Calnexin, calreticulin, and ERp57: teammates in glycoprotein folding. *Cell Biochem. Biophys.* 39, 223–247.

36. Endre, G., Kereszt, A., Kevei, Z., Mihacea, S., Kaló, P., and Kiss, G. B. (2002). A receptor kinase gene regulating symbiotic nodule development. *Nature* 417, 962–966.

37. Esen, A., and Blanchard, D. J. (2000). A specific β-glucosidase-aggregating factor is responsible for the β-glucosidase null phenotype in maize. *Plant Physiol.* 122, 563–572.

38. Feng, H., Xu, W.-Z., Lin, H.-H., and Chong, K. (2009). Transcriptional regulation of wheat VER2 promoter in rice in response to abscisic acid, jasmonate, and light. *J. Genet. Genomics* 36, 371–377.

39. Fouquaert, E., Peumans, W. J., Smith, D. F., Proost, P., Savvides, S. N., and Van Damme, E. J. M. (2008). The "old" *Euonymus europaeus* agglutinin represents a novel family of ubiquitous plant proteins. *Plant Physiol.* 147, 1316–1324.

40. Fouquaert, E., and Van Damme, E. J. M. (2012). Promiscuity of the euonymus carbohydrate-binding domain. *Biomolecules* 2, 415–434.

41. Fradin, E. F., Zhang, Z., Juarez Ayala, J. C., Castroverde, C. D., Nazar, R. N., Robb, J.,et al. (2009). Genetic dissection of *Verticillium* wilt resistance mediated by tomato Ve1.*Plant Physiol.* 150, 320–332.

42. Gilardoni, P. A., Hettenhausen, C., Baldwin, I. T., and Bonaventure, G. (2011).*Nicotiana attenuata* LECTIN RECEPTOR KINASE 1 suppresses the insect-mediated inhibition of induced defense responses during *Manduca sexta* herbivory. *Plant Cell*23, 3512–3532.

43. Gómez-Gómez, L., Bauer, Z., and Boller, T. (2001). Both the extracellular leucine-rich repeat domain and the kinase activity of FLS2 are required for flagellin binding and signaling in *Arabidopsis*. *Plant Cell* 13, 1155–1163.

44. Grant, S. R., Fisher, E. J., Chang, J. H., Mole, B. M., and Dangl, J. L. (2006). Subterfuge and manipulation: type III effector proteins of phytopathogenic bacteria. *Annu. Rev. Microbiol.* 60, 425–449.

45. Gudesblat, G. E., Torres, P. S., and Vojnov, A. A. (2009). *Xanthomonas campestris*overcomes *Arabidopsis* stomatal innate immunity through a DSF cell-to-cell signal-regulated virulence factor. *Plant Physiol.* 149, 1017–1027.

46. Gust, A. A., Willmann, R., Desaki, Y., Grabherr, H. M., and Nürnberger, T. (2012). Plant LysM proteins: modules mediating symbiosis and immunity. *Trends Plant Sci.* 17, 495–502.

47. Halter, T., Imkampe, J., Mazzotta, S., Wierzba, M., Postel, S., Bücherl, C., et al. (2014). The leucine-rich repeat receptor kinase BIR2 is a negative regulator of BAK1 in plant immunity. *Curr. Biol.* 24, 134–143.

48. Han, Z., Sun, Y., and Chai, J. (2014). Structural insight into the activation of plant receptor kinases. *Curr. Opin. Plant Biol.* 20, 55–63.

49. Häweker, H., Rips, S., Koiwa, H., Salomon, S., Saijo, Y., Chinchilla, D.,et al. (2010). Pattern recognition receptors require N-glycosylation to mediate plant immunity. *J. Biol. Chem.* 285, 4629–4636.

50. Hayafune, M., Berisio, R., Marchetti, R., Silipo, A., Kayama, M., Desaki, Y., et al. (2014). Chitin-induced activation of immune signaling by the rice receptor CEBiP relies on a unique sandwich-type dimerization. *Proc. Natl. Acad. Sci. U.S.A.* 111, E404–E413.

51. Howell, S. H. (2013). Endoplasmic reticulum stress responses in plants. *Annu. Rev. Plant Biol.* 64, 477–499.

52. Huang, P., Ju, H.-W., Min, J.-H., Zhang, X., Kim, S.-H., Yang, K.-Y.,et al. (2013). Overexpression of L-type lectin-like protein kinase 1 confers pathogen resistance and regulates salinity response in *Arabidopsis thaliana*. *Plant Sci.* 203–204, 98–106.

53. Huang, P.-Y., Yeh, Y.-H., Liu, A.-C., Cheng, C.-P., and Zimmerli, L. (2014). The*Arabidopsis* LecRK-VI.*2* associates with the pattern-recognition receptor FLS2 and primes *Nicotiana benthamiana* pattern-triggered immunity. *Plant J.* 79, 243–255.

54. Jiang, J. F., Han, Y., Xing, L. J., Xu, Y. Y., Xu, Z. H., and Chong, K. (2006). Cloning and expression of a novel cDNA encoding a mannose-specific jacalin-related lectin from*Oryza sativa*. *Toxicon* 47, 133–139.

55. Jiang, J. F., Xu, Y. Y., and Chong, K. (2007). Overexpression of OsJAC1, a lectin gene, suppresses the coleoptiles and stem elongation in rice. *J. Integr. Plant Biol.* 49, 230–237.

56. Jin, H., Hong, Z., Su, W., and Li, J. (2009). A plant-specific calreticulin is a key retention factor for a defective brassinosteroid receptor in the endoplasmic reticulum.*Proc. Natl. Acad. Sci. U.S.A.* 106, 13612–13617.

57. Kaku, H., Nishizawa, Y., Ishii-Minami, N., Akimoto-Tomiyama, C., Dohmae, N., Takio, K., et al. (2006). Plant cells recognize chitin fragments for defense signaling through a plasma membrane receptor. *Proc. Natl. Acad. Sci. U.S.A.* 103, 11086–11091.

58. Kanagawa, M., Liu, Y., Hanashima, S., Ikeda, A., Chai, W., Nakano, Y.,et al. (2014). Structural basis for multiple sugar recognition of jacalin-related human ZG16p lectin. *J. Biol. Chem.* 289, 16954–16965.

59. Kanzaki, H., Saitoh, H., Takahashi, Y., Berberich, T., Ito, A., Kamoun, S.,et al. (2008). NbLRK1, a lectin-like receptor kinase protein of *Nicotiana benthamiana*, interacts with*Phytophthora infestans* INF1 elicitin and mediates INF1-induced cell death. *Planta*228, 977–987.

60. Kapoor, M., Ellgaard, L., Gopalakrishnapai, J., Schirra, C., Gemma, E., Oscarson, S., et al. (2004). Mutational analysis provides molecular insight into the carbohydrate-binding region of calreticulin: pivotal roles of tyrosine-109 and aspartate-135 in carbohydrate recognition. *Biochemistry* 43, 97–106.

61. Kim, Y. T., Oh, J., Kim, K. H., Uhm, J. Y., and Lee, B. M. (2009). Isolation and characterization of NgRLK1, a receptor-like kinase of *Nicotiana glutinosa* that interacts with the elicitin of *Phytophthora capsici*. *Mol. Biol. Rep.* 37, 717–727.

62. Kittur, F. S., Yu, H. Y., Bevan, D. R., and Esen, A. (2009). Homolog of the maize β-glucosidase aggregating factor from sorghum is a jacalin-related GalNAc-specific lectin but lacks protein aggregating activity. *Glycobiology* 19, 277–287.

63. Knogge, W., and Scheel, D. (2006). LysM receptors recognize friend and foe. *Proc. Natl. Acad. Sci. U.S.A.* 103, 10829–10830.

64. Kouzai, Y., Nakajima, K., Hayafune, M., Ozawa, K., Kaku, H., Shibuya, N.,et al. (2014). CEBiP is the major chitin oligomers-binding protein in rice and plays a main role in the perception of chitin oligomers. *Plant Mol. Biol.* 84, 519–528.

65. Krol, E., Mentzel, T., Chinchilla, D., Boller, T., Felix, G., Kemmerling, B., et al. (2010). Perception of the *Arabidopsis* danger signal peptide 1 involves the pattern recognition receptor AtPEPR1 and its close homologue AtPEPR2. *J. Biol. Chem.* 285, 13471–13479.

66. Lannoo, N., Peumans, W. J., and Van Damme, E. J. M. (2008). Do F-box proteins with a C-terminal domain homologous with the tobacco lectin play a role in protein degradation in plants? *Biochem. Soc. Trans.* 36, 843–847.

67. Lannoo, N., Peumans, W. J., Van Pamel, E., Alvarez, R., Xiong, T., Hause, G.,et al. (2006). Localization and in vitro binding studies suggest that the cytoplasmic/nuclear tobacco lectin can interact in situ with high-mannose and complex N-glycans. *FEBS Lett.* 580, 6329–6337.

68. Lannoo, N., and Van Damme, E. J. (2010). Nucleocytoplasmic plant lectins. *Biochem. Biophys. Acta* 1800, 190–201.

69. Lannoo, N., Vandenborre, G., Miersch, O., Smagghe, G., Wasternack, C., Peumans, W. J., et al. (2007). The jasmonate-induced expression of the *Nicotiana tabacum* leaf lectin. *Plant Cell Physiol.* 48, 1207–1218.

70. Lee, J. R., Boltz, K. A., and Lee, S. Y. (2014). Molecular chaperone function of*Arabidopsis thaliana* phloem protein 2-A1 encodes a protein similar to phloem lectin.*Biochem. Biophys. Res. Commun.* 443, 18–21.

71. Lee, R. H., Wang, C. H., Huang, L. T., and Chen, S. C. (2001). Leaf senescence in rice plants: cloning and characterization of senescence up-regulated genes. *J. Exp. Bot.* 52, 1117–1121.

72. Lee, S. W., Han, S. W., Sririyanum, M., Park, C. J., Seo, Y. S., and Ronald, P. C. (2009). A type I-secreted, sulfated peptide triggers XA21-mediated innate immunity. *Science*326, 850–853.

73. Li, D., Wang, X., Yuan, D., Zhang, L., Jiang, X., Tao, Z.,et al. (2014). Over-expression of ArathEULS3 confers ABA sensitivity and drought tolerance in *Arabidopsis. Plant Cell Tissue Organ Cult.* 117, 431–442.

74. Li, J., Zhao-Hui, C., Batoux, M., Nekrasov, V., Roux, M., Chinchilla, D.,et al. (2009). Specific ER quality control components required for biogenesis of the plant innate immune receptor EFR. *Proc. Natl. Acad. Sci. U.S.A.* 106, 15973–15978.

75. Limpens, E., Franken, C., Smit, P., Willemse, J., Bisseling, T., and Geurts, R. (2003). LysM domain receptor kinases regulating rhizobial Nod factor-induced infection.*Science* 302, 630–633.

76. Liu, B., Li, J. F., Ao, Y., Qu, J., Li, Z., Su, J.,et al. (2012a). Lysin motif-containing proteins LYP4 and LYP6 play dual roles in peptidoglycan and chitin perception in rice innate immunity. *Plant Cell* 24, 3406–3419.

77. Liu, T., Liu, Z., Song, C., Hu, Y., Han, Z., She, J.,et al. (2012b). Chitin-induced dimerization activates a plant immune receptor. *Science* 336, 1160–1164.

78. Liu, Y., and Li, J. (2014). Endoplasmic reticulum-mediated protein quality control in*Arabidopsis*. *Front. Plant Sci.* 5:162.

79. Ma, Q. H., Tian, B., and Li, Y. L. (2010). Overexpression of a wheat jasmonate-regulated lectin increases pathogen resistance. *Biochimie* 92, 187–193.

80. Ma, Q. H., Zhen, W. B., and Liu, Y. C. (2013). Jacalin domain in wheat jasmonate-regulated protein Ta-JA1 confers agglutinating activity and pathogen resistance.*Biochimie* 95, 359–365.

81. Macho, A. P., and Zipfel, C. (2014). Plant PRRs and the activation of innate immune signaling. *Mol. Cell* 54, 263–272.

82. Madsen, E. B., Madsen, L. H., Radutoiu, S., Olbryt, M., Rakwalska, M., Szczyglowski, K.,et al. (2003). A receptor kinase gene of the LysM type is involved in legume perception of rhizobial signals. *Nature* 425, 637–640.

83. Malinovsky, F. G., Fangel, J. U., and Willats, W. G. T. (2014). The role of the cell wall in plant immunity. *Front. Plant Sci.* 5:178.

84. Melotto, M., Underwood, W., Koczan, J., Nomura, K., and He, S. Y. (2006). Plant stomata function in innate immunity against bacterial invasion. *Cell* 126, 969–980.

85. Miya, A., Albert, P., Shinya, T., Desaki, Y., Ichimura, K., Shirasu, K., et al. (2007). CERK1, a LysM receptor kinase, is essential for chitin elicitor signaling in *Arabidopsis.Proc. Natl. Acad. Sci. U.S.A.* 104, 19613–19618.

86. Mulder, L., Lefebvre, B., Cullimore, J., and Imberty, A. (2006). LysM domains of*Medicago truncatulaaa* NFP protein involved in Nod factor perception. Glycosylation state, molecular modeling and docking of chitooligosaccharides and Nod factors.*Glycobiology* 16, 801–809.

87. Muthamilarasan, M., and Prasad, M. (2013). Plant innate immunity: an updated insight into defense mechanism. *J. Biosci.* 38, 433–449.

88. Nagano, A. J., Fukao, Y., Fujiwara, M., Nishimura, M., and Hara-Nishimura, I. (2008). Antagonistic jacalin-related lectins regulate the size of ER body-type β-glucosidase complexes in *Arabidopsis thaliana. Plant Cell Physiol.* 49, 969–980.

89. Navarro-Gochicoa, M. T., Camut, S., Timmers, A. C. J., Niebel, A., Herye, C., Boutet, E., et al. (2003). Characterization of four lectin-like receptor kinases expressed in roots of *Medicago truncatula* structure, location, regulation of expression, and potential role in the symbiosis with *Sinorhizobium meliloti. Plant Physiol.* 133, 1893–1910.

90. Newman, M.-A., Sundelin, T., Nielsen, J. T., and Erbs, G. (2013). MAMP (microbe-associated molecular pattern) triggered immunity in plants. *Front. Plant Sci.* 4:139.

91. Nühse, T. S. (2012). Cell wall integrity signaling and innate immunity in plants. *Front. Plant Sci.* 3:280.

92. Paparella, C., Savatin, D. V., Marti, L., De Lorenzo, G., and Ferrari, S. (2014). The*Arabidopsis* LYSIN MOTIF-CONTAINING RECEPTOR-LIKE KINASE 3 regulates the cross talk between immunity and abscisic acid responses. *Plant Physiol.* 165, 262–276.

93. Park, C. J., and Ronald, P. C. (2012). Cleavage and nuclear localization of the rice XA21 immune receptor. *Nat. Commun.* 3:920.

94. Petutschnig, E. K., Jones, A. M., Serazetdinova, L., Lipka, U., and Lipka, V. (2010). The lysin motif receptor-like kinase (LysM-RLK) CERK1 is a major chitin-binding protein in*Arabidopsis thaliana* and subject to chitin-induced phosphorylation. *J. Biol. Chem.*285, 28902–28911.

95. Peumans, W. J., and Van Damme, E. J. M. (1995). Lectins as plant defense proteins.*Plant Physiol.* 109, 347–352.

96. Puthoff, D. P., Sardesai, N., Subramanyam, S., Nemacheck, J. A., and Williams, C. E. (2005). Hfr-2, a wheat cytolytic toxin-like gene, is up-regulated by virulent Hessian fly larval feeding. *Mol. Plant Pathol.* 6, 411–423.

97. Qin, Q. M., Zhang, Q., Zhao, W. S., Wang, Y. Y., and Peng, Y. L. (2003). Identification of a lectin gene induced in rice in response to *Magnaporthe grisea* infection. *Acta Bot. Sin.* 45, 76–81.

98. Qiu, Y., Xi, J., Du, L., Roje, S., and Poovaiah, B. W. (2012). A dual regulatory role of*Arabidopsis* calreticulin-2 in plant innate immunity. *Plant J.* 69, 489–500.

99. Radutoiu, S., Madsen, L. H., Madsen, E. B., Felle, H. H., Umehara, Y., Gronlund, M.,et al. (2003). Plant recognition of symbiotic bacteria requires two LysM receptor-like kinases. *Nature* 425, 585–592.

100. Regente, M., Taveira, G. B., Pinedo, M., Elizalde, M. M., Ticchi, A. J., Diz, M. S. S.,et al. (2014). A sunflower lectin with antifungal properties and putative medical mycology applications. *Curr. Microbiol.* 69, 88–95.

101. Roberts, D. D., and Goldstein, I. J. (1983). Adenine binding sites of the lectin from lima beans (*Phaseolus lunatus*). *J. Biol. Chem.* 258, 13820–13824.

102. Ron, M., and Avni, A. (2004). The receptor for the fungal elicitor ethylene-inducing xylanase is a member of a resistance-like gene family in tomato. *Plant Cell* 16, 1604–1615.

103. Rotblat, B., Enshell-Seijffers, D., Gershoni, J. M., Schuster, S., and Avni, A. (2002). Identification of an essential component of the elicitation active site of the EIX protein elicitor. *Plant J.* 32, 1049–1055.

104. Ryan, C. A. (2000). The systemin signaling pathway: differential activation of plant defensive genes. *Biochim. Biophys. Acta* 1477, 112–121.

105. Scheer, J. M., and Ryan, C. A. Jr. (2002). The systemin receptor SR160 from*Lycopersicon peruvianum* is a member of the LRR receptor kinase family. *Proc. Natl. Acad. Sci. U.S.A.* 99, 9585–9590.

106. Schouppe, D., Ghesquière, B., Menschaert, G., De Vos, W. H., Bourque, S., Trooskens, G., et al. (2011). Interaction of the tobacco lectin with histone proteins. *Plant Physiol.*155, 1091–1102.

107. Schulze, B., Mentzel, T., Jehle, A. K., Mueller, K., Beeler, S., Boller, T.,et al. (2010). Rapid heteromerization and phosphorylation of ligand-activated plant transmembrane receptors and their associated kinase BAK1. *J. Biol. Chem.* 285, 9444–9451.

108. Shahidi-Noghabi, S., Van Damme, E. J. M., De Vos, W. H., and Smagghe, G. (2011). Internalization of *Sambucus nigra* agglutinins I and II in insect midgut cells. *Arch. Insect Biochem. Physiol.* 76, 211–222.

109. Shahidi-Noghabi, S., Van Damme, E. J. M., and Smagghe, G. (2008). Carbohydrate-binding activity of the type-2 ribosome-inactivating protein SNA-I from elderberry (*Sambucus nigra*) is a determining factor for its insecticidal activity. *Phytochemistry*69, 2972–2978.

110. Shahidi-Noghabi, S., Van Damme, E. J. M., and Smagghe, G. (2009). Expression of*Sambucus nigra* agglutinin (SNA-I') from elderberry bark in transgenic tobacco plants results in enhanced resistance to different insect species. *Transgenic Res.* 18, 249–259.

111. Shang, C., and Van Damme, E. J. M. (2014). Comparative analysis of carbohydrate binding properties of *Sambucus nigra* lectins and ribosome-inactivating proteins.*Glycoconj. J.* 31, 345–354.

112. Sherman-Broyles, S., Boggs, N., Farkas, A., Liu, P., Vrebalov, J., Nasrallah, M. E., et al. (2007). S locus genes and the evolution of self-fertility in *Arabidopsis thaliana*. *Plant Cell* 19, 94–106.

113. Shimizu, T., Nakano, T., Takamizawa, D., Desaki, Y., Ishii-Minami, N., Nishizawa, Y., et al. (2010). Two LysM receptor molecules, CEBiP and OsCERK1, cooperatively regulate chitin elicitor signaling in rice. *Plant J.* 64, 204–214.

114. Shinya, T., Motoyama, N., Ikeda, A., Wada, M., Kamiya, K., Hayafune, M., et al. (2012). Functional characterization of CEBiP and CERK1 homologs in *Arabidopsis* and rice reveals the presence of different chitin receptor systems in plants. *Plant Cell Physiol.*53, 1696–1706.

115. Singh, P., Kuo, Y. C., Mishra, S., Tsai, C. H., Chien, C. C., Chen, C. W.,et al. (2012). The lectin-receptor kinase-VI.2 is required for priming and positively regulates *Arabidopsis*pattern-triggered immunity.*Plant Cell* 24, 1256–1270.

116. Singh, P., and Zimmerli, L. (2013). Lectin receptor kinases in plant innate immunity.*Front. Plant Sci.* 4:124.

117. Smith, J. M., Salamango, D. J., Leslie, M. E., Collins, C. A., and Heese, A. (2014). Sensitvity to Flg22 is modulated by ligand-induced degradation and de novo synthesis of the endogenous flagellin-receptor FLAGELLIN-SENSING2. *Plant Physiol.* 164, 440–454.

118. Song, M., Xu, W., Xiang, Y., Jia, H., Zhang, L., and Ma, Z. (2014). Association of jacalin-related lectins with wheat responses to stresses revealed by transcriptional profiling.*Plant Mol. Biol.* 84, 95–110.

119. Stefanowicz, K., Lannoo, N., Proost, P., and Van Damme, E. J. M. (2012). *Arabidopsis*F-box protein containing a Nictaba-related lectin domain interacts with N-acetyllactosamine structures. *FEBS Open Bio* 2, 151–158.

120. Stirpe, F., and Battelli, M. G. (2006). Ribosome-inactivating proteins: progress and problems. *Cell. Mol. Life Sci.* 63, 1850–1866.

121. Stracke, S., Kistner, C., Yoshida, S., Mulder, L., Sato, S., Kaneko, T., et al. (2002). A plant receptor-like kinase required for both bacterial and fungal symbiosis. *Nature* 417, 959–962.

122. Subramanyam, S., Smith, D. F., Clemens, J. C., Webb, M. A., Sardesai, N., and Williams, C. E. (2008). Functional characterization of HFR1, a high-mannose N-glycan-specific wheat lectin induced by Hessian fly larvae. *Plant Physiol.* 147, 1412–1426.

123. Sun, X., Cao, Y., Yang, Z., Xu, C., Li, X., Wang, S., et al. (2004). Xa26, a gene conferring resistance to *Xanthomonas oryzae* pv. oryzae in rice, encodes an LRR receptor kinase-like protein. *Plant J.* 37, 517–527.

124. Sun, X.-L., Yu, Q.-Y., Tang, L.-L., Ji, W., Bai, X., Cai, H., et al. (2013). GsSRK, a G-type lectin S-receptor-like serine/threonine protein kinase, is a positive regulator of plant tolerance to salt stress. *J. Plant Physiol.* 170, 505–515.

125. Takahashi, N., Kuroda, H., Kuromori, T., Hirayama, T., Seki, M., Shinozaki, K.,et al. (2004). Expression and interaction analysis of *Arabidopsis* Skp1-related genes. *Plant Cell Physiol.* 45, 83–91.

126. Tanaka, K., Nguyen, C. T., Liang, Y., Cao, Y., and Stacey, G. (2013). Role of LysM receptors in chitin-triggered plant innate immunity. *Plant Signal. Behav.* 8:e22598.

127. Thelin, L., Mutwil, M., Sommarin, M., and Persson, S. (2011). Diverging functions among calreticulin isoforms in higher plants. *Plant Signal. Behav.* 6, 905–910.

128. Thomma, B. P., Nurnberger, T., and Joosten, M. H. (2011). Of PAMPs and effectors: the blurred PTI-ETI dichotomy. *Plant Cell* 23, 4–15.

129. Tintor, N., and Saijo, Y. (2014). ER-mediated control for abundance, quality, and signaling of transmembrane immune receptors iplants. *Front. Plant Sci.* 5:65.

130. Vaid, N., Macovei, A., and Tuteja, N. (2013). Knights in action: lectin receptor-like kinases in plant development and stress responses. *Mol. Plant* 6, 1405–1418.

131. Vaid, N., Pandey, P. K., and Tuteja, N. (2012). Genome-wide analysis of lectin receptor-like kinase family from *Arabidopsis* and rice. *Plant Mol. Biol.* 80, 365–388.

132. Van Damme, E. J. M., Fouquaert, E., Lannoo, N., Vandenborre, G., Schouppe, D., and Peumans, W. J. (2011). "Novel concepts about the role of lectins in the plant cell," in*The Molecular Immunology of Complex Carbohydrates-3*, Vol. 705, ed. A. M. Wu (New York: Springer), 295–324.

133. Van Damme, E. J. M., Hao, Q., Chen, Y., Barre, A., Vandenbussche, F., Desmyter, S., et al. (2001). Ribosome-inactivating proteins: a family of plant proteins that do more than inactivate ribosomes. *Crit. Rev. Plant Sci.* 20, 395–465.

134. Van Damme, E. J. M., Lannoo, N., and Peumans, W. J. (2008). Plant lectins. *Adv. Bot. Res.* 48, 107–209.

135. Vandenborre, G., Groten, K., Smagghe, G., Lannoo, N., Baldwin, I. T., and Van Damme, E. J. M. (2010). *Nicotiana tabacum* agglutinin is active against Lepidopteran pest insects. *J. Exp. Bot.* 61, 1003–1014.

136. Vandenborre, G., Smagghe, G., and Van Damme, E. J. M. (2011). Plant lectins as defense proteins against phytophagous insects. *Phytochemistry* 72, 1538–1550.

137. Vandenbussche, F., Desmyter, S., Ciani, M., Proost, P., Peumans, W. J., and Van Damme, E. J. M. (2004a). Analysis of the in planta antiviral activity of elderberry ribosome-inactivating proteins. *Eur. J. Biochem.* 271, 1508–1515.

138. Vandenbussche, F., Peumans, W. J., Desmyter, S., Proost, P., Ciani, M., and Van Damme, E. J. M. (2004b). The type-1 and type-2 ribosome-inactivating proteins from Iris confer transgenic tobacco plants local but not systemic protection against viruses.*Planta* 220, 211–221.

139. Van Hove, J., Fouquaert, E., Smith, D. F., Proost, P., and Van Damme, E. J. M. (2011). Lectin activity of the nucleocytoplasmic EUL protein from *Arabidopsis thaliana.Biochem. Biophys. Res. Commun.* 414, 101–105.

140. Wan, J., Tanaka, K., Zhang, X. C., Son, G. H., Brechenmacher, L., Nguyen, T. H.,et al. (2012). LYK4, a lysine motif receptor-like kinase, is important for chitin signaling and plant innate immunity in *Arabidopsis*. *Plant Physiol.* 160, 396–406.

141. Wei, G. Q., Liu, R. S., Wang, Q., and Liu, W. Y. (2004). Toxicity of two type II ribosome-inactivating proteins (cinnamomin and ricin) to domestic silkworm larvae.*Arch. Insect Biochem. Physiol.* 57, 160–165.

142. Williams, C. E., Collier, C. C., Nemacheck, J. A., Liang, C., and Cambron, S. E. (2002). A lectin-like wheat gene responds systemically to attempted feeding by avirulent first-instar Hessian fly larvae. *J. Chem. Ecol.* 28, 1411–1428.

143. Williams, D. B. (2006). Beyond lectins: the calnexin/calreticulin chaperone system of the endoplasmic reticulum. *J. Cell Sci.* 119, 615–623.

144. Willmann, R., Lajunen, H. M., Erbs, G., Newman, M. A., Kolb, D., Tsuda, K.,et al. (2011). *Arabidopsis* lysine-motif proteins LYM1 LYM3 CERK1 mediate bacterial peptidoglycan sensing and immunity to bacterial infection. *Proc. Natl. Acad. Sci. U.S.A.*108, 19824–19829.

145. Wirthmueller, L., Maqbool, A., and Banfield, M. J. (2013). On the front line: structural insights into plant–pathogen interactions. *Nat. Rev. Microbiol.* 11, 761–776.

146. Wu, J., Luo, X., Guo, H., Xiao, J., and Tian, Y. (2006). Transgenic cotton, expressing*Amaranthus caudatus* agglutinin, confers enhanced resistance to aphids. *Plant Breed.*125, 390–394.

147. Xiang, Y., Song, M., Wei, Z., Tong, J., Zhang, L., Xiao, L., et al. (2011). A jacalin-related lectin-like gene in wheat is a component of the plant defence system. *J. Exp. Bot.* 62, 5471–5483.

148. Xin, Y., Xiangrong, Z., Mingju, Z., Wenchao, G., Yingchuan, T., Qizhong, X.,et al. (2011). Transgenic potato overexpressing the *Amaranthus caudatus* agglutinin gene to confer aphid resistance. *Crop Sci.* 51, 2119–2124.

149. Yamaji, Y., Maejima, K., Ozeki, J., Komatsu, K., Shiraishi, T., Okano, Y., et al. (2012). Lectin-mediated resistance impairs plant virus infection at the cellular level. *Plant Cell*24, 778–793.

150. Yong, W. D., Xu, Y. Y., Xu, W. Z., Wang, X., Li, N., Wu, J. S.,et al. (2003). Vernalization-induced flowering in wheat is mediated by a lectin-like gene VER2.*Planta* 217, 261–270.

151. Zhang, C., Shi, H., Chen, L., Wang, X., Lu, B., Zhang, S., et al. (2011). Harpin-induced expression and transgenic overexpression of the phloem protein gene AtPP2-A1 in*Arabidopsis* repress phloem feeding of the green peach aphid *Myzus persicae*. *BMC Plant Biol.* 11:11.

152. Zipfel, C., Kunze, G., Chinchilla, D., Caniard, A., Jones, J. D., Boller, T., et al. (2006). Perception of the bacterial PAMP EF-Tu by the receptor EFR restricts *Agrobacterium*-mediated transformation. *Cell* 125, 749–760.

153. Zhang, W., Peumans, W. J., Barre, A., Astoul, C. H., Rovira, P., Rougé, P.,et al. (2000). Isolation and characterization of a jacalin-related mannose-binding lectin from salt-stressed rice (*Oryza sativa*) plants. *Planta* 210, 970–978.

Index